Building Pathology and Rehabilitation

Volume 3

Series Editors

Vasco Peixoto de Freitas
Aníbal Costa
J. M. P. Q. Delgado

For further volumes:
http://www.springer.com/series/10019

Vasco Peixoto de Freitas
J. M. P. Q. Delgado
Editors

Durability of Building Materials and Components

Springer

Editors
Vasco Peixoto de Freitas
J. M. P. Q. Delgado
Building Physics Laboratory
Department of Civil Engineering
Faculdade de Engenharia
Universidade do Porto
Porto
Portugal

ISSN 2194-9832 ISSN 2194-9840 (electronic)
ISBN 978-3-642-44055-7 ISBN 978-3-642-37475-3 (eBook)
DOI 10.1007/978-3-642-37475-3
Springer Heidelberg New York Dordrecht London

Printed on acid-free paper

Springer is part of Springer Science+Business Media (www.springer.com)

Preface

In the distant past, construction solutions were validated empirically through several years of experience, whereas with the onset of further industrialization of the construction process, it was recognized within the construction community that a performance-based selection of materials, components, and systems was required if innovation was to be fostered and progress in the construction domain achieved. However, it was equally apparent to those promoting such novel approaches that the selection on the basis of understanding of performance requirements could only be met if the results of research and development were made available and indeed exploitable by practitioners.

Given the availability of new construction solutions, these days ever evolving in the construction domain, and these offering greater levels of construction complexity, there is perhaps a presupposition that their performance over time can be readily estimated on the basis of a knowledge of material properties and the ability to simulate the performance of the individuals parts of a complex system. In fact, only an in-depth knowledge of the area of durability allows the prediction of performance over time and from this, the designation of suitable solutions for the most complex problems affecting the built environment. In addition, the adoption of innovative technological solutions can only truly be undertaken with assurance to provide the expected performance over time with appropriate and focused research; the costs of a non-durable alternatives are simply too high.

The purpose of this book, *Durability of Building Materials and Components* is to provide a collection of recent research works to contribute to the systematization and dissemination of knowledge related to the long-term performance and durability of construction and, simultaneously, to show the most recent advances in this domain. It includes a set of new developments in the field of durability, service life prediction methodologies, the durability approach for historical and old buildings, asset and maintenance management, and on the durability of materials, systems, and components. The book is divided into several chapters that intend to be a resume of the current state of knowledge for the benefit of professional colleagues.

Vasco Peixoto de Freitas
J. M. P. Q. Delgado

Contents

A Probabilistic Framework for Performance-Based Durability Engineering

Madeleine Flint, Jack Baker and Sarah Billington

Abstract A proposed probabilistic framework for performance-based durability engineering links uncertainty in exposure, deterioration, and repair to estimations of decision information based on economic, environmental and social sustainability. This information includes costs, environmental impacts, downtime, and safety. The assessment of a new or existing structure or structural element is completed in three separate stages of analysis, allowing flexibility in choice of models and assumptions. An example simulation analyzes the energy required for repair of a reinforced concrete structure subjected to carbonation and corrosion. It is shown that the framework can be used to determine the strategy minimizing the lifetime repair energy of the structure.

Keywords Durability · Reliability · Corrosion · Reinforced concrete

1 Introduction

The design and rehabilitation of sustainable buildings and infrastructure poses a great challenge for structural engineers in a world of funding constraints, increasing demands for performance, and uncertain future environmental

---

M. Flint (✉)
Blume Earthquake Engineering Center, Stanford University,
Bldg. 540 MC: 3037, Stanford, CA 94305-3037, USA
e-mail: flint@stanford.edu

J. Baker · S. Billington
Department of Civil and Environmental Engineering, Stanford University,
MC: 4020, Stanford, CA 94305-4020, USA
e-mail: bakerjw@stanford.edu

S. Billington
e-mail: billington@stanford.edu

V. P. de Freitas and J. M. P. Q. Delgado (eds.), *Durability of Building Materials and Components*, Building Pathology and Rehabilitation 3,
DOI: 10.1007/978-3-642-37475-3_1, © Springer-Verlag Berlin Heidelberg 2013

conditions. Accurate consideration of economic, environmental, and social sustainability requires full life cycle modeling from initial construction to subsequent deterioration and repair. A robust estimate of impacts and sustainability should also explicitly consider the uncertainty inherent to construction, repair and impacts, and the environmental conditions that lead to structural deterioration. The probabilistic framework for performance-based durability engineering (PBDE) under development by the authors aims to characterize the durability of structures and structural elements through detailed analysis of exposure, deterioration, repair, and economic, environmental and social impacts with incorporation of uncertainty in all stages.

The PBDE approach is expected to result in benefits for researchers, engineers, and decision-makers. Foremost is the generality of the approach, which allows the assessment of any type of structure undergoing any type of exposure and deterioration so long as models exist for the exposure, degradation and impact phenomena of interest. Additionally, the separation of the phases of analysis allows flexibility in choice of models used, which in turns allows the analyst to assess how sensitive the end results are to assumptions and models in each stage. This separation of phases of analysis can allow, for example, comparison of the accuracy of different research and engineering models for deterioration. With validated and accurate models in place, the incorporation of uncertainty allows the calculation of information useful to decision- and policy-makers. In addition to data such as expected service life it is possible to obtain e.g., the distribution of possible repair costs over the life of the structure, or environmental impacts such as greenhouse gas emissions and energy consumption. Social impacts are included both through the calculation of closure time (downtime) for repair and through time-variant structural reliability analysis.

The background of the methodology, additional motivations, a review of existing durability frameworks and an overview of the approach is presented in this section. The probabilistic methodology used is detailed in Sect. 2 and a description of general stages of analysis is presented in Sect. 3. The methodology can be applied to a variety of deterioration causes, including moisture and temperature-induced degradation of buildings and façades, and an example of a reinforced concrete building suffering from carbonation-induced corrosion damage is presented in Sect. 4. Finally future work and expected challenges are discussed and conclusions drawn in Sect. 5.

1.1 Background

The presented methodology draws on the experience gained in the development of analogous approaches to performance-based risk management in the nuclear, seismic, and natural hazard engineering communities. The cumulative nature of deterioration damage makes the development of performance-based risk management for problems of long-term durability particularly challenging. Both risk simulation (e.g., [1]) and performance-based earthquake engineering (PBEE) [2] use separate phases of analysis linked together at a limited number of "pinch points" where a few variables are passed from one analysis to another. Uncertainty is incorporated in each stage, and at the end of the analysis an integral (convolution) equation is used to calculate the probability of exceeding a value of a decision variable, such as cost. In addition to similarities in the probabilistic approach and general framework, the methods make use of advanced scientific and engineering models for structural performance. Whereas PBEE requires simulation of the nonlinear response of a structure to a large number of seismic records, PBDE requires the simulation of the nonlinear response of the structure to a large number of records of environmental exposure. Due to the intensive computation requirements of these approaches parallelization or use of optimization schemes to reduce the number of records required are frequently desired.

1.2 Motivation

In addition to the factors previously stated (generality, flexibility, assessment of sensitivity, accuracy, and usefulness of information), the successful adoption of the cited PBEE framework [2] by the research, practicing, and policymaking communities is a motivation for a new effort towards PBDE. With PBEE, a large cooperative research effort aimed at validating the probabilistic approach and developing the required models, has been used in large-scale analyses for regional policymaking, and is being codified in simpler methods for use in practice [3].

Recent interest within the broader engineering and policymaking communities in both sustainability and performance metrics provides motivation for the development of the PBDE framework. In Europe the introduction of new construction product regulations [4] requires estimation of life cycle impacts for all construction products, and must therefore include an accurate assessment of the service life of a new structure. The US transportation community will also need to demonstrate the effectiveness of bridge rehabilitation and repair actions to address the findings of the US Governmental Accountability Office [5]. As developers and infrastructure owners push for the design of structures having longer service lives [6, 7], including those located in challenging environments, methods of ensuring that a design will meet performance-based specifications will be required. Finally, as the global engineering community works to address climate change by reducing energy consumption through initiatives such as the International Energy Agency Annex 55 [8], fully probabilistic tools for predicting the deterioration of buildings over time are needed. Taken together these developments and needs call for new and innovative methods of assessing and designing for structural durability.

PBDE has the potential to aid and complement many approaches to durability design and assessment currently in use and under development, e.g., those described in Sect. 1.3. These include: design of new structures to meet performance-based specifications; selection of new designs amongst alternative designs using decision information; selection of repair strategies (see example in Sect. 4); assessment of the safety of deteriorating structures; structural repair planning to minimize social impacts, such as closure time; assessment of networks of structures for policy-making, and; design code calibration.

1.3 Existing Durability, Deterioration, and Sustainability Frameworks

Many approaches for durability design and modeling of structural deterioration and service life exist. These approaches generally fall into three categories: (1) service life estimation methods and durability design standards, (2) maintenance optimization methods, and (3) time-variant structural reliability. A brief description of these methods, focusing on their relation to PBDE assessment, is given here.

1.3.1 Service Life Estimation and Design for Durability

Various standards and methods focus on predicting the service life of a deteriorating structure such as those from CIB W80 and RILEM TC 140 [9, 10]. Methods include the factor method [9], dose–response, damage, or performance functions, survival analysis or discrete event simulation [11], and Markov models (Sect. 1.3.2). For reinforced concrete structures, approaches include those in *fib* Bulletin 34 [12], RILEM's DuraCrete model [13], and the performance-based design approach from JSCE-331, Technical Committee on Structural Performance of Deteriorated Concrete Structures [14]. Many of these frameworks use empirical degradation models, or incorporate data taken from the inspection of a large number of structures. The degree of uncertainty modeled varies widely, but many frameworks output a probabilistic distribution of service life.

In these frameworks meeting a design service life with a selected probability is taken as an indicator for acceptable predicted performance of the structure. PBDE takes an alternate approach, in which service life is not modeled directly, and instead the economic, environmental and social impacts are modeled over a set assessment time. Similar to the factor, dose–response, and survival methods, PBDE is capable of assessing the performance of a wide variety of structures, but can be adapted by analysts to use their own preferred models. These models could include empirical or survey-derived models as well as numerical models. The flexibility of PBDE comes with the tradeoff of increased complexity and data requirements. It is envisioned that a simplified framework for design engineers could be developed as has been the case within the PBEE community [3].

1.3.2 Maintenance Optimization Methods

The well-developed field of maintenance optimization uses a variety of techniques to determine optimal repair timing for a single structure or a network of structures within various performance constraints. These constraints might include available funding or desired level of service. Structural façades or envelopes and bridges, types of structures for which maintenance is semi-regular and inventories are large, have been the focus of most optimization methods. Envelope modeling has generally used probabilistic deterioration models, derived using either survival analysis [11] or from Markovian approaches [15, 16]. Bridge Management Systems, such as Pontis [17], track inspections of common bridge elements, using Markov models to predict deterioration based on environmental classes and current damage state [18]. These results can be fed into various single and multi-objective optimization schemes to select optimal repair timing, e.g., [19].

The proposed PBDE framework presented herein does not yet approach the level of sophistication of the various maintenance optimization schemes, though it shares the concept of using multiple objectives to assess the durability or sustainability of a structure or structural element. The optimization methods generally require probabilities of transitioning from one level of damage to another, and these probabilities can be computed using a PBDE assessment rather than from inverse analyses of inspection data sets. It is therefore possible to make use of the already-developed, advanced optimization methods after completing a PBDE assessment. This combination offers the additional benefits of being able to predict the deterioration of new elements or elements in which sufficient survey data is not available, and adds transparency to the approach through the output of intermediate data.

1.3.3 Time-Variant Structural Reliability

Many researchers have used the time-variant structural reliability methods to predict the performance of deteriorating structures. As these approaches focus on the probability of overload or threats to safety, they have so far tended to focus on deterioration mechanisms with the potential to cause structural failure. Enright and Frangopol [20], Vu and Stewart [21], and Li and Melchers [22], among many others, performed reliability assessment of corroding concrete structures. Matsuki et al. [23], Rao et al. [24], and Choe et al. [25] have performed combined assessments of the seismic risk of deteriorating reinforced concrete structures.

Although the type of analysis performed differs (FORM, Monte Carlo), reliability methods may calculate sensitivity to different uncertain parameters in addition to predictions of failure rates. This information can be used to identify where additional detailed information may improve precision, or to determine which variables might be modeled as deterministic to limit the complexity of the analysis. On the other hand, the complexity of the general reliability methods themselves tends to limit the complexity of the models that may be adopted for deterioration modeling, and the results of the analysis are only as robust as the closed-form deterioration models used. Monte Carlo approaches allow the use of more sophisticated deterioration models, but the number of simulations required when exposure and decision information uncertainty is to be included can be very large. The proposed PBDE framework offers an approach to combining the sophisticated models for deterioration used in Monte Carlo simulations with reliability analysis.

1.4 Overview of Proposed Methodology

The PBDE methodology utilizes three distinct stages of analysis separated by pinch-points, shown in Fig. 1: (1) exposure analysis to determine exposure conditions (*EC*) acting on the structure; (2a) deterioration and (2b) repair analysis to track accumulated damage measures (*DM*) and subsequent repair actions (*RA*), and; (3) impact analysis to calculate resulting decision information (*DI*) including costs, downtime, environmental impacts, and safety indicators [26].

Stages (1), (2) and (3) are computed separately, with only a few variables being passed from one stage to the next, i.e., it is not necessary to track a single exposure record from the prediction of exposure all the way through to costs. Because of the important interplay between deterioration and repair these two phases must be computed together for a given record, as a repair action will change the course of deterioration at points in time after the repair. A more detailed description of the analysis stages is presented in Sect. 2.

The analysis format shown in Fig. 1 is used to calculate the probability distributions of interest. Conditional exceedence curves, also known as conditional complementary cumulative distributions (*G*), obtained from the stages are convolved together at the pinch-points to obtain intermediate results, e.g., damage measure exceedence curves, and final results, e.g., costs, at the end of the analysis timeline, shown in Eq. 1. The probability of exceeding a certain characteristic exposure condition level *ec* is denoted G_{EC}, and is obtained from exposure analysis. Exposure condition records are simulated based on a the level of the characteristic value and are passed to deterioration and repair analysis to obtain the

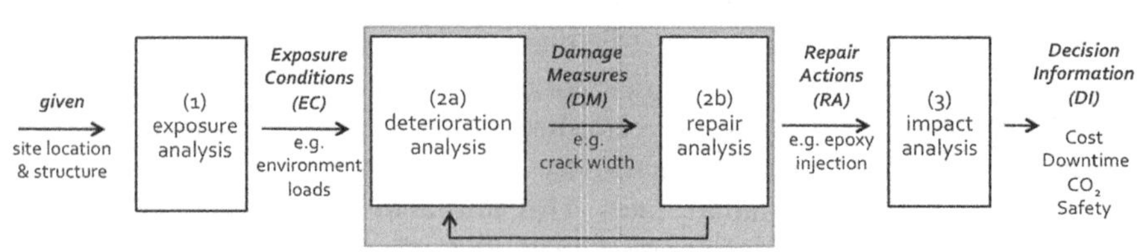

Fig. 1 Stages and pinch-points of the PBDE assessment. Simulations are performed at analysis stages (*1*) exposure analysis, (*2*) deterioration and repair analysis, and (*3*) impact analysis based on a limited number of pinch-point variables passed on from the preceding stage

joint conditional distribution of damage measure and repair timing, $G_{tRA},\ _{DM\ |\ EC}$. Impact distributions are then computed in an impact analysis based on the possible combinations of repair action timing, $G_{DI\ |\ tRA}$. A distribution for the damage as a function time can also be obtained from the convolution Eq. 2.

$$G_{DI,t\,>\,0}(di) = \int\int G_{DI|t_{RA}}(di|t_{RA})dG_{t_{RA},DM|EC}(t_{RA}, dm|ec)dG_{EC}(ec) \quad (1)$$

$$G_{DM}(dm, t) = ||G_{DM|EC}(dm, t|ec)dG_{EC}(ec) \quad (2)$$

In some cases, for example for new structures, it may be important to include initial impacts, e.g., original construction costs, in the PBDE assessment. In this case the impacts become the sum of impacts caused by repairs, calculated through the three stages, and initial impacts described by the distribution $G_{DI,t=0}$. Initial impacts are discussed in more detail in Sects. 2.1.1 and 2.2.1.

2 Probabilistic Methodology

Several of the potential benefits of the proposed PBDE methodology are evident in the use of Eq. 1, namely flexibility and ability to assess outcome sensitivity to various models. However, under some circumstances the use of convolution is not possible. The requirements for valid use of convolution and an alternative approach, the use of various Monte Carlo methods, are discussed in Sects. 2.1 and 2.2, respectively. Initial impacts, correlation within analysis stages, and methods to reduce the number of simulations required are also discussed for each approach.

2.1 Convolution Approach

The integral form of Eq. 1 is only valid given certain assumptions about correlation and independence of the analysis stages. Essentially, there must be conditional independence at the pinch-points: the deterioration and repairs must not affect the exposure, and the decision information must be completely described by the repair action and timing and may not be additionally dependent on either damage or exposure. Each stage may only conditionally depend on the stage immediately preceding it. Exposure may influence decision information via it its impact on deterioration and repair, but the distribution for decision information may not be directly dependent on the value of the exposure condition. Should impacts be conditionally dependent on exposure or should the damage or repair affect the exposure another approach must be used, detailed in Sect. 2.2.

If the above-mentioned assumptions are taken as valid and then verified, the analyst still has choices to make regarding the probabilistic distributions.

As frequently occurs, the tradeoffs involved in obtaining the probabilistic distributions are between accuracy and computation time. To obtain high confidence in the conditional exceedence distributions many simulations must be performed. When computationally-intensive models are used for the simulations the performance of thousands of simulations may not be practical. Thus, it may be desirable to perform fewer simulations and to fit a parametric distribution to the results, rather than to use an "empirical" distribution function. In some cases this approach may result in only a small loss in accuracy. For example, Weibull distributions have been found to fit the results of Monte Carlo simulation for time-to-cracking in a corroding reinforced concrete beam well [27]. In other cases, such as when the results diverge into two or more characteristic behaviors (e.g., for intact and cracked reinforced concrete members), it may be difficult to choose an appropriate single parametric distribution. In these cases the combination of multiple parametric distributions or a technique such as kernel smoothing may be appropriate. Sheskin provides a thorough description of the fitting of probabilistic distributions to non-parametric data [28].

2.1.1 Incorporation of Initial Impacts

Incorporation of initial impacts such as initial construction cost complicates the computation of the final impact distribution because the initial and lifetime impact distributions cannot be simply summed to obtain the final result. Should the decision information distributions at $t = 0$ and $t > 0$ be reasonably modeled by normal distributions then the resulting sum will be normal. If either distribution is not normal then an additional convolution must be performed, according to

$$G_{DM}(di) = \int G_{dt,t>0}(di - z) f_{Dt,t>0}(z) dz \tag{3}$$

where z is a dummy variable for integration. Similar to Eqs. 1, 3 also takes the form of a convolution integral. As described in Sect. 2.1, the distribution $G_{DI,\ t>0}$ may be parametric or non-parametric. For the remainder of this chapter the subscript $t > 0$ will be omitted.

2.1.2 Correlation of Pinch-Point Variables

While correlation within one of the three analysis stages (Fig. 1) does not invalidate the convolution approach, it does add complexity. There are three main sources of correlation: correlation in output exposure records, spatial correlation of adjacent areas in deterioration and repair analysis, and correlation of impacts.

The degree to which correlation must be explicitly accounted for in exposure analysis depends on the sophistication of the models used. It is important that an exposure record (i.e., a time series of weather data used to perform a simulation in deterioration and repair analysis) has consistency between temperature, humidity, solar exposure, etc. A physical model that directly simulates how the local climate acts on the structure will produce consistent exposure records without any adjustments required. However, if multiple, independent models are used to determine the distributions of the exposure conditions over time, then the simulation of a record should include correlation. For example, unusually warm temperature, rainfall, and chloride deposition likely have a joint cause (e.g., a tropical storm), and the resulting domain hygrothermal state will be different from the domain hygrothermal state produced by an exposure record with non-simultaneous high values of temperature, moisture, and chlorides.

Another source of correlation occurs because of spatial discretization of the structural elements necessary for modeling. If only one segment or area of a structure is modeled explicitly, then the damage state of other segments must be correlated with the modeled segment. This is necessary in order to perform a realistic repair analysis (i.e., repairs decisions are not made independently for each square meter of a surface) and for structural reliability analysis. Examples of spatially correlated analyses for corrosion in reinforced concrete structures are given in [29] and [30].

Finally, the impacts analyzed in the third stage of analysis may be correlated. For example, a repair action with an unusually high downtime will likely result in a larger cost as well. Similarly, low repair energy is likely to occur together with low greenhouse gas emissions.

Where correlated variables must be simulated, methods are given by, for example, Kelton and Law [31]. In general, the simplest approach is the use of multivariate normal distributions, which require only correlation coefficients between all correlated variables. Obtaining these coefficients may be challenging. Weather data sets can be analyzed to obtain the necessary correlation coefficients for exposure records. Guidance for spatial correlation coefficients is contained in [29] and [30]. Correlation of impacts requires the selection of a coefficient based on engineering judgment. When the use of normal distributions is not possible various transformations from the parametric distribution of choice may be possible. Finally, in some cases the development of a joint distribution from correlated data may be required.

2.1.3 Methods to Reduce Computational Demand

Any of the stages in a PBDE assessment may be computationally intensive, and a method of limiting the number of simulations may be desired. In addition to fitting a parametric distribution to pinch-point conditional distributions, a response surface may be created to allow consideration of for example, uncertainty in structural parameters [32]. The use of a response surface facilitates performing more simulations within a stage fewer computationally-intensive analyses.

2.2 Monte Carlo Approach

In some cases the assumption of independence at pinch-points adopted in the convolution approach will not be valid. One such situation would be the option of replacing rather than repairing a deteriorated structure. Unless the new structure uses exactly the same materials and geometry as the old, a new exposure analysis would need to be performed if replacement were ever triggered during a deterioration/repair simulation. A more complex scenario could invalidate the independence between impacts and environment, e.g., if changes in global climate conditions influenced material costs. In these cases the analysis cannot be performed as a series of conditionally-independent stages, but rather as a collection of full-lifetime simulations, i.e., Monte Carlo analysis, on which statistics are later performed to calculate any distributions of interest.

Performing the PBDE assessment as a Monte Carlo analysis has some benefits: no assumptions regarding the distributions at pinch points are required, and complex interactions within a record can be more easily handled. While using Monte Carlo is valid in all cases current research aims to preserve the integral format when possible. This is desired for two practical reasons: (a) it allows the analyst to swap out different models within each stage without affecting the other stages, and (b) the results of different computer programs can be linked together at the pinch-points at the end of the analysis stages, rather than several times during each record. Nonetheless, properly-performed Monte Carlo analyses will always be probabilistically valid, and for this reason the results of the convolution approach may be compared to a parallel Monte Carlo simulation for validation of the probabilistic methodology.

2.2.1 Incorporation of Initial Impacts

Given a distribution for initial impacts it is very simple to incorporate these in the Monte Carlo approach. A value of the initial impact is simulated at the beginning of the analysis and added to the impact value obtained over the lifetime simulation. The impact distribution fitted to the results will then contain both initial and lifetime impacts.

2.2.2 Correlation of Pinch-Point Variables

As described in Sect. 2.1.2, correlation within an analysis record requires the simulation of correlated data. The same methods used for the convolution approach may be used for Monte Carlo. Some additional complications arise for the incorporation of spatial correlation because distributions are generally not calculated until all of the simulations are performed. Without a damage measure or repair action distribution to draw from it is not possible to simulate the state of other segments during one record before continuing to impact analysis. This can be overcome by performing an initial batch of analyses to determine deterioration distributions to be used for later simulation or for re-sampling.

2.2.3 Methods to Reduce Computational Demand

Depending on the level of uncertainty in input parameters, a full Monte Carlo simulation might require thousands or millions of simulations to achieve a high confidence level. This may not be practical when one or more of the models used is computationally intensive, which suggests the use of methods to reduce computational demand. One previously mentioned method is re-sampling or bootstrapping, in which simulations used to build a distribution are later re-used as simulations of the distribution. Response surfaces may also be used to reduce computation time. For uncertain structural parameters Monte Carlo has several available methods of reducing the number of required simulations, including Latin hypercube. An effective approach fits pinch-point distributions after a number of simulations have been performed, and then further simulations are performed from the distributions. This approach is logical when only one stage is computationally intensive (e.g., deterioration/repair) but when other stages have high uncertainty (impacts) and require a large number of simulations to capture the underlying uncertainty.

3 Description of Analysis Stages and Models

This section discusses general requirements for modeling in each of the analysis stages. Depending on the cause(s) of deterioration chosen for the analyses the requirements can vary broadly, so the description here focuses on essential information required, with less discussion of mechanism-specific requirements. A more detailed description of the models used for carbonation-induced corrosion in reinforced concrete is given in Sect. 4.

3.1 Exposure Analysis

Exposure analysis (stage (1) in Fig. 1) uses basic information about the structure and location to determine the distribution of exposure adjacent to the structure. Figure 2 summarizes the approach. The exposure at the site is necessary to model boundary conditions including heat and moisture, chemical concentrations, and forces or tractions. Required input information, analysis format, output distributions, suggested data sources, and commentary are discussed in this section.

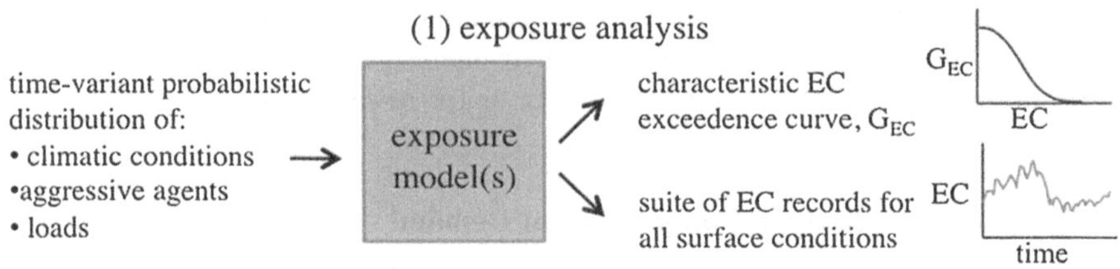

Fig. 2 Exposure analysis schematic. Given basic information about the site and structure exposure analysis computes hygrothermal, chemical, and traction conditions on or adjacent to structural surfaces. Uncertainty in these exposure conditions (*EC*) is accounted for in a characteristic exceedence curve and through the simulation of a series of records given a value of the characteristic *EC*

3.1.1 Input Information

The structure's geometry and materials and its location are the fundamental inputs to exposure analysis. From this basic information it is possible to obtain the time-varying distributions of regional climatic conditions and loads required for the exposure analysis. Necessary inputs vary depending on the exposure and deterioration models chosen, but in general distributions may be required for temperature, solar concentration, air humidity, precipitation, wind velocity and direction, and chemical concentrations in air and water. Additionally, dead and live loads may be required.

3.1.2 Analysis Format

The function of exposure analysis is to use these given distributions to determine an exceedence curve for the exposure conditions acting on the structure during its lifetime. Thus, uncertainty in the inputs is translated by the deterministic exposure model to uncertainty in the surface conditions. Multiple models may be required to compute a set of time-series data (exposure records). Local air temperature and relative humidity might be obtained from a weather station, or from the simulations of a global climate model. A computational fluid dynamics model might be used to translate wind and precipitation measurements at a weather station to rain driven onto a structural surface. Or the computational fluid dynamics model might be used to model where marine aerosol chlorides are deposited on the structure, and their washing away by rainfall. Other models might be used for the deposition of acids on a building envelope. Tractions might be simulated based on empirical models for the impact loads caused by truck traffic.

A variety of model types may be used in exposure analysis. Some models may lend themselves to direct statistical analysis, whereas in other cases Monte Carlo simulation may be used within the exposure analysis. Thus, the form of this stage may be as simple as a single calculation or may require thousands of complex simulations. In very simple cases an exposure model may not be necessary to translate input information into structural boundary conditions, as the input information itself can serve as the boundary condition. This case might occur in, for example, the modeling of a submerged reinforced concrete column using recorded water temperature and salinity as input information.

3.1.3 Output Information

There are two main types of output from exposure analysis, shown in Fig. 2. The first is the exceedence curve (G_{EC}) for the characteristic exposure condition, used in the convolution equation. The second is a series of exposure records to be used in deterioration and repair analysis. Depending on the exposure models used, these exposure records could be (a) direct outputs from exposure simulations, or (b) new stochastic simulations created from the distributions of exposure conditions. In (a) the direct output of the exposure model is used to determine the value of the characteristic exposure condition, and is then passed on to deterioration analysis directly. In (b) a statistical evaluation of all of the outputs of the deterioration model is performed, time-varying distributions for each condition are fit, and then random (consistent, correlated) records are simulated from these distributions. Option (a) is more adapted to complex exposure models, whereas (b) is more appropriate for simple models. In either case the temporal discretization of the output must match that used in the deterioration analysis.

3.1.4 Data Sources and Models

Much of the necessary input data can generally be obtained from databases of climate and weather information, e.g., [33]. Other data (e.g., salinity of coastal water, carbon dioxide exposure, soil contamination concentration) can be obtained from literature. Dead loads and a distribution of live loads can generally be obtained from the design documents and related codes, or from surveys and other load studies. If any monitored or sensed data is available greater accuracy in exposure condition prediction should be possible.

Because the level of detail required in exposure analysis varies quite greatly it is not possible to suggest specific models to be used. Textbooks on the fundamental types of models will provide more detailed reference lists. Examples of models to determine these hygrothermal boundary conditions may be found in [34]. For airborne deposition a computational fluid dynamics model may be desired, and these models and other advanced simulation tools used for structure and buildings are described in [35]. A review of methods of simulating load conditions from design codes or from other models can be found in [36].

3.1.5 Discussion

As several different types of exposure condition may be required, a decision needs to be made regarding which "characteristic" condition will be used for the exceedence curve required for the integration of Eq. 1. Two main approaches are proposed: one is to use the condition most directly tied to the deterioration; the second is to use a meta-parameter. In the first case the exceedence curve might represent the maximum value expected over the lifetime, or uncertainty in the distribution parameters (e.g., the mean) where this value is of more importance. Alternatively, the characteristic value might be related to scenario uncertainty, e.g., the annual increase in atmospheric greenhouse gas levels. In the second approach, a meta-parameter would represent the level of severity of the exposure. In using a meta-parameter care must be taken not to introduce a false degree of correlation in the analysis, i.e., it is unlikely that the structure will concurrently undergo severe hygrothermal conditions, chemical attack, and loading. Thus, this definition may be of more use when the outputs of the exposure model are used directly for deterioration analysis (case (a) in Sect. 3.1.3). The affects of various assumptions about the characteristic exposure condition are currently being explored.

The descriptions provided in Sects. 3.1.1–3.1.4 have assumed that that modeling from input statistical distributions of climatic and loading data will sufficiently capture the exposure of the structure. Such time-averaged data may not accurately characterize the sorts of rare events that could greatly impact the actual structural deterioration. While it is impossible to predict some unusual load cases (e.g., vehicle collisions) if such information is recorded for an existing structure it should be incorporated into the exposure analysis. It is noted that other durability assessment methods similarly neglect the affects of these very rare events.

3.2 Deterioration Analysis

The goal of the deterioration analysis phase (2a) is to model the deterioration resulting from the exposure of the structure and to link the deterioration to observable or detectable damage measures, as shown in Fig. 3. Both damage and repairs are tracked simultaneously for each record or simulation in stage (2), as indicated in Fig. 1. For clarity, inputs, analysis format, outputs, suggested models and commentary are discussed here for the deterioration sub-phase only.

3.2.1 Input Information

Input information includes that already required for exposure analysis (structural materials, geometry), the output records of exposure analysis, and any additional data required for the selected deterioration models. If uncertainty in the structure

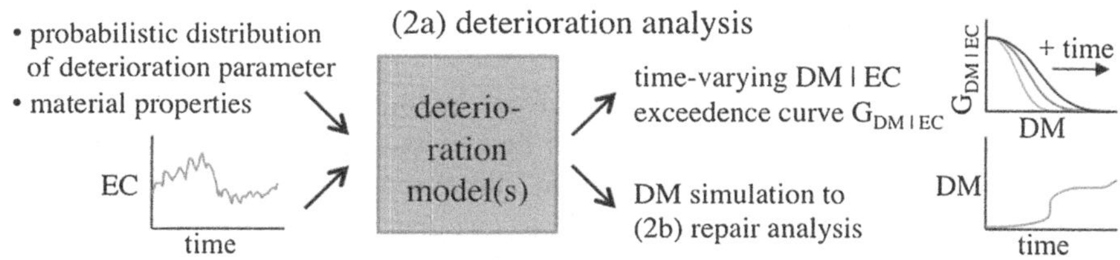

Fig. 3 Deterioration analysis schematic. Given a level of exposure and a suite of exposure condition (*EC*) records, deterioration analysis computes boundary conditions and hygrothermal, chemical, electrochemical, stress, and strain states in the domain over the analysis timeline. Some of these states are tracked as damage measures (*DM*) conditional on the *EC* level to create the time-varying conditional *DM* curves. The state of the structure is passed to repair analysis at inspection times

itself is to be considered, probabilistic distributions will be required for e.g., elastic modulus, diffusion coefficients, and cover depth. Many deterioration models require other parameters related to the material and construction that are intrinsically uncertain. Chloride concentration required to initiate active corrosion is one such parameter.

3.2.2 Analysis Format

Deterioration analysis begins by determining surface boundary conditions given the exposure analysis data record adjacent to the surface. For example, a hygrothermal model may be tasked with converting air, precipitation, and solar information to surface temperature and surface relative humidity or moisture flux. From the boundary conditions the deterioration analysis models, among other phenomena, ingress of moisture and chemicals, changes in material properties, and damage measures such as cracking. The damage measures should be selected based on what is observable or detectable from the inspection planned for the structure or element. In reinforced concrete this might include crack widths and delamination for routine inspections, and potential mapping or chloride profiles for special investigations. For façades cracking, staining or efflorescence might be chosen. Visible corrosion or flaking of paint might be important in an exposed steel structure.

Uncertainty within this stage is a result of uncertain material parameters and deterioration and repair thresholds. In many cases the deterioration models and repair decision trees will be deterministic given these input parameters. In order to introduce uncertainty, important non-exposure input parameters are stochastically simulated for each exposure record. For each incremental degree of exposure condition obtained from exposure analysis a number of deterioration and repair simulations are performed. The number of simulations required is determined by the number of stochastic deterioration parameters and by their uncertainty.

3.2.3 Output Information

As in exposure analysis, there are two types of output from the deterioration model. The damage measure simulations themselves are passed back and forth from stage (2a) to (2b) at each "inspection" time step. Once the required number of simulations is performed it is possible to calculate the distribution of damage measures given the input level of exposure conditions (i.e., the characteristic value of the exposure record). This conditional distribution, $G_{DM \mid EC}$, for a particular damage measure can be integrated with the exposure condition curve to get a prediction of the course of damage over time using Eq. 2. This intermediate result is used to verify the accuracy of the deterioration models and various assumptions if inspection records are available. A similar curve is output for the structural capacity of the element at each point in time if a structural reliability analysis is to be performed.

3.2.4 Data Sources and Models

Data is required to develop distributions for the uncertain parameters required by the selected deterioration model. Guidelines for these values can be found in literature, design codes such as those described in Sect. 1.3.1, and in textbooks for the structural material. For existing structures material properties may be determined through testing.

Many types of models are available in literature to predict deterioration, some of which were discussed in Sect. 1.3.1. Models developed from empirical studies, surveys of actual deterioration, and physio-chemical analytical or numerical models may be used within the PBDE framework (see for example, [9, 10]).

3.2.5 Discussion

Models that attempt to capture the physics and chemistry of the deterioration process are preferred, as it is believed that these models best take advantage of the full description of exposure and incorporation of probabilistic information. Empirical models may also be used provided that the input information is appropriate. Thus, an analyst using a finite element model for moisture and chemical ingress must use "true" diffusion coefficients and boundary conditions, whereas an analyst using a model based on effective Fickian diffusion should use "apparent" diffusion coefficients and surface concentrations. A major goal of the current research on PBDE by the authors is to assess the difference in decision information outcomes with the use of various types of deterioration models.

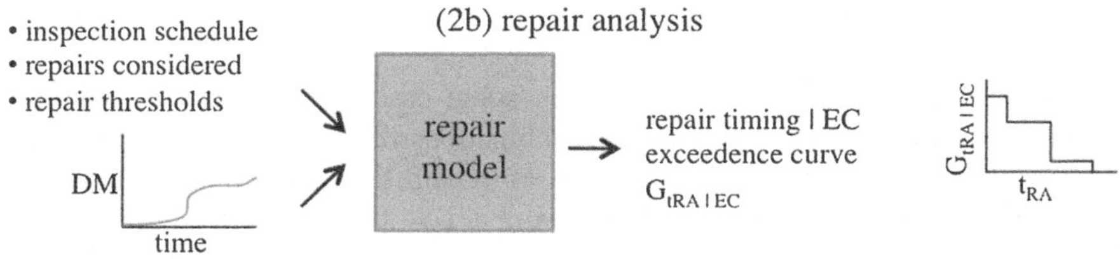

Fig. 4 Repair analysis schematic. The damage measures (*DM*) for a given record are passed to repair analysis at inspection times. Distributions for repair action (*RA*) timing (t_{RA}) given exposure condition (*EC*) level are computed from the results of a suite of damage and repair simulations

3.3 Repair Analysis

Given the damage caused by a certain level of exposure, repair analysis (2b) models the decision-making process of the structure operator or maintenance manager in repairing the structure, shown in Fig. 4.

3.3.1 Input Information

Besides the simulations of damage measures from deterioration analysis, the repair model requires information about maintenance and repair practices and preferences. A limited set of practical repairs must be selected before the assessment is performed, and the circumstances (thresholds) under which each repair will be applied must be described. If stochastic thresholds for repair are to be used then these distributions are also required.

3.3.2 Analysis Format

One relatively simple way of linking damage measures to specific repair actions is through the use of a decision tree. Certain repair actions are triggered based on the component type, and the level and spatial extent of damage. At specified time steps in the deterioration analysis an inspection is "performed" and the degree and extent of each damage measure is recorded. A series of subsequently smaller branches on the tree are traced to determine whether or not a repair is triggered. If triggered, the time at which the repair occurs is recorded (t_{RA}) and any appropriate changes to materials, geometry, boundary conditions, or damage levels are applied within the simulation. All damage measures do not need to link to all repair actions, and there may be different inspection timing for each damage measure. For example, it might be specified that regular visual inspections occur every two years, but that a test for corrosion initiation only be performed every 20 years or if the preceding visual inspection finds cracking. When distributions for thresholds are used a critical value for each record must be simulated.

3.3.3 Output Information

Unlike for exposure and deterioration, in some cases repair analysis will only output the probability of the possible combinations of repair action timing given exposure, $G_{tRA \mid EC}$. Each combination of repair timing will have a discrete probability of occurring given the exposure level. If desired, the repair action timing results can be convolved as an intermediate result to check that the decision tree has captured the decision-making of the maintenance manager or operator. If there are many theoretically possible repair combinations but relatively few that occur in repair analysis, the repair timing combinations with non-zero probability of occurring can be passed to impact analysis to reduce the number of impact calculations required.

3.3.4 Data Sources and Models

While some repair decision information is available from literature, the analyst will more likely have to develop a new decision tree for the structure and operator analyzed. These trees can be obtained through two general methods. The first is through solicitation of expert opinion and review of maintenance and inspection standards for the operator. The second is by performing an inverse analysis of a large number of inspections and repairs for a certain component. This analysis is somewhat comparable to the analysis performed in the development of transition matrices for assessment using Markov chains. Both methods have advantages and disadvantages. Among other possible issues, experts may have unconscious bias in their judgments [37]. When inverse analysis is performed the thresholds computed might not reflect the actual damage-based decision-making. For example, the repair thresholds obtained might be more reflective of the availability of funding to the operator rather than the need for repair. Thus, a combination of the two approaches is suggested, where the decision tree is developed based on expert opinion and review of maintenance standards and then validated using existing inspection records.

3.3.5 Discussion

As implied in the previous section, the decision of when and how to repair a damaged structure is complex, and generally a function of more than the degree and extent of damage. It is easy to think of scenarios in which a repair action is applied when not justified by the level of damage, or where its application is delayed beyond the crossing of a damage threshold. Some maintenance operators have preventative maintenance preferences, and may, for example, apply sealants to all bridge decks if funding is available, regardless of the level of damage. In other cases a repair such as replacement of part of a façade might occur early because of other work being performed on the structure, or for aesthetic reasons.

A needed repair could be delayed due to lack of funds, or so that it might be combined with other repairs due to occur at a later time. These external uncertainties in repair timing are not captured directly in PBDE assessment. However, presenting a rational repair scheme will allow more consistent comparison across repair methods and structural designs than one that attempts to take into account these external conditions.

3.4 Impact Analysis

Impact analysis, stage (3) shown in Fig. 5, takes a given combination of repair timing and uses an inventory of the materials, equipment, energy, and time taken for the repairs to predict the economic, environmental and social sustainability. Decision information (cost, downtime, environmental impacts) both at the time of the initial repair and any ongoing impacts are considered. Impact analysis can additionally consider the probability of structural failure if this information is desired. Costs, downtime, and environmental impacts are the focus in this section, as reliability methods have been well reviewed elsewhere, e.g., in Melchers [36], and no fundamental changes to reliability methodology are proposed.

3.4.1 Input Information

Impact analysis requires an inventory for each repair in addition to which repair timing combinations are possible. Each repair is broken down into tasks until the steps are sufficiently simple that their impacts can be determined. For environmental impact, re-painting a façade might be broken down into: removal of old paint by hydrojetting, application of primer, application of paint, transportation of materials, and use of an diesel generator. From a cost and downtime perspective it might not be necessary to break down the tasks as finely. The inventory quantities are all uncertain, however the analyst can choose to incorporate quantity uncertainty separately or combine it with uncertainty in the impacts themselves.

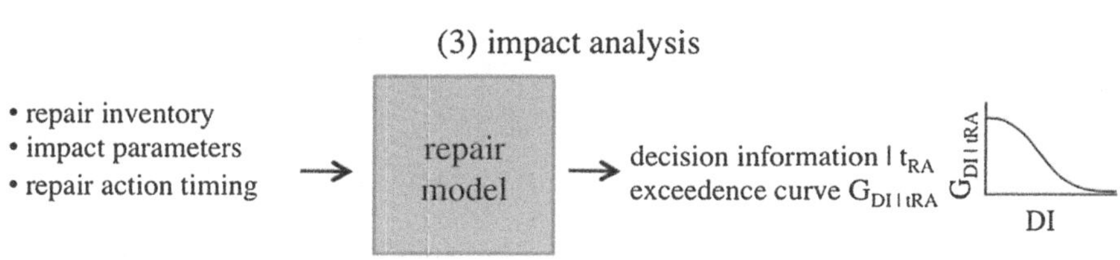

Fig. 5 Impact analysis schematic. Given possible combinations of repair actions and timings the impact analysis computes distributions of decision information (*DI*) including costs, downtime, environmental impacts, and structural safety

The calculation of the inventory quantities leads to the other type of required information, which are the impacts (cost, environmental, time) of the quantity of material, equipment usage, energy, etc. or of the repair as a whole. A distribution is required both for the initial impact (at the time of the repair) and any ongoing impacts. In the simplest case these distributions can be considered stationary, for example, the toxic emissions of sealing a bridge deck are the same in year 10 as in year 30. Time-variant information may be required for other impacts, e.g., to incorporate cost inflation. Obtaining the input information is generally the most difficult part of impact analysis.

3.4.2 Analysis Format

Performing impact analysis is quite simple once the conditional curves have been developed. For each possible repair combination the cumulative lifetime impact distribution is obtained by combining the contributions of the individual repairs, again including both initial and ongoing impacts. In the very simple case where the individual repair impacts are normally distributed the lifetime impacts can be directly obtained using the statistics of normal variables. If the impacts are not normally distributed, or if they are correlated, other methods, such as Monte Carlo simulation, must be used.

This description suggests that unlike exposure and deterioration/repair analysis, there is no true "model" in impact analysis. At this point in the development of the PBDE framework, this is an accurate description. This does not preclude, however, the future use for example, of an economic model to predict costs.

3.4.3 Output Information

The distributions created by combining the impact distributions for the possible repair timings are exactly the $G_{DI \mid tRA}$ required for the convolution. The number of curves computed is determined by the number of possible repair timings multiplied by the number of types of decision information. In theory it is possible to develop these curves without any information from repair analysis by computing the curves for all possible repair timings. In practice it is simpler to only compute the curves for repair timing combinations that have non-zero probability of occurring, which tends to reduce the number of curves required.

3.4.4 Data Sources and Models

Local contract cost data, construction bids, and engineering judgment are used to create a distribution of the costs of the materials, equipment, and energy required for the possible repair tasks or the full repairs. A life cycle impact assessment tool such as SimaPro [38], among others, can be used to model the environmental impacts of the tasks. Alternately, other databases (e.g., economic input–output sources such as Open IO [39]), that contain environmental impact data may also be used. The required information for impact analysis is theoretically the same as that required for life-cycle costing (LCC) and life-cycle impact assessment (LCIA). As LCC and LCIA become more commonly used tools for engineers it is expected that obtaining the necessary impact data will become easier. Modeling downtime is a challenge, and using a combination of construction schedules and engineering judgment is the suggested method of obtaining the necessary information.

3.4.5 Discussion

As with developing the repair decision tree, there are several anticipated difficulties in obtaining the data needed for the impact analyses. Besides the variance in construction practices and materials between regions it is also necessary to break down the impacts to some degree on the area or volume of the structure on which the repair takes place. In some cases the repair may have an inventory that consists of a "per-repair" value (e.g., for setup and breakdown) as well as a value based on the extent of the repair required. It is therefore essential that the repair actions themselves are well defined. It may be beneficial to break down a repair group into, for example, "concrete patching 0.5–1.0 m^2," "concrete patching 1.0–10 m^2," and "concrete patching 10–20 m^2." These estimates will never precisely match the values used in the actual repair but should still allow a general comparison of impacts between alternatives.

It is recognized that this data will never be exact. LCC and LCCA data comes from sets that are averaged among producers or are valid only for certain regions. Different contractors bid jobs differently, and apparent unit material prices may include many non-material costs. As with the quantities calculation, the results may be imprecise but should suffice to assist in decision-making.

4 Example Assessment: Carbonation of a Reinforced Concrete Wall

The exterior wall of a reinforced concrete structure was designed as a case study for the application of the proposed methodology to choice of repair strategy. This fictitious example was chosen as a simple demonstration of how the PBDE framework may be implemented. In the example, carbonation-induced corrosion is of concern for a 60-year-old building with poor cover and concrete durability located within an industrial zone in a tropical rainforest region. The deterioration of such a structure is most likely caused by diffusion of carbon dioxide into the cement mortar, carbonation and neutralization of the bulk cement electrolyte, initiation of active corrosion and subsequent uniform corrosion propagation.

Simple models and assumptions are chosen, and indeed the example itself has been designed to allow the use of simple models with limited loss of accuracy. The goal of the analysis is to determine which two repair options, re-alkalization of carbonated cement (Option A), or cathodic protection (Option B) offers the optimal strategy in terms of the amount of energy used for the repairs over the remaining 90 years of a planned 150-year service life. At the time of the assessment the state of damage in the structure is unknown. The given information (location and structure), models used, implementation, results, and a discussion of the results are presented here.

4.1 Location and Structure

The tropical rainforest climate undergoes very small variation in average daily temperature and air humidity over the course of the year. Freeze–thaw, chloride contamination, alkali-silica reaction and other sources of degradation are not expected to affect the structure, leaving carbonation as the likely source of deterioration. The exterior walls, with low cover to the light tie reinforcement are of particular concern, as corrosion-induced cracking and spalling have a deleterious aesthetic effect and are possibly hazardous. For the analysis a one-meter square representative area is modeled. Due to the uniform nature of carbonation-induced corrosion the segment length or width can be considered arbitrary.

Geometric properties are considered deterministic with the exception of cover depth, which is modeled as lognormally distributed with mean of 15 mm and standard deviation of 5 mm. The diameter of the #3 exterior ties is 9.5 mm. The concrete is assumed to have a water to cement ratio (w/c) of 0.5 using 350 kg of ordinary Portland cement per cubic meter of concrete and with a 28 day compressive strength of 28 MPa. This concrete is expected to yield paste porosity, ε_p, of 0.20, and post-hydration concentrations of C–S–H and CH in the concrete of 210 and 90 kg/m^3, respectively. Resistivity of carbonated concrete is expected to be considerably higher than for uncarbonated concrete [40] and is modeled as a function of concrete relative humidity. The tensile splitting strength of the concrete is taken as 2.8 MPa. Uncertainty in material properties is not included in the assessment.

4.2 Models

Only those exposure conditions that are needed to predict deterioration are computed, and only those damage measures that are likely to be found during routine inspection are tracked. The environmental impact chosen, energy use, does not require discounting of future impacts.

4.2.1 Exposure Analysis

Modeling carbonation-induced corrosion in an area of stable temperature and air humidity requires the calculation of the ingress of carbon dioxide, which is dependent on three environmental parameters: the amount of carbon dioxide present on the surface of the structure, the temperature of the concrete, and the relative humidity inside the concrete.

In this example the carbon dioxide content of the air in the industrial zone (450 ppm or approximately 0.00075 kg/m^3) and the temperature (34 °C) are modeled as stationary in time, as the effects of any seasonal variations are expected to be overwhelmed by average conditions, especially considering the slow process of CO_2 diffusion. The uncertainty in the relative humidity of the concrete is the main exposure condition controlling the deterioration, as it affects both the diffusion coefficient and resistivity of the concrete. Outside of the "convection zone" the concrete relative humidity will also be stable, but unknown without testing [41]. Thus, the exposure analysis results in an exceedence curve for concrete relative humidity, shown in Fig. 6. The relative humidity is modeled as $N(70, 2)$.

4.2.2 Deterioration Analysis

The deterioration model must be capable of accurately capturing both the diffusion of carbon dioxide and resulting corrosion for the concrete type and exposure conditions studied. Küter [42] contains a very thorough description of the thermodynamics of carbonation-induced corrosion in oxygen-rich concrete, and the assumptions made in this analysis generally follow that description. The

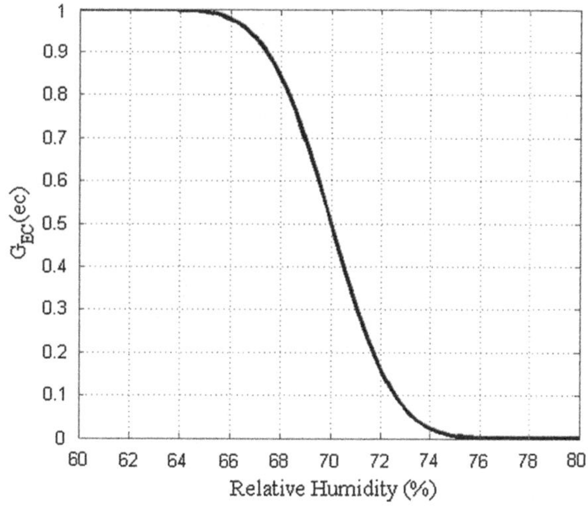

Fig. 6 Characteristic exceedence curve for concrete relative humidity. 50 data points were used in the analysis and passed on to deterioration analysis

stationarity of environmental parameters suggests the use of a carbonation model of the form:

$$x_c = \left(2[\mathrm{CO_2}]^0 D_{\mathrm{e,CO_2}} / \left([\mathrm{CH}]^0 + [\mathrm{C-S-H}]^0\right)\right)^{1/2} t^{1/2} \quad (4)$$

where x_c is the depth of carbonation, $D_{e,CO2}$ an effective diffusion coefficient and the concentration superscript '0' denoting the concentrations at the end of hydration [43]. The effective diffusion coefficient is related to the relative humidity (RH) in the concrete by an empirical equation,

$$D_e = \left(B_{\varepsilon_p}(1 - \mathrm{RH}/100)\right)^2 \quad (5)$$

where B is an empirical constant taken as $(1.2 \pm 0.1) \times 10^{-3}$ m/s$^{1/2}$, which is modeled as a normally distributed variable with standard deviation equal to the error term [43]. Equation 5 was developed from experiments at 30 °C, and so the coefficient must be adjusted for exposure temperature by an Arrhenius equation,

$$D_{e,CO_2}(T) = \exp(E_a/R(1/T_S - 1/T))D_e \quad (6)$$

[44], where the activation energy, E_a, is taken as 45 kJ/mol. The combination of (5) and (6) results in a stochastic function for the calculation of the effective diffusion coefficient given relative humidity.

A one-dimensional model can be used to predict the carbonation ingress in the wall representative area. The initiation of microcell corrosion once the carbonation front has reached the rebar is a stochastic process, but will here be assumed to occur as soon as the carbonation front reaches the reinforcing bar. The analysis assumes that corrosion propagation (corrosion rate, i_{corr}) in carbonated concrete tends to be governed by concrete resistivity, which varies greatly depending on the relative humidity. The rate is found by combining a power-law relationship between relative humidity and concrete resistivity for carbonated concrete [45]:

$$\rho_{conc} = \left((100/\mathrm{RH} - \mathrm{a}) \,/\, \mathrm{c}\right)^{1/b} \quad (7)$$

where $a = 0.999$, $b = 3.22$, and $c = 7.55 \times 10^{-11}$, with a simplified relationship for corrosion rate given resistivity:

$$\log i_{corr} = \log\left(57.8 \times 10^{-3} \rho_{conc}^{-0.8125}\right) - 1.0057 \quad (8)$$

A discussion of possible thermodynamic inaccuracies in these assumptions can be found in Gulikers [46]. For this example simulation, these models are sufficient in that they are simple and reasonably accurate.

Based on the electrochemistry of the system it is likely that the rust products formed will be a mix of Fe_2O_3 and Fe_3O_4, which have a relative density of 2.1 and 2.0, respectively [42]. Due to the small difference in these numbers the relative density is taken as 2.0. These expansive products will eventually cause cracking of the concrete cover, which is modeled according to the equations found in DuraCrete [47]:

$$x_{crack} = 86.8 + 7.4d_c/D_b - 22.4f_{sp} \quad (9)$$

$$w_{crack} = 0.05 + 0.0125(x - x_{crack}); \quad 0 \leq w_{crack} \leq 1.0 \text{ mm} \quad (10)$$

where x_{crack} is the corrosion penetration (μm) to cause cracking, and the tensile splitting strength f_{sp} is given in MPa. This simple equation has been shown to predict the cracking behavior in accelerated testing reasonably well [48].

The main damage measure to be extracted from the analyses is surface cracking as it is easily detectable in regular inspection and will likely trigger repair actions. It is assumed that inspections at 10 year intervals past the current structure age of 60 years would capture any cracks with width greater than 0.1 mm. Additionally, a special inspection is also planned at the current time (structure age 60 years) regardless of the presence of surface cracks. Regular inspections will occur every 10 years until year 140 in the life of the structure.

4.2.3 Repair Analysis

Two repair strategies are investigated in the analysis. Option A uses repair action 1 (ra_1), which involves applying an alkali-rich electrolyte solution to the surface of the concrete and applying direct current to electrochemically re-alkalize the concrete, which in turn halts corrosion. ra_1 may be triggered by the presence of surface cracks or determination of corrosion initiation. ra_1 is assumed to be effective on 80 % of the first applications, and another round of re-alkalization is planned if testing suggests that the first repair has not been successful. Option B uses repair action 2(ra_2), in which an impressed current cathodic protection system is installed, which is expected to completely eliminate the possibility of future corrosion with 100 % effectiveness. After 10 years the cathodic protection system may be turned off for half of the year without re-initiation of corrosion.

The threshold relating crack width to repair initiation is modeled as $N(0.1, 0.01)$ mm, and each analysis must only consider one repair action. Thus, two damage measures (corrosion initiation and cracking) are mapped to one repair action in each analysis, shown in Fig. 7. More generally multiple repair actions must be considered, and the thresholds for their application may be stochastic. Consideration of multiple repairs and uncertain thresholds adds complexity to the analysis but does not require any adjustment of the underlying probabilistic approach.

Because only one repair is considered for each analysis there are relatively few combinations required for repair timing. ra_1 and ra_2 may not be triggered until the first inspection at year 60, and then may be triggered in any inspection year thereafter, giving 9 possible times to the application of ra_1 or ra_2.

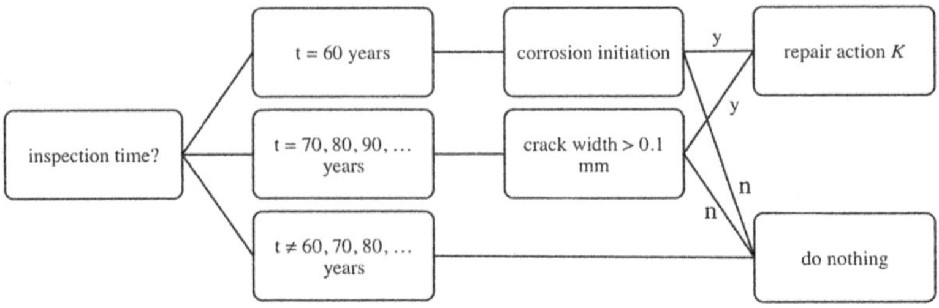

Fig. 7 Decision tree for repair actions (*RA*) re-alkalization or cathodic protection. Inspections begin in year 60 and continue at 10 year intervals until year 140. Decisions to repair are based on the damage measures (*DM*) of corrosion initiation and crack width

4.2.4 Environmental Impact Analysis

To perform the environmental impact analysis inventories of both repair actions are created. Only the energy (direct current) used during the repair is considered, i.e., the energy required for the materials, transport, and equipment is not considered. The required energy is modeled as a normal variable taking into account variation in usage based on engineering judgment. For repair option A, re-alkalization, only initial energy use is considered, modeled as $N(10, 4)$ kWh. Repair option B, cathodic protection, has only a continued energy use, $N(1.8, 0.4)$ kWh/yr for the first 10 years and $N(0.9, 0.2)$ kWh/yr in subsequent years. The assumption of normal impacts allows simple calculation of repair-combination distributions following rules for sums of normal variables.

4.3 Implementation

The example simulation is implemented in a series of Matlab R2011b functions, utilizing the Statistics toolbox. The general structure is shown in Fig. 8. The convolution method of assessment is used, and only the relative humidity and repair timing combinations are passed between the analyses. A bounded kernel-smoothed distribution for the damage measures and repair timing is implemented using the "fitdist" function within the Statistics toolbox. Numerical integration (convolution) of the conditional curves uses the trapezoidal rule. The assessment used 50 exposure condition points, 1,500 time steps, 500 records per exposure, 1,000 damage points and 1,000 decision information points. In general the number of records and discretization required should be assessed to ensure robust results. Use of the Parallel Computing toolbox allows parallelization of the damage/repair records at the conditional exposure level (i.e., simulations were performed in parallel at $ec = ec_1, ec_2,\ldots$).

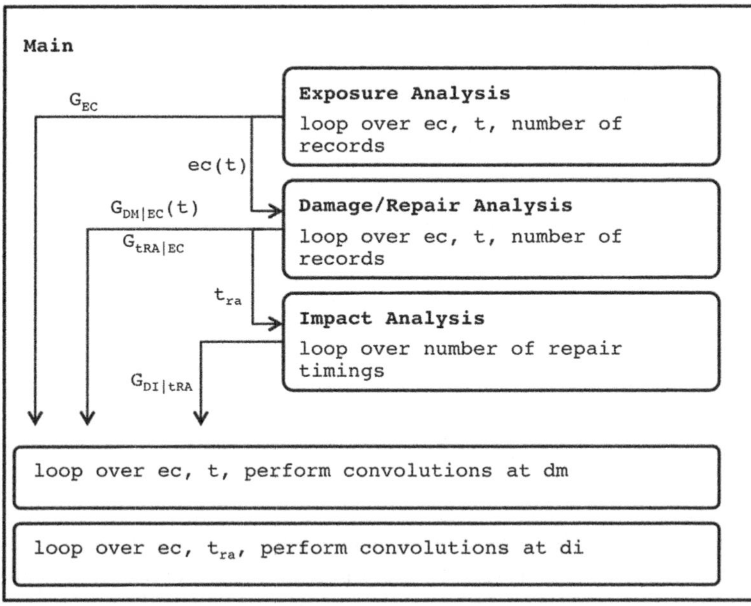

Fig. 8 Implementation of the PBDE assessment in Matlab. The assessment runs as a series of three analysis loops that pass on a limited number of pinch-point variables and two loops to perform the numerical integration of damage over time and decision information. The analyses of the suites of exposure condition (*EC*) records at a given value are performed in parallel

Fig. 9 Evolution of damage measure (*DM*), crack width, over time. The median and 25th exceedence percentile crack widths are very small, whereas the 10th percentile value demonstrates the larger crack widths present in the upper tail of the distribution

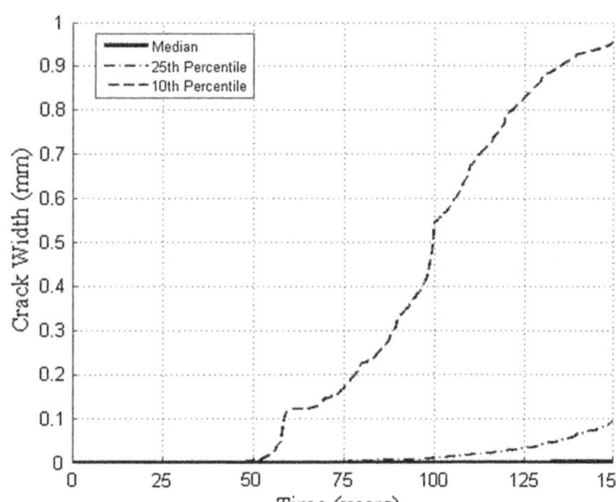

4.4 Results

Two main categories of results are presented: intermediate results (convolved damage measures and repair timing) and final results (convolved decision

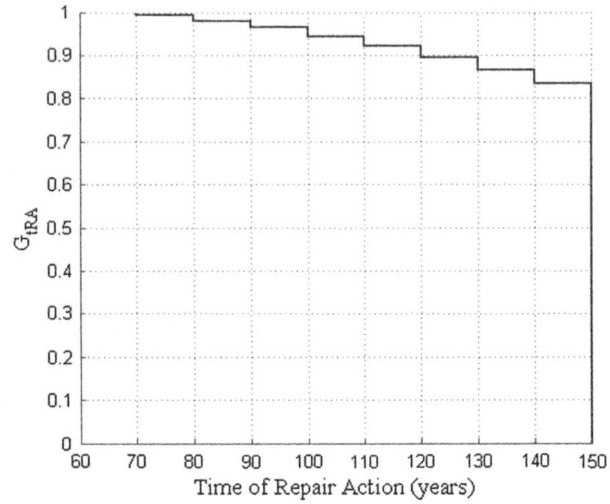

Fig. 10 Repair action timing exceedence distribution. Repairs (either re-alkalization or cathodic protection) occur with 18 % likelihood during the lifetime of the structure

information). Figure 9 shows an intermediate result for the evolution of crack width over time. Because damage is relatively unlikely in this example crack widths at the median (50 % exceedence) and 25 % exceedence are small. However, the 10 % exceedence curve has much higher crack widths, as some less likely scenarios cause very early corrosion initiation. Figure 10 shows the exceedence curve for repair timing. Due to the low levels of damage very few repairs occur during the simulations, leading to an 82 % probability that the repair time exceeds the 150-year service life. Figures 11 and 12 contain the final results for both repair options.

For option A the expected energy over the lifetime is 1.7 kWh, whereas for option B the expected energy is 7.6 kWh. Another assessment was performed using only mean values in a single simulation. In this case no repair is ever triggered during the lifetime of the structure.

4.5 Discussion

There are several methods of making decisions using the PBDE results. By the criterion of minimum expected value, option A, re-alkalization, is optimal, as its mean energy use is four times lower than the mean energy use of option B. A characteristic exceedence probability criterion might be used if the operator wanted to select the repair scheme minimizing the probability of exceeding, for example, 20 kWh. By this criterion option A would also be preferred. Similarly, option A would be preferred if minimizing the amount of energy consumed in unlikely scenarios was important, for example selecting the scheme with the smallest energy having 10 % probability of exceedence. In this example assessment three different decision criteria yield the same result (option A is preferred), but in other PBDE assessments different decision criteria may not give the same result.

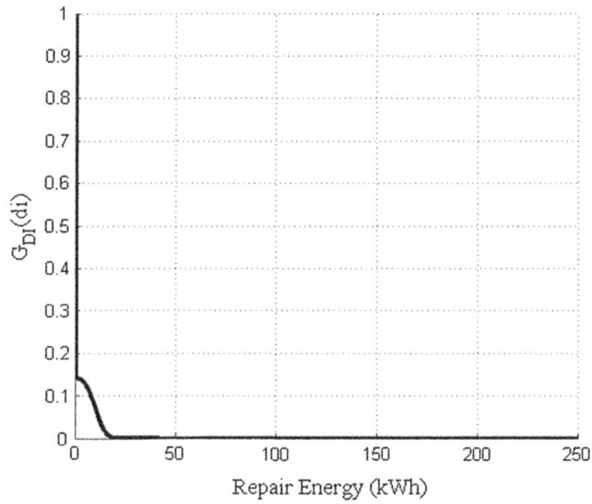

Fig. 11 Option A (re-alkalization) repair energy exceedence distribution. Due to the low probability of repair a zero-impact result is very likely. As re-alkalization is modeled with only an initial, time-of-repair impact the rest of the distribution follows a normal curve

The assumption of deterministic material properties, temperature, and carbon dioxide levels reduces the uncertainty in repair timing and therefore also the repair energy. Additionally, the assumption of constant (mean-annualized) exposure conditions affects the final results by reducing the uncertainty in the exposure and thus deterioration process. This reduction in uncertainty would tend to result in fewer simulations of rare behavior (e.g., extremely fast deterioration). The effect of these assumptions will be explored in future work.

The deterministic assessment performed as a comparison to the PBDE assessment yields no information regarding which repair option is preferred for this particular example. In the mean-value analysis no repairs are ever triggered, which makes it impossible to compare the repair schemes using any criteria. Thus, this very simple example demonstrates that performing the PBDE assessment and incorporating uncertainty offers improved decision-making information. A Monte Carlo assessment is also capable of incorporating uncertainty and may have been acceptable for this simulation. However, when more complex models are linked, assumptions on probability distributions are to be changed after the analysis, or spatial variation is to be taken into account (as discussed in Sect. 2) the modular, integral format of PBDE may be preferred.

5 Future Work

While the foundation for a probabilistic PBDE methodology has been laid in this and in other works, e.g., [26], much work remains to validate the approach and to demonstrate its usefulness. Validation must occur both for the models used in the analysis stages, against laboratory and survey data, and for the probabilistic methodology against the results of a Monte Carlo analysis using the same models.

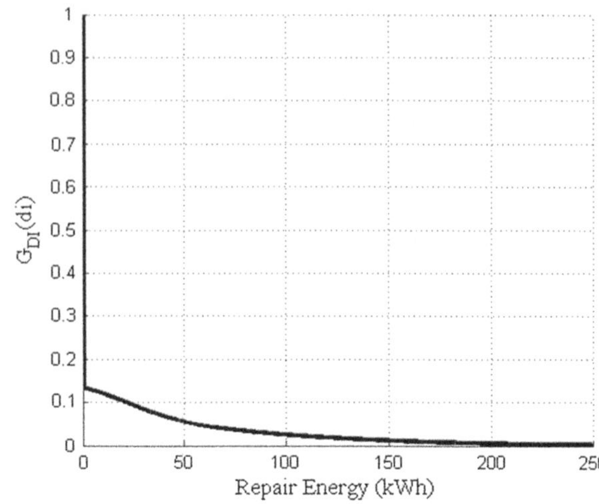

Fig. 12 Option B (cathodic protection) repair energy exceedence distribution. No impact is the most likely result, but the ongoing energy requirements of impressed-current cathodic protection combined with uncertain repair timing result in a smoothed distribution for lifetime energy use

This will be achieved by performing a "testbed" simulation of a single structure, which will also serve to demonstrate the usefulness of the methodology. A reinforced concrete bridge undergoing chloride-induced corrosion has been selected as the first testbed structure.

Additionally, assessing the sensitivity of the decision information results to assumptions made and models used in each analysis stages is currently underway. Comparison of the results of a large number and type of deterioration models on the same problem appears to be lacking in literature and could provide helpful information for researchers and practitioners trying to model deterioration. While some findings on the level of exposure condition detail needed for accurate deterioration results have been presented, it is thought that confirmation of these results for more sophisticated deterioration models is wanting. Finally, sensitivity assessment will allow the development of general guidelines for analysts using the PBDE methodology on which sort of variables are essential to model stochastically and which variables may be modeled deterministically. The choice of which variables may be modeled deterministically is likely to vary depending on what sort of deterioration is modeled, but the sensitivity assessment performed on the testbed structure should provide an example for how to perform similar assessments for other types of structure and conditions.

6 Conclusions

The presented probabilistic methodology for performance-based durability engineering (PBDE) offers a comprehensive, modular framework for the assessment of the impacts of structural durability. By combining different stages of analysis at pinch-points the framework allows flexibility in modeling choices, and through convolution allows incorporation of uncertainty at each stage. Current efforts are focused on applying the methodology to corrosion in reinforced concrete structures, although the type of deterioration and structure is not limited to this case.

An example simulation for an existing structure undergoing carbonation served to demonstrate how the methodology is applied. A repair scheme was selected based on the minimization of lifetime energy used in repair. Had only a deterministic simulation been used to predict the course of deterioration and repair it would not have been possible to offer decision information regarding the optimal repair scheme.

Due to the number of analysis stages, data requirements, and sophistication of models used, performing a PBDE assessment can be complex and computationally intensive. The development of frameworks for nuclear and seismic performance-based design and assessment suggests that these limitations can be overcome with further development and collaboration. As decision-makers increase their demands for quantitative estimates of economic, environmental, and social sustainability measures, methodologies such as PBDE will be valuable assets.

References

1. Kaplan, S., Garrick, B.J.: On the quantitative definition of risk. Risk Anal. **1**(1), 11–27 (1981)
2. Deierlein, G., Krawinkler, H., Cornell, C.A.: A framework for performance-based earthquake engineering. (2003)
3. Hamburger, R.: The ATC 58 project: Development of next generation performance based earthquake engineering design criteria for buildings. Structures Congress, ASCE (2006)
4. Construction Products Regulation. EC, vol 305. EU (2011)
5. GAO.: Highway bridge program: Clearer goals and performance measures needed for a more focused and sustainable program (2008)
6. LTBP: Long-Term Bridge Performance Program. http://www.fhwa.dot.gov/research/tfhrc/programs/infrastructure/structures/ltbp/ (2009). Accessed 12 Oct 2011
7. Gjørv, O.E.: Durability design of concrete structures in severe environments. Taylor & Francis Group, London (2009)
8. Annex55: Reliability of energy efficient building retrofitting—probability assessment of performance and cost (RAP-RETRO). IEA energy conservation in buildings and community systems (2009)
9. Hovde, P.J., Moser, K.: Performance based methods for service life prediction. State of the art reports, CIB Report: Publication 294 (2004)
10. Jernberg, P.: 140-TSL RTC, W80 CWC Guide and bibliography to service life and durability research for building materials and components. CIB (2004)
11. Juan, A.A., Ferrer, A., Serrat, C., Monteforte, A., Faulin, J.: Applications of discrete-event simulation to reliability and availability assessment in civil engineering structures. IEEE, pp. 2759–2767 (2009)
12. Fib: Bulletin 34: Model code for service life design. Fib, Lausanne (2006)
13. Siemes, T., Edvardsen, C.: Duracrete: Service life design for concrete structures, pp. 1343–1356 1999
14. Shimomura, T., Miyazato, S., Yamamoto, T., Kobayashi, K., Sato, T., Saito, S., Kamiharako, A., Akiyama, M.: Report of research project on structural performance of deteriorated concrete structures by JSCE-331, (2006)

15. Van Winden, C.: Rationalisation of building maintenance by Markov decision models: A pilot case study. J. Operational Res. Soc. **49**(9), 928–935 (1998)
16. Lacasse, M.A., Kyle, B., Talon, A., Boissier, D., Hilly, T., Abdulghani, K.: Optimization of the building maintenance management process using a markovian model, pp. 1–9 (2010)
17. Robert, W., Marshall, A., Hwang, S., Aldayuz, J.: ÑThe next generation of the Pontis Bridge management system ì. Bridge maintenance, safety management, life cycle performance and cost, IABMAS (2006)
18. Guerre, J., Robert, W., Fraher, M.: Framework for using Pontis bridge management system to support development of bridge programs. Paper presented at the transportation research board 86th annual meeting, Washington, D.C (2007)
19. Patidar, V., Labi, S., Sinha, K., Thompson, P.: Multi-objective optimization for bridge management systems, NCHRP Report 590. Transportation Research Board (2007)
20. Enright, M., Frangopol, D.: Probabilistic analysis of resistance degradation of reinforced concrete bridge beams under corrosion. Eng. Struct. **20**(11), 960–971 (1998)
21. Vu, K.A.T., Stewart, M.G.: Predicting the likelihood and extent of reinforced concrete corrosion-induced cracking. J. Struct. Eng. **131**, 1681–1689 (2005)
22. Li, C.Q., Melchers, R.E.: Time-dependent reliability analysis of corrosion-induced concrete cracking. ACI Struct. J. **102**(4), 543–549 (2005)
23. Matsuki, S., Billington, S.L., Bakerm J.W.: Impact of long-term material degradation on seismic performance of a reinforced concrete bridge (2006)
24. Rao, A.S., Lepech, M.D., Kiremidjian, A.S., Sun, X.Y.: Time varying risk modeling of deteriorating bridge infrastructure for sustainable infrastructure design. Time (Stage 3) pp. 2494–2501 (2011)
25. Choe, D.E., Gardoni, P., Rosowsky, D., Haukaas, T.: Seismic fragility estimates for reinforced concrete bridges subject to corrosion. Struct. Saf. **31**(4), 275–283 (2009)
26. Flint, M., Billington, S.: A probabilistic framework for performance-based durability engineering. Paper presented at the DBMC-12, Porto, Portugal (2011)
27. Thoft-Christensen, P.: Modelling of the deterioration of reinforced concrete structures. Department of Building Technology and Structural Engineering, Aalborg University (2000)
28. Sheskin, D.: Handbook of parametric and nonparametric statistical procedures. CRC Pr I Llc (2007)
29. Vu, K., Stewart, M.: Cracking and spalling reliability analysis considering spatial variability for reinforced concrete structures. Structural Safety and Reliability: ICOSSAR'01 (2001)
30. Ying, L., Vrouwenvelder, T., Wijnants, G.H.: Spatial variability of concrete degradation. In: Frangopol, D. (ed.) Life-cycle Performance of Deteriorating Structures: Assessment Design and Management, pp. 49–58. ASCE-SEI, Virginia, USA (2004)
31. Kelton, W.D., Law, A.M.: Simulation Modeling and Analysis, vol. 2. McGraw Hill, New York (2000)
32. Guan, X., Melchers, R.: Effect of response surface parameter variation on structural reliability estimates. Struct. Saf. **23**(4), 429–444 (2001)
33. Rutledge, G.K., Alpert, J., Ebuisaki, W.: NOMADS: A climate and weather model archive at the national oceanic and atmospheric administration. Bull. Am. Meteorol. Soc. (2006)
34. Hagentoft, C.E.: Introduction to building physics. External organization (2001)
35. Malkawi, A., Augenbroe, G.: Advanced Building Simulation. Taylor & Francis, London (2004)
36. Melchers, R.E.: Structural Reliability: Analysis and Prediction, vol. 2. Wiley, New York (1999)
37. Chhibber, S., Apostolakis, G., Okrent, D.: A taxonomy of issues related to the use of expert judgments in probabilistic safety studies. Reliab. Eng. Syst. Saf. **38**(1–2), 27–45 (1992)
38. Consultants, P.: SimaPro 7 life-cycle assessment software package, Version 7. Plotter 12, 3821 BB Amersfoort, The Netherlands (2007)
39. Open, IO: The sustainability consortium. http://www.sustainabilityconsortium.org/open-io/ (2011)

40. Papadakis, V.G., Vayenas, C.G., Fardis, M.N.: Fundamental modeling and experimental investigation of concrete carbonation. ACI Mater. J. 88 (4) , 363–373 (1991)
41. Relling, R.H., Sellevold, E.J.: In situ moisture state of coastal concrete bridges. In: International Conference on Concrete Repair, Rehabilitation and Retrofitting (ICCRRR), Cape Town, South Africa, 21–23 November, 2005. Taylor & Francis, p 191 (2006)
42. Küter, A.: Management of Reinforcement Corrosion: A Thermodynamic Approach. DTU, Kongens Lyngby, Denmark (2009)
43. Papadakis, V.G., Vayenas, C.G., Fardis, M.N.: Experimental investigation and mathematical modeling of the concrete carbonation problem. Chem. Eng. Sci. **46**(5–6), 1333–1338 (1991)
44. Allen RTL: Concrete in coastal structures. Thomas Telford (1998)
45. Michel, A., Geiker, M.R., Stang, H., Olesen, J.F.: Numerical modelling of reinforcement corrosion in concrete structures. In: 8th fib PhD Symposium, Kgs. Lyngby, Denmark, 20–23 June (2010)
46. Gulikers, J.: Theoretical considerations on the supposed linear relationship between concrete resistivity and corrosion rate of steel reinforcement. Mater. Corros. **56**(6), 393–403 (2005)
47. DuraCrete: Modelling of deterioration. BRITE EURAM, Probabilistic performance based durability design of concrete structures. (1998)
48. Markeset, I.G., Myrdal. R.: Modelling of reinforcement corrosion in concrete-State of the art. COIN Project Report. SINTEF Building and Infrastructure, Trondheim (2008)

Planned Preventive Maintenance Activities: Analysis of Guidance Documents

Sónia Raposo, Jorge de Brito and Manuel Fonseca

Abstract Nowadays the management of public built assets gains a significant importance owing to their size, social relevance and the corresponding working, operational and maintenance costs. Since it is economically relevant to the total cost of a building, the activity of maintenance management cannot be done in an improvised and casual way. In this document some of the reasons for the increased interest and development of several researches works in the area of building maintenance in Portugal are presented. Documents that provide guidance to the planned preventive maintenance activities are analysed in a general way and four of them in detail. The information contained in these four documents was applied to the study of planned activities for some current components existing in primary school buildings, in Lisbon. The analysis revealed that these documents do not address all the construction elements, systems and equipments and exterior spaces equally. These discrepancies between documents can raise difficulties in carrying out benchmarking actions for LCC or LCA assessment purposes. It is important to continue the work under development to standardize this information and create national, European or international references databases.

Keywords Planned maintenance · Guidance documents · Databases · Elementary public schools

S. Raposo (✉) · M. Fonseca
Buildings Department, National Laboratory of Civil Engineering (LNEC), Lisbon, Portugal
e-mail: sraposo@lnec.pt

M. Fonseca
e-mail: mfonseca@lnec.pt

J. de Brito
Civil Engineering and Architecture Department, Instituto Superior Técnico (IST), Lisbon, Portugal
e-mail: jb@civil.ist.utl.pt

V. P. de Freitas and J. M. P. Q. Delgado (eds.), *Durability of Building Materials and Components*, Building Pathology and Rehabilitation 3,
DOI: 10.1007/978-3-642-37475-3_2, © Springer-Verlag Berlin Heidelberg 2013

1 Introduction

Nowadays the management of public built assets gains a significant importance owing to their size, social relevance and the corresponding working, operational and maintenance costs. Since it is economically relevant to the total cost of a building, the activity of maintenance management cannot be done in an improvised and casual way.

In this document some of the reasons for the increased interest and development of several researches works in the area of building maintenance in Portugal are presented. Documents that provide guidance to the planned preventive maintenance activities are analysed in a general way and four of them in detail: a French source "La maintenance des bâtiments en 250 fiches pratiques", a Spanish source "Fitxes rehabilitació of ITeC", an English source "Housing component life manual" and a North American source "The Whitestone facility maintenance and repair cost reference 2009–2010".

The information contained in these four documents was applied to the study of planned activities for some current components existing in primary school buildings, in Lisbon.

2 Building Maintenance in Portugal

The maintenance activity in Portuguese public buildings and infrastructures has been, over the years, underestimated. The shortage of available resources assigned to the governmental and municipal buildings' maintenance has led to an accelerated degradation of both physical and functional spaces, with negative impacts on services and user satisfaction, to property devaluation and to the declining image of the institutions themselves.

Some reasons for this traditional attitude are related to the absence of legislation requiring long-term consideration of maintenance, as for example, the Italian law 109/94 which requires the development of maintenance plans for every public work's design, providing the scheduled maintenance activities for guaranteeing over time the wanted functionalities, characteristics of quality, efficiency and economic value [1].

The new Portuguese building regulation, still under discussion, introduces requirements for durability and maintenance purposes. According to it, the development of an inspection and maintenance building manual as part of the execution project is mandatory. Three definitions are introduced in this new regulation: building service life, component service life and service life required by the owner/promoter. The building service life corresponds to the period during which the structure shows no degradation of materials. The building component's service life must be set by the manufacturer. The building's service life is a requirement set by the owner/promoter and, when it is not defined, it is considered, by default, 50 years.

Legislation relating to safety and health on temporary or mobile construction site forces the owner/promoter to provide a building technical compilation that includes all the necessary information to take into account in the use, conservation and maintenance activities, taking into consideration the safety and health of those who deliver them. This document shall include in particular:

- Complete identification of all actors;
- Technical information on the overall project and specialities, including written technical specifications, as built drawings and structural aspects, services, systems and materials that are relevant to the prevention of occupational hazards;
- Technical information relating to installed equipment that is relevant to risk prevention concerning its use, conservation and maintenance;
- Necessary information to plan the maintenance activities involving access and circulation with particular risks for the workers.

Although this legislation was published in 2003, the preparation of the building technical compilation is far from current practice. The existence of other priorities related with lack of time, meeting deadlines already short, the intervention of many actors with various scattered information, are presented as an obstacle to the development of this important document.

2.1 Reasons for the Turning Point

Nowadays there are more reasons for an increased attention to buildings operation and maintenance phases in Portugal. There is, in general, a very sharp decrease in the construction of new buildings, especially for housing purposes. Small construction companies have to adapt to this new market reality to survive.

In the public sector, the economic considerations associated with the management of property assets, are on the agenda in face of the present severe financial constraints. The public owners have been aware of the costs associated with the working, operational and maintenance activities. Since it is economically important to the building total cost, the maintenance management cannot be done in an improvised and casual way. It is necessary to start adopting practices aiming at a sustainable approach by rationalizing and optimizing the available resources through the implementation of integrated and cost-effective solutions, allowing an acceptable performance of buildings throughout its life cycle.

Finally, with the emergence of public–private partnerships for the health area in 2002 and the modernization program for the secondary school buildings in 2007 there was an increased need of knowledge on best practices in building maintenance.

2.1.1 Public–Private Partnerships

In 2002, a program of public–private partnerships (PPP) for the health area was created. The PPP model, established by Decree-Law No. 185/2002, provides for the hiring of a private entity to perform a global mission, covering design, construction, financing, maintenance and operation of building hospitals and also the general management of the hospital including the provision to clinical services.

This new procurement model requires a clear specification of the Operation and Maintenance Manual (O&M) and the definition of life cycle programs for the facilities, before its construction. The private facility management is responsible for establishing all the procedures, activities and resources deemed necessary to maintain the good performance of the buildings and the quality of the services provided throughout the duration of the contract. After 30 years (maximum period allowed) the building shall revert to the public partner in perfect conditions.

This innovative approach, forced the study of maintenance requirements, during the design phase, considering the building life cycle and its related costs. However, the lack of reliable information within the Portuguese construction industry also raised some relevant issues such as what types of planned maintenance work must be performed in order to keep the components and systems in good condition, with which frequency it must be performed or when is the end of the service live (total replacement).

2.1.2 Modernization of Secondary Schools

In 2007, the modernization program of secondary schools was launched by the government. This program aims to restore and modernize school buildings. The procurement format included the provision of maintenance services for a period of 10 years. Services to be provided by the contractor include preventive maintenance, corrective maintenance, functional maintenance and large conservation.

The preventive maintenance includes works to be completed according to Preventive Maintenance Plan and necessary to prevent and reduce the degradation of operability functions and of school spaces. The maintenance plan is drawn covering all the actions on a monthly, quarterly, semiannual and annual base.

The corrective maintenance includes works that aim to correct any anomalies resulting from misuse or natural degradation of facilities and equipments.

In functional maintenance the work to be done at the request of the school buildings' management entity is included. The large conservation includes the operations, to be held 5 and 10 years after the end of the building rehabilitation, to restore the functional, safety or accessibility conditions, if appropriate.

3 Collecting Planned Preventive Maintenance Information

Information on planned maintenance activities, currently performed in public buildings, is scattered in several divisions and governmental and municipal entities, responsible for its management and use. It is necessary to gather and process available information to use it in buildings' life cycle analysis, under real conditions.

The implementation of a monitoring system applicable to a sample of buildings would also be desirable. In 1980, Urien [2] referred to the research work conducted by CSTB, aiming at the development of methodologies to determine the maintenance and operational costs of several types of buildings and to relate these costs with technical solutions adopted at the design phase. An observation system covering 30,000 dwellings was set up, with construction age ranging from 10 to 46 years and about 190 different maintenance activities operations were collected. The results for exterior finishes, roofs and exterior openings elements and some related data for life cycle costs purposes were presented.

On the other hand, there is a lot of technical information in the Portuguese construction sector, but it is scattered in several study areas, such as materials, components or systems used in buildings. Much of this documentation include, for specific components, the main maintenance activities to be performed and the main factors that may affect their performance. So it is important to collect the scientific information covering the national regulatory documents and national maintenance recommendations, as a starting point for a national reference database.

Neely and Neathammer [3] presented in 1991 the methodology used in preparing four databases with the information needed for determining life cycle costs of building facilities depending on functional use. These databases were developed using expert knowledge and practical and professional experience. The study follows the governmental decision to construct army related facilities with the lowest life cycle cost and not facilities that have the lowest initial construction cost. The absence of maintenance data to support this kind of analysis involved a great initial work, including:

- Identify all of the possible components that could exist in buildings of different functionalities;
- Identify all tasks that had to be performed to maintain the component in standard operating order. This work was developed by professional in each trade. Task resource information was developed for each task and recorded in a format containing: labour hours, material costs and indirect labour hours to cover job planning, material collection or breaks. Three frequencies of task occurrence were developed. The high frequency is defined as the earliest time a task would have to be performed. The low frequency is defined as the latest time a task would have to be performed, and the average frequency is defined as the most probable time of occurrence.

This last information was introduced in the most detailed database containing the labour, equipment and material hours per square foot. The second database contains both yearly component replacement and annual reoccurring maintenance cost per square foot by building age and building use, e.g. aviation unit maintenance hangars, hospital or child support center. The third database contains the labour, equipment and material hours per square foot by building age and building use. An average cost by facility type, in costs per square foot (4th database content) and maintenance resources requirements were obtained for the first 120 years of building life.

Knowing this information allows determining useful data, namely:

- Predict an annual cost when only the building floor is known, when the floor area and the current functional use is known (different level of information during design phase development);
- Predict an annual cost when the floor area, current use and age of the facilities are known (important for building stock annual financial management).

Finally it is worth noting that one objective of this study was to determine the high-cost components and tasks that can be assumed to be cost drivers in building maintenance—the major cost and replacement tasks. This information could be used to determine the least amount of data required to be known about a building to develop accurate resource prediction and could also be used to identify possible areas of future research to reduce the total building maintenance cost (resource reduction through technologies advances or productivity improvements) [3].

3.1 Some Key Documents

In 1993, the Architectural Institute of Japan published the English edition of the "Principal Guide for Service Life Planning of Buildings" [4]. It concerns the fundamental concept of durability within each stage of the life cycle of buildings, such as planning, design, contract, construction, utilization, maintenance and modernization and demolition. Responsibilities are assigned to all stakeholders: owner, designer, constructor and user. It is stated that the maintenance planning to ensure the appropriate level of performance and the durability of a building is to be provided by the designer, and the characteristics of this planning are preferably notified to the client and the constructor.

The guide provides recommended classes of planned service life for whole building and different parts of building elements and components. This document takes into consideration not only the physical deterioration, but also aspects concerning the flexibility and obsolescence of the building [4, 5].

Since the prediction of the service life is an important item in the design for durability, it is stated that it involves difficult problems, to be solved in the future, and therefore at present there is no choice but to continue the efforts for estimating the service life as correctly as possible based on the present available knowledge and experience.

The guide provides some examples of the method for predicting service life of some building elements and components determined by deterioration factors: factors relating to the inherent durability characteristics (performance of materials, design level, works execution level, maintenance level) and factors relating to deterioration (site and environment conditions and building conditions).

The Canadian Standard S478-95 (R2007) [6] provides guideline on durability of buildings. Durability requirements and quality assurance are emphasized as an essential consideration in every stage in the design service life of any building as, for example, design (detail, specifications), tendering, construction, handover, operation and maintenance and renovation.

It contains generic advice on the environmental agents and other mechanisms that have an impact on the durability of building components and materials, and provides advice for incorporating requirements for durability in the design, operation, and maintenance provisions for buildings and their components. It identifies the need to consider long term costs.

The British Standard BS 7543:2003 [7] gives guidance on durability, required and predicted service life and design life of buildings and their components. It also gives guidance on presenting information on the service life and design life of buildings and their components and/or parts when a detailed brief is being developed. The various conditions that influence the durability prediction and building life include workmanship and maintenance and the practical problems of handling, installing and inspection of components in a building site. Annex A gives information on the prediction of durability for buildings and parts of buildings and guidance on the way agents can affect service life. Agents that can affect the service life of building components and materials: temperature, radiation (solar radiation, thermal radiation), water, normal air constituents, air contaminants, freeze/thaw, wind, biological factors, stresses factors, chemical incompatibility factors and use factors.

This standard says that the maintenance, repair and replacement of buildings and parts of buildings provide a wealth of experience on durability. However, in all but a few cases, there is a lack of systematically collated data that can form a basis from which durability can be predicted with any certainty.

The design life is defined by the designer in discussion with the client and aims to support the client's performance specifications and having the following information is essential:

- Time (or a time measure such as running time or cycles of use) against which durability is to be assessed;
- Conditions in which the item will have to perform (i.e. environmental conditions and levels of maintenance and use);
- Performance point at which functions become unsatisfactory.

The standards' family ISO 15686 provides the actual framework and guidance for buildings and constructed assets service life planning. It consists on:

- Part 1: General principles and framework. Identifies and establishes general principles for service life planning and a systematic framework for undertaking service life planning of a planned building on construction work throughout its life cycle. Reference service live (RSL) is defined as a documented period in years that the component or assembly can be expected to last in a reference case under certain well-defined service conditions. RSL is not defined and it is anticipated that national and international guidance will be developed through collaboration between owners, suppliers, materials specialists and constructors;
- Part 2: Service life prediction procedures. Describes generic procedures for testing the performance of components, materials and assemblies to provide service life predictions;
- Part 3: Performance audits and reviews. Deals with measures to ensure that the life care of a constructed asset is considered through each stage of decision making from initial briefing, through design and construction, to occupancy and eventual disposal and reinstatement of the site;
- Part 4: (under development) Service life planning using IFC based Building Information Modeling;
- Part 5: Life-cycle costing. Gives guidelines for performing LCC analysis of buildings and constructed assets and their parts;
- Part 6: Procedures for considering environmental impacts. Describes how to assess, at the design stage, the potential environmental impacts of alternative designs of a constructed asset. It identifies the interface between environmental life cycle assessment and service life planning;
- Part 7: Performance evaluation for feedback of service life data from practice. It aims to describe a generic methodology, including the terms to be used, that provide guidance on the planning, documentation and inspection phases, as well as on analysis and interpretation of performance evaluations, both on the object (single building) and network (stock of buildings);
- Part 8: Reference service life. Provides guidance on the provision, selection and formatting of reference service-life data and on the application of these data for the purposes of calculating estimated service life using the factor method. The standard gives the following criteria to the data records [8]:
  - General information;
  - Scope (including purpose);
  - Material/component;
  - Methodology;
  - Reference in-use conditions;
  - Degradation agents;
  - Critical properties and performance requirements;
  - Reference service life;
  - Data quality;
  - Reliability of data;

  - Additional information considered;
  - References;

    The standard also gives the rules for validation of data sources that are not fully in accordance with the standard. Depending on the quality of the data source, a laborious process with more extensive research and validation by experts has to take place [8];

- Part 9: Guidance on assessment of service life data. Gives guidance for the derivation and presentation of reference service-life data. It is applicable to manufacturers or producers that provide reference service life data for use in service life planning in accordance with ISO 15686-1; ISO 15686-2, ISO 15686-3, ISO 15686-5 and ISO 15686-6;
- Part 10: When to assess functional performance. Establishes when to specify or verify functional performance requirements during the service life of buildings and buildings—related facilities, and when to check the capability of buildings and facilities to meet identified requirements;
- Part 11: Terminology.

3.2 Some Available Databases

In 2003, Chew [9] presented a research carried out by the National University of Singapore (NUS) and the Building Construction Authority (BCA) in Singapore [10]. Some of the outputs of this study were:

- Database called "Defect library" with information about types of defects and their causes, maintenance and diagnostic methods, good practices and guidelines which can be applied during the design and the construction stage to control repetitive problems;
- Database called "Material manual" with information on the performances and durability of materials under tropical conditions;

This "Material manual" offers information about building materials used in façade, internal wet areas, basements and roofs. It focuses on the various performances including durability, sustainability and clean ability of materials under tropical conditions [9]. These two databases are available on the internet [11].

In the UK, The Building Life Plans (BLP), presented by Mayer and Bourke in 2005 [11], have made available on the internet a construction durability database comprising:

- Durability data on over 800 component types, listed for fabric and building services;
- Durability rankings for component types described generically by criteria which determine durability;
- Adjustment factors which may increase or decrease durability;
- Maintenance and inspection requirements;

- Design, installation, commissioning assumptions;
- Key failure modes and key durability issues;
- Notes about durability issues specific to components and related references;
- Detailed methodology for using durability rankings to determine expected service lives.

The BLP durability assessment was developed based on BLP's 15 years research and practices experience in assessing component durability for latent defect warranty purposes. Durability data which was already in a published format was inputted into a database and an opportunity was taken to cross link durability with construction industry standard codes. The information sources used was based on HAPM methodology and in information generally available in the public domain. Broadly listed in order of reliability [11]:

- International, European and British Standards;
- Authoritative publications and independent certifications;
- Trade associations;
- Manufacturers;
- Practical and professional experience.

For building services systems the "Building services component life manual" [12] presents a total of 20 components, divided into a number of subtypes which represent the most common types, materials and/or configurations available in the UK. For each component subtype, a series of distinct specifications benchmarks, or quality levels, is provided and each description is assigned a life assessment (ranging from 5 to 35+ years). A further life class (designated "U" for unclassified) is assigned to components that fail to comply with relevant British or European standards or are unsuitable for the purpose specified, or where there is insufficient information provided to enable a life to be allocated.

The life assessments assume compliance with good practice in design and installation, a normal level of maintenance, and typical exposure and use conditions. When the service life is likely to be affected by changes in these factors, positive and negative adjustment factors are provided to enable lives to be amended. This methodology is similar to the followed by HAPM and BPG Component Life Manuals.

For each component the expected service lives for a set of related durability factors and assumptions which are made explicit are presented. For each component, material or assembly the key failure modes and durability issues are presented, which provides a framework for failure mode effect and criticality analysis which underlies the durability rankings assessed for each component (see Table 1).

In the Netherlands, the Dutch Building Research Institute (SBR) published in 1995 a catalogue of the reference service lives of building components. It gives reference service lives of roughly 600 common building components and building services. Data was gathered from various sources and judged by experts. Only service lives on which consensus was reached by these experts were included [8].

Table 1 Extract of cold water pipe work of distribution pipe work table information [12]

Cold water pipe work	
Years	35+
Description	Stainless steel pipe work for water applications to BS 4127 (Specification for light gauge stainless steel tubes, primarily for water applications). Fittings to BS 4825 (Stainless steel tubes and fittings for the food industry and other hygienic applications)
Inspection	Long cycle inspection for external corrosion/damage and joint/fixing integrity
Maintenance	Generally, pipe systems require little or no maintenance if correctly specified. Any maintenance will be in the form of periodic repainting where materials such as cast iron and some types of plastic are exposed to the elements or are installed in a corrosive setting
Key failure modes	External corrosion; internal fouling, scaling, corrosion and vibration, leading to tube leakage and cracking; damage due to excessively high liquid velocity, temperature and/or pressure; reduced flow rate due to partially or fully blocked pipe work
Key durability issues	Corrosion resistance of base materials; overall water quality and suitability of water treatment used, quality of handling, installation and commissioning; quality of protection and/or protective coating on materials to prevent corrosion from the environment in which they are located
Adjustments factors	Installed in adverse (but not severely corrosive) environments: 5 years; Not sleeved through walls: 5 years (not cumulative: the factor that is the largest should be applied)
Assumptions list for design and installation and commissioning are provided	

A review of this service life catalogue is being made by expert judgements in order to publish a new Dutch service life database. This new publication provides service lives of 600 general buildings components, covering: substructure and frame, external walls, upper floors and floor finishes, roofs and roof finishes and window and external doors. Paints are not included as separate products as paintwork is an activity necessary to maintain the life of a large number of construction groups in external walls, and windows and external doors. Internal components, fittings, sanitary appliances and building services were left out of this document. Table 2 gives an extract of the example, provided by Straub [8], for flat roof coverings.

Table 2 Extract of an example of service lives of building components [8]

Flat roof coverings		RSL
Metals	Steel, trapezium, galvanized, coated	50
	Steel, trapezium, galvanized	15
	Aluminium, folded, enamelled	60
	Aluminium, folded, coated	40
	Copper, folded	75
	Zinc plates, 10 % overlapping	25
	Zinc	40

The CSTB and Politecnico di Milano are structuring an international RSL database-proposed tool for building materials and components data collection. This RSL database has been developed to collect a series of grids in which set of RSL are stored and indexed. The set of RSL consists of [1]:

- The duration in years, choosing among different type of RSL distributions;
- The failure mode;
- The selection of the several levels of factors in the grid, according to ISO 15686 factor method.

Complementary data such as year, place, sources, data quality and observations are also provided. RSL according to collected data can have various origins [1]:

- Experience;
- Ageing tests in natural environment;
- Accelerated ageing tests;
- Numerical simulation;
- Bibliographical studies.

3.3 Detailed Description of Four Guidance Documents

Four databases with data on maintenance activities were consulted: a French source "La maintenance des bâtiments en 250 fiches pratiques" [13], a Spanish source "Fitxes rehabilitació of ITeC" [14–16], an English source "Housing component life manual" [17] and a North American source "The Whitestone facility maintenance and repair cost reference 2009–2010" [18].

3.3.1 Spanish Source

The *Institut de Tecnologia de la Construcció de Catalunya—IteC* published in 1991 three books in the areas of building [14], services [15] and urban spaces maintenance [16], respectively. These documents are made available on line by the IteC website and its main goal is to provide in a simple and practical way, the basic information related with buildings maintenance in the Spanish province of Catalonia.

In the first volume, the building is divided into exterior elements, roof covering elements and exterior finishes, windows and doors, and interior elements, covering the finishes, the interior doors and the sanitary and kitchen equipments. A total of 67 elements information are presented in an individual sheet format containing four sections:

- List of the most common anomalies;
- The principal maintenance activities and their frequencies;
- The service life;
- The main factors that causes degradation of the element.

The list of the most common anomalies attempts, according to the authors, to give an idea of the common problems that can cause degradation of each element. This section has an informative function but it can also help detect the possible pathology of the element and find the best way to face the task of maintenance.

The exposure to specific environmental conditions or use factors, defines the risk of degradation of the element which can be high (H), medium (M) and low (L). The level of maintenance required (frequency of execution of maintenance activities) and the element service life, are related with the risk of degradation. Table 3 presents the main factors considered for the classification of the risk of degradation for different kind of elements.

In Fig. 1 an example of the information sheet for an anodized aluminium window [14] is presented. 0 in the periodicity column of a maintenance operation means that this is done regardless of the degree of maintenance.

The volume related to services maintenance includes: smoke evacuation installations, ventilation and waste disposal facilities, HVAC, kitchen facilities, water supply and sewer facilities, valves, pumps and pressure groups, electric and lighting systems, transportation facilities, fire protection facilities and installations of audio and video.

The volume of urban areas includes: fences and gates, paving elements, equipments, services and landscaping.

IteC developed two software packages, one of them for maintenance management of one or few buildings and the other for a large building stock. These informatics applications create all the information related with the buildings, such as the contractual information, stakeholders list, execution development report, products, systems and equipments installed. All technical specifications, containing service life information and use and maintenance instructions, and "as built" project must be included. The maintenance plan is generated according to the constructive solutions adopted, containing the schedule of all the maintenance, technical and human resources, costs, regulatory obligations, requirements for

Table 3 Type of factors considered in risk definition "Manteniment de l'édifici. Fitxes" [14]

Element	Factors defining the risk of degradation
Pitched roof claddings	Rainfall (high, normal or low), temperature range (sudden or smooth), risk of accumulation of leaves and debris (high or low)
Exterior finishes—mortar	Humidity and risk of erosion (high and low) and accessibility to people and cars
Doors/windows: anodized aluminium	Humidity (high, low), exposure to wind and strong winds (low exposure to wind) accessibility to people and cars and sun exposure (high or low)
Interior ceramic floor finishes	Stairs, public use, corridors, work areas and overloads (private use and particular house)
Interior wall finishes stone	Harsh environments for the stone composition, temperature range, humidity and light-coloured stones
Sanitary equipment	Public areas, private areas and occasional use (parking, stores)

WINDOWS. ALUMINIUM
ANODIZED ALUMINIUM WINDOWS

POSSIBLE ANOMALIES: Surface dirt
Small deterioration in the finish
Joints sealing problems in opening elements
Defect in the functioning of mechanisms

MAINTENANCE ACTIVITIES:

PERIODICITY	H	M	L
Cleaning the frames and glass with appropriate product	0	1	0
Inspection of the state of the finish	0	1	0
Minor repairs in the finish	0	1	0
Inspection of the condition of ironworks	2	4	6
Inspection and lubrication of mechanisms	0	1	0
Inspection of the tightness of the joints	0	1	0
Inspection of drainage	0	1	0

SERVICE LIFE:

	H	M	L
Replacement of window	40	50	60

H High level of
High humidity
High exposure to wind
Acessble to people
High exposure to sun

M Medium level of maintenance

L Low level of maintenance
Low humidity
Low exposure to wind
Not acessible
Low exposure to sun

Fig. 1 Information for the anodized aluminium exterior windows [14]

certification of products, personnel or entities and conditions for hiring of maintenance.

3.3.2 French Source

The French book "La maintenance des bâtiments en 250 fiches pratiques" [13], published for the first time in 1995, presents maintenance activities for 270 elements, distributed in 9 major groups: structure, exterior envelope, roofing and waterproofing, interior construction, plumbing, HVAC, electrical equipments components, other electrical installations, basements and other services and external works and landscaping. Each sheet presents the following information:

- The maintenance activities and frequencies, broken down into five levels: inspection, cleaning and preventive maintenance, light intervention, heavy intervention and replacement;

- The cost of preventive and corrective interventions expressed in percentage of the cost of new construction;
- The regulations and normative references.

In Table 4 an example of the information sheet for an aluminium window [13] is presented.

This guide also presents an evaluation grid to use in building condition assessment. Knowing the building physical condition is an important indicator (among others) to be used in the implementation of a maintenance management system.

Ten summary tables are also provided in this book to help in the analysis of buildings with regulations, covering: asbestos, disabled person, *legionella*, plumbing and painting with lead, fire safety and electrical installations. This analysis gives an important feedback for the definition of maintenance activities, requiring a periodic monitoring by the building stock manager.

Table 4 Aluminium windows sheet information adaptation from Albano [13]

Maintenance operations	Periodicity	Ratio (%)
Inspection	Semester	1
Detailed examination of the windows		
Mechanical resistance and finish window (frame, openings and glazing) inspection		
Control the proper functioning of all mechanisms and hinges		
Open the windows to check its functionality		
Check the water drainage mechanisms		
Preventive maintenance	Annual	1
Cleaning the frames and openings with proper product		
Mechanisms and hinges should be lubricated according to manufacturers recommendations		
If the opening is difficult adjustments should be made in mechanisms (hinges, rollers, sliding components)		
Clean and unclog drains and ducts		
Minor intervention	Whenever necessary	5
Diagnosis: poor orthogonality and difficult in opening movement		
Intervention: verification and if necessary replacement hinges		
Heavy intervention	10 years	35
Diagnosis: cracks between the frame and masonry; infiltration (frame and sill interface); poor orthogonality and difficult in opening movement		
Intervention: treatment of joint between frame and masonry and frame and sill and mechanisms treatment/replacement if necessary		
Replacement	25 years	100
References DTU 36.1 and DTU 37.1		

3.3.3 UK Source

The HAPM (Housing Association Property Mutual Ltd) manual was produced by Construction Audit Limited and published in 1999 in the UK [17].

Although this document has been prepared specifically for insurance purposes its use goes far beyond them. The information is structured into the seven following groups of components: flooring components, walling and cladding components, roofing components, doors, windows and ironworks, mechanical equipment components, electrical equipment components and external works.

In general, each group is divided into types and sub-types that are assigned a score corresponding to one of the following insured lives classes: A = 35+ years, B = 35 years, C = 30 years, D = 25 years, E = 20 years, F = 15 years, G = 10 years, and H up to 5 years. There is also a life class designated "U" (meaning uninsured) which is used where the component does not comply with British Standards or is unsuitable for the purpose specified, or where there is insufficient information presented to allocate an insurance life class.

The life classes embrace good practice, a normal amount of maintenance and typical exposure. General guidelines also include installation in accordance with manufactures' direction, relevant code of practice and British standards and the use of appropriate design details. When non-typical or extreme conditions prevail, adjustments factors are provided. These may be negative or positive depending on whether the deviation from the norm is detrimental or beneficial. Examples of such conditions are indicated in Table 5. Where the life of a component may vary depending on its location in the building (e.g. ground floor, intermediate floor) alternative component life columns are given for different locations.

Data sheets on each component group include a description of each component type, adjustment factors, assumptions, locations, typical maintenance, and specific notes. Table 6 presents an extract of the data in HAPM for aluminium windows.

Table 5 Definitions for the adjustments factors [9]

Conditions	
Normal environment	Inland, with normal urban atmospheric pollution only
Polluted/industrial environment	With airborne sulphur dioxide, acid or alkali pollution, normally from an industrial source
Marine environment	Coastal areas subject to salt spray and/or sea water splashes. It may extend up to 3 km from the coast or tidal estuary depending on prevailing wind and topography
High risk frost locations	Coastal areas subject to salt spray and/or sea water splashes. It may extend up to 3 km from the coast or tidal estuary depending on prevailing wind and topography
With 3rd party assurance	A product with a certificate indicating that ongoing testing and assessment of the product's suitability and/or adherence to claimed standards has been carried out by an independent 3rd party such as the BBA or others

Table 6 Aluminium windows sheet information adaptation from HAPM [17]

Type: windows	Sub types: aluminium
Description	Aluminium windows kite marked to BS 4873 (Specification for aluminium alloy windows); Anodised or liquid organic coated or polyester powder coated. Windows supplied under LHC (London Housing Consortium) bulk quotation arrangement A1
Years	35 (Location: external masonry walls)
Years	35 (Location: external walls timber framed)
Maintenance	If painted, repeat every 5 years. Renew weather-stripping and gaskets every 10 years. Renew hardware every 10 years

3.3.4 USA Source

The "Whitestone Facility Maintenance and Repair Cost Reference 2009–2010" [18] provides maintenance and repair costs from various sources of information and various types of North American establishments. This reference represents the result of 14 years of continuous work from the Whitestone Research Company with several US government agencies and consulting firms that work in this area. This collaboration yielded the information needed to calculate the cost of maintaining a building over its service life, the amount and type of resources associated with this maintenance and the lifespan of the various constituents of the building.

The document uses the North American classification system UNIFORMAT II (classification of asset elements), published by *ASTM*—American Society of Testing and Materials (ASTM E1557 2009 [19]), which includes the following items: substructure (foundation and basement construction), shell (super structure, exterior envelope and roofing), interiors (interior construction, stairs and interior finishes), services (lift, plumbing, HVAC, fire protection and electrical) and equipment and furnishings.

The data presented refer to maintenance and repair activities (M&R) and are divided into: preventive maintenance (PM), unscheduled maintenance (UM) and replacement (R). PM and minor repair consists on scheduled tasks that sustain a component's level of service during a prescribed service life. The UM consists of service calls, emergency response, and other tasks that cannot be individually anticipated and replacement consists of component overhaul or major replacement tasks. These tasks extend a component's service life, and reset the schedule of PM and minor repair tasks. Activities related to facilities operation, such as landscape maintenance, are not included in this book.

The book features 72 maintenance profiles for various types of facilities and equipment, for a study period of 50 years, with the costs of maintenance and repair, per m^2 of building area and as a percentage of replacement value. An example of a maintenance profile of a North American primary school is given in Table 7. The school, of 3.7 m height, has a gross internal floor area of 4,360 m^2, built of reinforced concrete, screed, carpet and vinyl floor tiles and ceilings with

Table 7 50-Year M&R cost summary for an elementary school [18]

Task type	50 year total cost	Annual cost per GSFT	Annual cost as percentage of replacement (%)
PM & minor repair	$1,389,231	$0.59	0.29
Unscheduled maintenance	$1,070,522	$0.46	0.22
Renewal & replacement	$5,470,564	$2.33	1.13
Total	$7,930,317	$3.37	1.63

plaster finish. The building has a replacement cost of $ 9,701,531 [18]. Distribution of maintenance and replacement cost is presented in Fig. 2.

Table 8 present an extract of the information contained in the USA source for aluminium windows.

3.3.5 Summary of the Collected Data and the Main Application Fields

In Table 9 a summary of the data and the main application fields for planned maintenance activities are presented.

4 Preventive Maintenance Activities: Case Study

A study was carried out by Raposo [20] to develop an analysis tool that integrates technical, economic and organizational factors in the implementation of a Maintenance Management System (see Fig. 3).

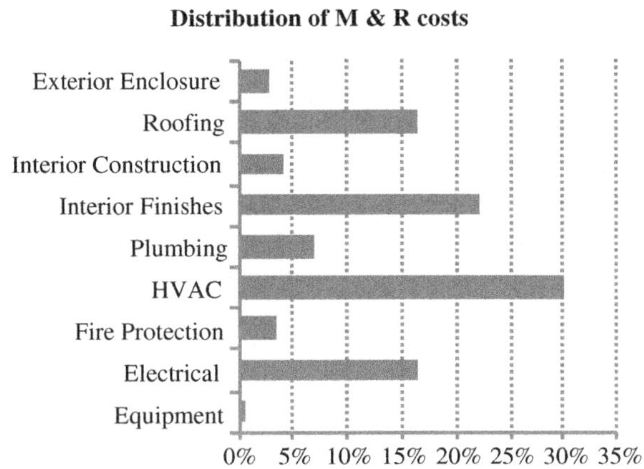

Fig. 2 Distribution of maintenance and replacement cost [18]

Table 8 Exterior aluminium windows sheet information

Aluminium operable window, 24 square feet					
Labour hours	Material costs	Equipment type	Frequency	Task type	Replacement task
Repair (replace) 1st floor—each—trade: carpenter—labour type: contract					
0.330 (2.540)	4.99 (909.66)	None	15 (75)	Major	No (yes)
Repair (replace) 2nd floor—each—trade: carpenter—labour type: contract					
0.330 (3.240)	4.99 (909.66)	None	15 (75)	Major	No (yes)
Repair (replace) 3rd floor—each—trade: carpenter—labour type: contract					
0.330 (3.940)	4.99 (909.66)	None	15 (75)	Major	No (yes)

Adaptation from Whitestone [18]

Table 9 Type of data available in the references listed and application fields in the research study

Type of information available	[18]	[13]	[17]	[14–16]
Service life	Yes	Yes	Yes	Yes
Factors influencing the service life	No	No	Yes	Yes
Description of maintenance activities and their periodicity	Yes	Yes	Yes	Yes
Factors that influence the maintenance activity	Yes	Yes	No	No
Human resources and labour hours per maintenance activity	Yes	No	No	No
Data to calculate the maintenance activity cost	Yes	Yes	No	No
Most frequent anomalies in components and building systems	No	No	No	Yes

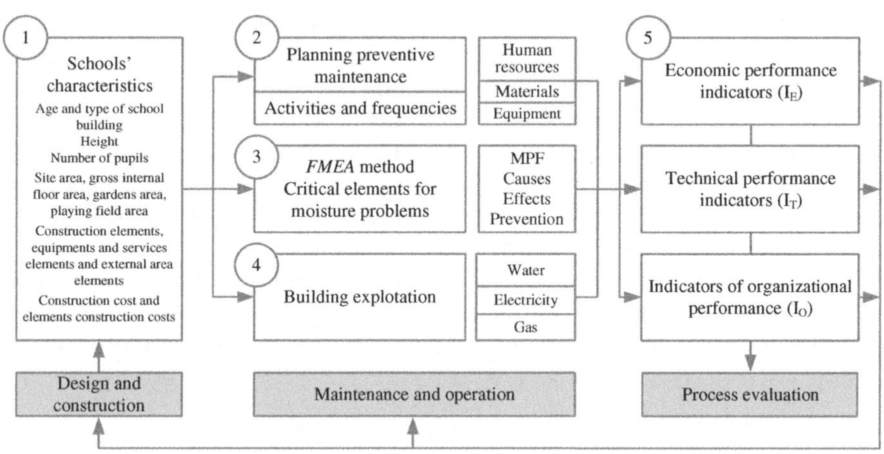

Fig. 3 Maintenance management system (MMS)

In total 17 key performance indicators were defined covering economic and technical performance (IEE, ITT), organizational development (IOO). This methodology was tested on a sample of five primary schools (EB1) in Lisbon.

The planned preventive maintenance activities and the statutory planned preventive activities were established through international references and national regulations consultation and national manufacturer's opinion.

The four databases referred to in detail were consulted and information on preventive maintenance activities was collected, including routine inspections (I), maintenance and cleaning routine (M), and component replacement (R).

4.1 Building Envelope Components

In the five schools studied the exterior envelope is made of double walls with hollow ceramic brick masonry and thermal insulation and exterior windows/doors with anodized aluminium frame and colourless single-pane glass (Fig. 4).

The walls should be subjected to regular inspections every three years and the windows and doors every six months. The condition state of the finishing of the frame, operation and state-setting mechanisms for openings and fittings, condition of gaskets and mastics and verification of clearance at outlets for water flow should be followed [13].

Preventive maintenance of the exterior walls consists in their cleaning and eventual closing of cracks and joints, whenever the conservation status of the walls requires it or once every 10 years [18].

The exterior walls service life exceeds 50 years of the research study, and according to Abate et al. [18] it exceeds 75 years, whereas Albano [13] does not mention any period. The outer openings require a yearly maintenance of hardware replacement, lubrication of door locks and joint replacement. The replacement of windows and doors is indicated every 25 years in Albano [13], with replacement of mastics to occur between 10 and 15 and the fittings around 15 years. The HAPM [17] and Abate et al. [18] differ in terms of the service life of window elements, indicating 35 and 75 years respectively, and of the service life of exterior doors, proposing 30 and 50 years respectively.

Roofs are predominantly of sheet steel and reversed flat roof types. The former must be annually inspected, performing at the same time preventive maintenance activities, such as cleaning the surface and checking the tightness of fasteners;

Fig. 4 **a** São Bartolomeu school and **b** Windows and façade

unclogging the rainwater plumbing and checking the proximity of tree branches. For the latter, inspections should be carried out every six months with maintenance activities to be held annually.

The metal roof's service life is 25 years [13], whereas IteC [14] makes a distinction between the roof metal covering's service life (30 years) and the support structure repair (50 years). The service life of flat roofs depends primarily on the conservation status of the layers of support, thermal insulation and waterproofing; Albano [13] suggests 20 years.

Skylights should be inspected annually by verifying the conservation status of the window frames, glass, gaskets condition and mechanisms for opening and closing the windows. Maintenance activities, to be held twice a year, consist of cleaning, lubrication and adjustment of all moving parts.

The rain water drainage system should be inspected and cleaned every six months, checking the condition of connections and pipes and all system performance, namely the water flow. Albano [13] refers the entire system replacement after 15 years, whereas Abate et al. [18] mentions the need to replace up to 20 % of pipe after 30 years and total replacement after 40 years.

4.2 Building Interior Components

The interior walls are in general made of single pane perforated ceramic brick masonry (in thicknesses ranging 0.20–0.35 m) and should be subjected to regular inspections every three years checking the cracking widths, deformations and moisture stains.

Ceilings in gypsum board are an uncommon solution in school buildings, appearing occasionally in recent schools in circulation areas or in the lobbies of the classrooms. Annually one should proceed with a visual inspection/maintenance, to check the aspect and the attachment to the support structure. Replacement occurs after 20 years, during which one should be aware of its possible deformation or the existence of water originating from the underlying services.

The interior openings are mostly in wood finish, with sliding doors in the separation between the classrooms and areas of artistic expression. Control inspections should be performed annually, in general checking all moving elements. Maintenance should be held biannually with the lubrication and adjustment of the moving parts and hardware. The service life of the doors was considered to be in general 20 years and that of sliding doors 15 years.

4.3 Exterior and Interior Finishes

The exterior cladding of the facades' schools is predominantly painted plaster. The control inspections should take place every three years, checking the general appearance of walls, in particular the aspect of painting and the existence of cracks and spots [13]. The current maintenance activities include cleaning stains and dirt in localized areas. IteC [14] refers a periodicity of 15 years for repainting walls and ceilings and at 40 years a more in-depth intervention on the plaster must be performed before repainting.

In external walls, ceramic tiles, varnished concrete and wood panels are used less frequently. In ceramic tiles the superficial appearance of walls, joints and the flatness of the surfaces should be checked. The main maintenance activity consists on cleaning surface dirt and efflorescence stains. The replacement of this type of coating is usually not considered.

Walls in exposed varnished concrete require regular monitoring at 10 years intervals, to determine the need for surface treatment with waterproofing products, and cleaning maintenance every 20 years. Normally their replacement is not considered [14]. Wood panelling demands an annual monitoring inspection to check the surface condition of the panels, the performance of the support elements and the integrity of joints. It should be cleaned periodically [14]. Replacement of the support elements occurs around 40 years.

The most common coating for walls and ceilings is painted render. There should be an inspection every year to check for cracks or other damage and the adherence condition to the support. For paint coating, the surface appearance and the presence of moisture, dirt stains or detachment of material must be verified. Washing the surfaces is the appropriate maintenance procedure to be carried out once every 3 years. Repainting should occur every 15 years, including the repair of the support. In wet areas and in circulation areas, ceramic wall cladding is used requiring an annual inspection control of the surface aspect of the wall and cleaning of surface dirt and efflorescence stains, where necessary.

The flooring currently adopted for the classrooms and corridors is linoleum (see Fig. 5). In the older schools ceramic or wood flooring can be found. The inspection

Fig. 5 School n.° 34—Alta de Lisboa. Floor finishes

Table 10 Summary of information on preventive maintenance (extract) [20]

Element		I	M	R (years)	NI
Building envelope					
	Exterior walls	3 year	10 year (W)	NC	I = 16; M = 5; R = 0
	Windows	6 m	1 year	25;35(H); 75(W)	I = 100; M = 50; R = 2; 1(H); 0(W)
		1–4 year(I)	1 year(I)	50(I)	I = 50/12; M = 50; R = 1(I)
	Doors	6 m	1 year	25; 30(H); 50(W)	I = 100; M = 50; R = 1(H); 1(W)
		1–4 year(I)	1 year(I)	50(I)	I = 50/12; M = 50; R = 1(I)
	Mastic			10/15	R = 5/3
	Ironworks			15	R = 3
	Joint			10	R = 5
Roofing					
	Metal roofing	1 year	1 year(I, H)	25; 30(I); 40(W)	I = 50; M = 50; R = 2; 1(I, W)
	Flat roof	6 m	1 year (A; I)	20	I = 100; M = 50; R = 2
		5 year(W)	1 year(W)	35(W)	I = 10; M = 50; R = 1(W)
	Skylights	1 year	6 m	20/30; 40(W)	I = 100; M = 100; R = 2/1; 1(W)
	Rain water system	6 m	6 m	15; 40(W)	I = 100; M = 100; R = 3; 1(W)
Interior elements					
	Walls	3 year	10 year(W)	30	I = 16; M = 5; R = 1
	Gypsum board	1 year	2 year	20	I = 50; M = 25; R = 2
	Interior doors	1 year	2 year	20; 50(I); 40(W)	I = 50; M = 100; R = 2; 1 (I;W)
	Ironmongery		5 year	10(W)	M = 10; R = 10(W)
	Sliding doors	1 year	2 year	15; 40(W)	I = 50; M = 25; R = 3; 1(W)
Finishes					
	Painted wall render	3 year	3 year	15(A; I); 10(W)	I = 16; M = 16; R = 3; 5(W)
	Ceramic wall cladding	1 year; 5(I)	20 year(I)	NC	I = 50; 10(I)M = 2; R = 0
	Concrete	10 year(I)	20 year(I)	NC(I); 75(W)	I = 5; M = 2; R = 0(I; W)
	Wood panelling	1 year		30(I)	I = 16; R = 1(I)
	Linoleum flooring	1 year		10; 18(W)	I = 50; R = 5; 2(W)
		3 year(I)		20(I)	I = 16; R = 2(I)
	Wood parquet flooring	1 year	2–5 year	40(A; W)	I = 50; M = 25/10; R = 1(A; W)

of the linoleum flooring is to be done annually (every three years according to IteC) focusing on the verification of the pavement condition in current areas (flatness) and localized areas (adherence and mechanical performance). Flooring requires periodic cleaning and its service life varies from 10 [13], 18 [18] to 20 years [14].

Wood flooring is a solution currently adopted for administrative areas and the monitoring inspection should be held annually, noting the state of conservation of surfaces and signs of moisture. The application of suitable surface products such as varnishes or wax is part of cleaning and routine maintenance. The referred service life of this pavement is 40 years [13, 18].

The floor finishing used in wet areas is ceramic tiles that have a service life of 50 years [14, 18]. It is a type of floor finishing that does not require great frequency of inspection but must be subjected to a weekly cleaning with products suited to its surface.

Table 10 presents a summary (extract) of the information collected on different elements preventive maintenance activities and frequencies and their impact in the number of intervention in building (technical performance indicator-NI). The Albano [13] source was chosen as a basis for the study and the Whitestone-W [18], IteC-I [14] and HAPM-H [17] sources were used to assess the former information.

5 Conclusions

There is a great need for reliable information in the area of building maintenance. Information on maintenance activities in Portugal is still scattered in different study areas that relate to building elements and components and in the various government agencies that manage building parks. In some regulated areas, such as the electrical equipment components or gas installations, Portuguese legislation is unclear on neither the type of maintenance activities nor their frequency of application. Lifts, fire protection and HVAC systems are subject to specific regulations, recently published, which require the existence of user manuals and maintenance plans.

Several countries have developed guides and documentation to support the maintenance activities in its various aspects, such as:

- Common buildings components and buildings services used;
- Performance and durability of materials in normal conditions;
- Service life;
- Factors influencing the service life;
- Maintenance activities description and related frequencies;
- Factors that influence the maintenance activity;
- Human, materials and equipment resources related with maintenance;
- Data to calculate the maintenance activity cost;
- Most frequent anomalies in components and buildings systems.

By consulting various databases, it was found that would be very helpful to have all this information concentrated in one document for a more comprehensive analysis and maintaining the consistence and integrity of the information.

It is known that the major maintenance costs that are incurred over the 50 years of a building's service life result from the renewal and replacement of components and elements at the end of their service life. This information is not always coincident in the different consulted sources and can considerably change the results of a life cycle cost analysis. It appears that there is a difficulty in carrying out benchmarking actions for LCC or LCA assessment purposes. It is important to continue the work under development to standardize this information and create national, European or international references databases.

At a national level, it was found that material specifications are generally poorly detailed (design specification and lack of as built information), which may reflect the adoption of traditional technical solutions with no durability or economic concerns.

References

1. Daniotti, B., Spagnolo, S.L., Chevalier, J.L., Hans, J., Chorier, J.: An international service life database: the grid definition for an actual implementation of factor methods and service life prediction. In: Barrett, P., Amaratunga, D., Haigh, R., Keraminiyage, K., Pathirage, C. (eds.) CIB 2010 World Congress (2010)
2. Urien, R.: Dépenses d'entretien et coût global des ouvrages de bâtiment. Research on maintenance and modernization, CIB/W70, Rotterdam, The Netherlands (1980)
3. Neely, E., Neathammer, R.: Life cycle maintenance costs by facility use. J. Constr. Eng. Manag. (1991)
4. Architectural Institute of Japan: The English edition of principal guide for service life planning of buildings. Architectural Institute of Japan, Tokyo (1993)
5. Soronis, G.: Standards for design life of buildings: utilization in the design process. Constr. Build. Mater. **S0950–0618**(96), 00018–00019 (1996)
6. Canadian Standards Association: S478–1995 Guideline on durability in buildings. Canadian Standards Association, Etobicoke (2001)
7. British Standards Institution: BS 7543 Guide to durability of buildings and building elements, products and components. British Standards Institution, London (2003)
8. Straub, A.: To a new Dutch service life database of building products. COBRA 2011: Royal Institution of Chartered Surveyors International Research Conference, Salford, UK (2011)
9. Chew, M.: Maintainability of buildings. In: 2nd International Symposium on Building Pathology, Durability and Rehabilitation, Learning from errors and defects in building, Lisbon, Portugal (2003)
10. National University of Singapore and Building Construction Authority. Maintainability under tropical conditions. http://www.hpbc.bdg.nus.edu.sg/
11. Mayer, P.D., Bourke, K.P.: Durability rankings for building component service life prediction. In: 10th International Conference on Durability of Building Materials and Components, Lyon, France (2005)
12. Building Life Plans: Building services component life manual. Blackwell Science, London (2001)
13. Albano, J.R.: La maintenance des bâtiments en 250 fiches pratiques. Le Moniteur, Paris (2005)

14. Institut de Tecnologia de la Construcció de Catalunya: IteCa: Manteniment de l'edifici: Fitxes. Barcelona, Spain (1991)
15. Institut de Tecnologia de la Construcció de Catalunya: IteCb: Manteniment installacions: Fitxes. Barcelona, Spain (1991)
16. Institut de Tecnologia de la Construcció de Catalunya: IteCc: Manteniment urbanització: Fitxes. Barcelona, Spain (1991)
17. Housing Association Property Mutual—HAPM.: Component life manual CD-Rom, E&FN Spon (1999)
18. Abate, D., Towers, M., Dotz, R., et al.: The Whitestone facility maintenance and repair cost reference 2009–2010. Whitestone Research, California (2009)
19. American Standard of testing materials: ASTM E1557–09 Standard classification for building elements and related site work: UNIFORMAT II. American Standard of Testing Materials, Philadelphia (2009)
20. Raposo, S.: The management of maintenance activity in public buildings. Model and strategies for a sustainable intervention. PhD thesis, Instituto Superior Técnico (IST), Technical University of Lisbon, Portugal (2011)

Further Reading

21. BRE Construction Division: Estimation of the need to spend on maintenance and management in the Local Authority housing stock. BRE Construction Division, London (2003)
22. Christian. J., Pandeya, A.: Cost prediction of facilities. J. Manag. Eng. **13**(1), ASCE (1997)
23. Damen, T.: Planning and budgeting maintenance activities, Systems of maintenance planning, CIB/W70, Edinburgh (1983)
24. Daniotti, B., Spagnolo, S.: Service life prediction for buildings' design to plan a sustainable building maintenance. In: 11th International Conference on Durability of Building Materials and Components, Istanbul, Turkey (2008)
25. Frohnsdorff, G.J., Martin, J.W.: Towards prediction of building service life: the standards imperative. In: 7th International Conference on Durability of Building Materials and Components, Stockholm (1996)
26. Inose, I.: Building maintenance and management in NTT, Research on maintenance and modernization, CIB/W70, Rotterdam (1980)
27. Lacasse, M.A., Sjostrom, C.: Methods for service life prediction of building materials and components: Recent activities of the CIB W80/RILEM 175-SLM, Workshop on Management of Durability in the Building Process, Milan (2003)
28. Masters, L.W.: Service life prediction. In: A state of the art 4th International Conference on Durability of Building Materials and Components, Lyon (1987)
29. Strand, S.M., Hovde, P.J.: Use of service life data in LCA of building materials. In: 8th International Conference on Durability of Building Materials and Components, Vancouver (1999)

Microstructure of Cement Paste Blended with Micronized Sand (MS)

Ying Wang, Guang Ye and K. van Breugel

Abstract In this article, micronized sand is selected as filler to partially replace Portland cement in order to reduce the global warming potential, i.e. CO_2 emission. It is necessary to figure out the effects of micronized sand on the properties of cement paste. Microstructure, especially the pore structure, influences the transport properties, which are widely recognized as the most important factors to determine the durability of cementitious materials. In this study, it focused on the microstructure of cement paste blended with micronized sand in comparison with the reference ordinary Portland cement paste. Non-evaporable water test was carried out to calculate the degree of hydration of cement. By means of mercury intrusion porosimetry measurements and electronic scanning electron microscope, the pore structure of cement pastes blended with micronized sand as well as interfacial transition zone between micronized sand and hydration products have been investigated. Parameters, including the Portland cement replacement percentages: 10, 20 and 30 % and fineness of micronized sand: M6, M300 and M600, have been taken into account. The results show that although micronized sand is considered as inert filler, it can indirectly modify the hydration kinetics by increasing the degree of hydration of cement and influence the porosity and pore size distribution of the paste. There is a clear interfacial zone between micronized sand and hydration products. This interfacial zone caused by the micronized sand is considered as the decisive factor which influences the transport properties of the material.

Y. Wang (✉) · G. Ye · K. van Breugel
Microlab, Delft University of Technology, Delft, The Netherlands
e-mail: ying.wang@tudelft.nl

G. Ye
e-mail: g.ye@tudelft.nl

K. van Breugel
e-mail: k.vanbreugel@tudelft.nl

V. P. de Freitas and J. M. P. Q. Delgado (eds.), *Durability of Building Materials and Components*, Building Pathology and Rehabilitation 3,
DOI: 10.1007/978-3-642-37475-3_3,

Keywords Micronized sand · Cement replacement · Microstructure · Pore structure · Interfacial transition zone

1 Introduction

Long term properties, i.e. durability of concrete structure are in the center of issues for several decades. It has been proved that durability of concrete is essentially influenced by the processes that involve the interconnectivity of ions or molecules in the form of liquids and gases [1–3]. The passage of these potentially aggressive agencies is primarily influenced by the permeability of the concrete [3], which is greatly controlled by the details of porous microstructure [4]. The most decisive characteristics in pore structure, including porosity, threshold pore diameter, pore size distribution and connectivity of capillary pores, play a significant role in microstructure of cement paste binder. In other words, the total volume, size, shape and connectivity of pores influence on the permeability and diffusivity [1]. Therefore, durability quality of concrete can be assessed indirectly from the knowledge of its pore system characteristics. As the pore structure might be different when cement is partially replaced by micronized sand (MS), some changes might occur in durability behavior. It is significant to study the paste microstructure and to see how it is affected by replacement of MS and to know the influence of MS on pore structure. In this study, different replacement levels and different fineness of MS are applied in the blended material in order to figure out its influence on the microstructure.

1.1 Pores in Cement Paste

Pores in cement paste can be divided into gel pores, capillary pores and air voids [1]. There are several definitions of classification of pores in technical articles. In this study, the classification proposed by [5] is going to be used. In Table 1, the general classification of pores in cement paste according to their average width is proposed. Capillary pores have highly irregular shape and their size could range from very small to large values, i.e. form 2 nm to 10 μm. They also change with time by the precipitation of hydration products, mainly C–S–H gel, in the originally water-filled space. Based on the classification described in Table 1, capillary pores are divided into three types: small, medium and large capillary pores. The borderlines between the different classes are not strict and depend on the shape of the pores. Pores having diameter of 2.5–10 nm are called gel/capillary. Since this small capillary pores have less influence on permeability, in this study capillary pores are defined as medium and large capillary pores, which affect the permeability and have a major effect on transport processes in cement paste [1].

Table 1 Classification of pores in hydrated cement paste, according to Mindess [5]

Name		*Diameter*	*Paste properties affected*
Micropores "inter layer"	Gel Pores	Up to 0.5 nm	Shrinkage, creep at all RH
Micropores "inter layer"		0.5 nm to 2.5 nm	Shrinkage, creep at all RH
Small (gel) capillaries		2.5 nm to 10 nm	Shrinkage between 50% and 80% RH
Medium capillaries	Capillary Pores	10 nm to 50 nm	Strength, permeability, shrinkage at high RH, >80%
Large capillaries		50 nm to 10 μm	Strength, permeability
Entrained air		0.01 mm to 1 mm	Strength

1.2 Definition of Pore Structure Parameters

To thoroughly describe the properties and performance of hardened cement pastes, several parameters are used to determine the pore structure. The parameters most commonly used to characterize the pore structure of the cement-based material are:

- Porosity;
- Threshold pore diameter;
- Pore size distribution;
- Connectivity of capillary pores.

1.2.1 Porosity

Total porosity is one of the most important terms used to define pore structure. It is the fractional volume of pores with respect to bulk volume of the material, which is expressed by Eq. 1. It includes both open and closed pores.

$$p = \frac{V_p}{V_b} \cdot 100\ \% \tag{1}$$

where p is the total porosity (%), V_p is the total pore volume in the bulk material, V_b is the bulk volume of the material.

1.2.2 Critical Pore Size

Critical pore size, d_{cr}, corresponds to the steepest slope of the cumulative porosity curve. This factor is frequently referred to the maximum continuous pore radius, i.e. it is the most frequently occurring pore size in interconnected pores [1]. This peak corresponds to the value of critical diameter which is one of the most important indicators to permeability property.

1.2.3 Pore Size Distribution

Pore size distribution represents the fraction of the pore volume in which the pores lie within a state size range. It is defined as:

$$D_v(r) = -\frac{dV}{dr} \tag{2}$$

where $D_v(r)$ is the pore size distribution function, dV is the change in the pore volume (mm^3), dr is the change in the pore radium (mm) [1].

The pore size distribution can be represented in two ways. A plot of the volume of voids that are smaller than a given size versus the pore size is the cumulative pores size distribution. The slope of this curve, plotted against pore size, is the differential pore size distribution. These two characters are used in concrete science commonly [1].

1.2.4 Connectivity of Pores

Connectivity of pores is the fraction of pores with respect to the bulk volume of the material constituted only by open and interconnecting pores [6]. When C = 1, means all pores are connected, C = 0, means that no pore path could go from one side to other side [6].

$$Connectivity\,of\,pores\,(C) = \frac{Connected\,pore\,volume}{Total\,pore\,volume} \tag{3}$$

Compared with closed pores, interconnecting pores provide a continuous channel of communication with the external surface of the material. Therefore, the connectivity of pores is a more important indication of permeability [1]. Effective porosity is alternative term used to describe connectivity of pores. Obviously, effective porosity is always less than or equal to the total porosity.

2 Materials and Methods

2.1 Materials

2.1.1 Material Information

The cement paste specimens are prepared using the same type of cement, CEM I 42.5 N Portland cement produced by ENCI company, the Netherlands. The basic chemical composition of the cement is stated in Table 2. Four main constituents of cement are listed in Table 3.

Three different types of MS, named M6, M300 and M600, are used in the experiments. The chemical composition and specification of MS are given in the Table 4 and 5, respectively www.sibelco.be. Table 4 shows that basically, MS is quartz with a very high SiO2 Content.

Particle size distributions of OPC and MS are demonstrated in Fig. 1 [7]. As can be seen in this figure, M6 is coarser and M600 is much finer than cement. The size range of particles in M6 or in M600 is larger and more different than in OPC and in M300. Therefore, the mixture can have wider diversity in particle size when cement mixed with M6 or M600.

Fineness of MS is also related to the price. The higher the fineness is, the more expensive the sand is. That should be taken into consideration regarding the economical aspect.

2.1.2 Mixture Design

The effect of fineness of micronized sand on the microstructure of the cement paste will be investigated based on the mixtures with M6, M300 and M600. Different replacement percentages, i.e. 10, 20 % and up to 30 %, are selected to study the

Table 2 Chemical composition of ordinary Portland cement

Composition	%
CaO	63.96
SiO_2	20
Al_2O_3	4.88
Fe_2O_3	3.36
SO_3	2.4

Table 3 Constituents of CEM 1 42.5 N cement

Phase	%
C_3S	62
C_2S	10.5
C_3A	7.3
C_4AF	10.2

Table 4 Chemical composition of micronized sand (MS)

Proportion (%)	SiO_2	Fe_2O_3	Al_2O_3	TiO_2	K_2O	CaO
M6	99.5	0.03	0.2	0.03	0.04	0.02
M300	99.5	0.03	0.2	0.03	0.05	0.02
M600	99.2	0.05	0.4	0.03	0.05	0.02

Table 5 Physical properties of MS

	Fineness (Blaine) (cm^2/g)	D_{10} (μm)	D_{50} (μm)	D_{90} (μm)	Density (g/cm^2)
M6	2,400	5	30	90	2.65
M300	4,000	3	17	40	2.65
M600	13,000	2	4	9	2.65

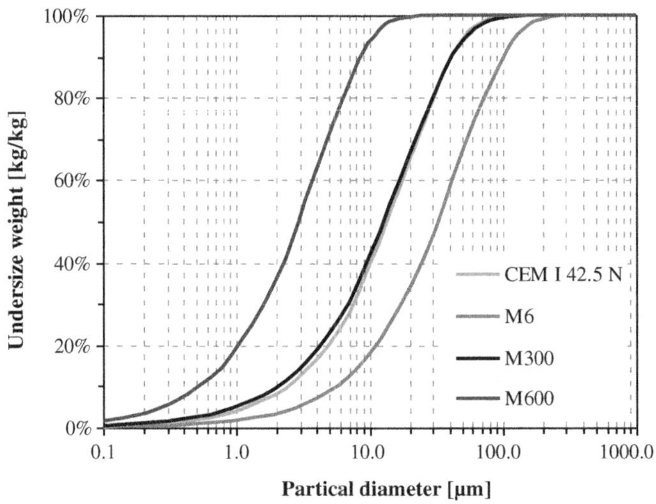

Fig. 1 Particle size distribution of OPC and micronized sand (MS) [7]

Table 6 Mix proportions

Name	OPC (%)	Micronized sand (%)	Sand/Binder[a] (s/b)	Water/Cement (w/c)	Water/Binder[a] (w/b)
OPC 0.40	100	0	0	0.40	0.40
M6_10 %	90	10	0.1	0.44	0.40
M6_20 %	80	20	0.2	0.50	0.40
M6_30 %	70	30	0.3	0.57	0.40
M300_20 %	80	20	0.2	0.50	0.40
M600_20 %	80	20	0.2	0.50	0.40

[a] *Powder* OPC + MS

effect of the replacement percentage of micronized sand on the transport properties of the blended material. The mix proportions are listed in Table 6.

2.2 Methods

2.2.1 Mercury Intrusion Porosimetry (MIP)

A wide variety of techniques has been used for the characteristics of the pore structure of cement paste. Mercury intrusion porosimetry (MIP) has proven over several decades to be a useful technique in characterizing pore structure parameters of cement-based material [1, 8, 9]. Such parameters include total pore volume, pore size distribution, density of solid and specific surface area of pores. Thus, MIP has become a convenient and fast technique for pore structure characterization. In addition, MIP is the only available technique that is supposed to cover nearly the whole range of sizes that must be analyzed in cement paste. Important characteristics, including total porosity, critical pore diameter and pore size distribution, can be determined from the results of MIP. The schematic drawing is demonstrated in Fig. 2. The total porosity obtained from the cumulative porosity curve corresponds to the point of highest pressure and the smallest equivalent pore size (c). The peak in (d) corresponds to the critical pore size [1].

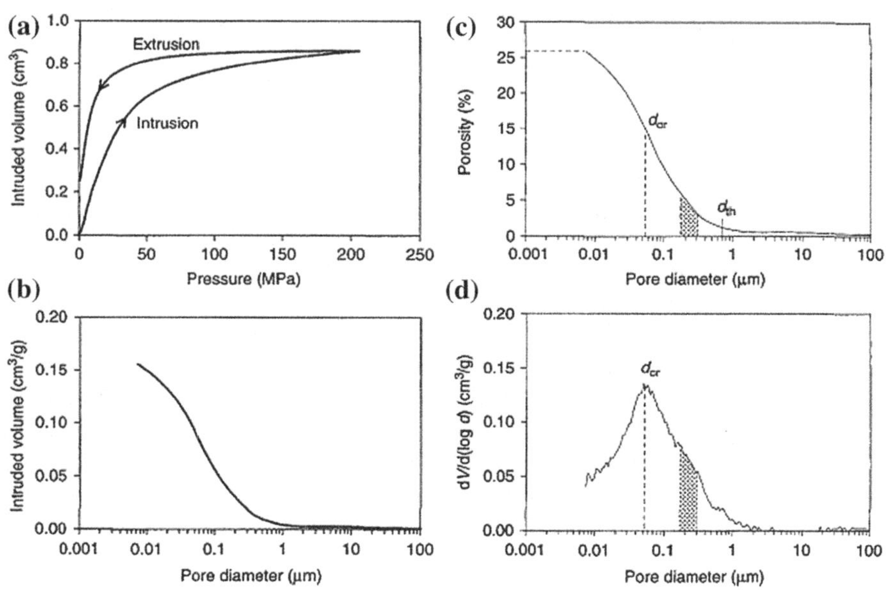

Fig. 2 Plots used most frequently in MIP to report experimental results [1]. **a** Pressurization curve. **b** Cumulative intruded volume curve. **c** Cumulative pore size distribution. **d** Differential pore size distribution

Although data from MIP experiments on cement paste have been used to represent the pore-size distribution for many years, there is still much debate about the validity of this method [8]. The assumption in MIP is that each pore is connected to the sample surface directly or through larger pores. However, by using microscopy technique, it has been found that the pore size of cement-based materials is randomly distributed and most pores are connected to the surface of the sample through a chain of pores with varying sizes and shapes [10]. With such a pore structure, mercury can not intrude into larger pores until the applied pressure is sufficient to force mercury to go through smaller throats, which is so called the ink-bottle effect. As a result, the volume of these larger pores is counted as the volume of smaller throats [10, 11]. Then MIP can not provide a real pore size distribution. The drawback of this technique will be taken into consideration when analyzing the results as well.

Samples are placed into a chamber, chamber is evacuated, the samples are surrounded by mercury, and pressure on the mercury is gradually increased [9]. When the pressure is increasing, mercury is forced into the pores on the surface of the sample. If the pore system is continuous, a pressure may be achieved at which mercury can penetrate the smallest pore necks of the system and penetrate the bulk sample volume. If the pore system is not continuous, mercury may penetrate the sample volume by breaking through the pore walls. After achieving the highest rate of intrusion, mercury has been shown to penetrate the interior of the sample [9].

MIP is a method which injects mercury into the porous material under very high pressure (several hundred MPa). The test is governed by the Washburn–Laplace equation in which the size of intruded pore accesses, assimilated to cylindrical capillaries, are inversely proportional to the applied pressure (Eq. 4):

$$P = -\frac{4\gamma \cos\theta}{d} \tag{4}$$

where P is the mercury injection pressure (Pa), γ is the surface tension of mercury (N/m), θ is the contact angle between solid and mercury (°), and d is the pore access diameter (m).

MIP tests were performed with a Micromeritics PoreSizer 9320. The PoreSizer 9320 is a 207 MPa mercury intrusion porosimeter, which determines pore sizes in the range of 7 nm to 500 μm. The measurement is conducted in two stages: a manual low pressure run from 0 to 0.170 MPa and an automated high pressure run from 0.170 to 205 MPa. For cementitious material, the surface tension value of 480 (N/m) and the contact angle of 139° were suggested [12] for the intrusion process and the contact angle of 106° was suggested [13] for the extrusion process. The MIP test procedure includes intrusion and extrusion.

2.2.2 Environmental Scanning Electron Microscope (ESEM)

Because of the so-called "ink-bottle" effect of MIP, the results from this technique sometimes can be misreported [8, 10, 11]. Therefore, direct observation of microstructure through optical and electron microscopy has the potential to provide insight into the spatial character of pore structure [14]. This technique has been developed, improved and applied for dozens of years [15–17]. A large amount of information can be interpreted from images, including pore structure, phase distribution characteristics of hydration products [18]. That can help understanding the development of hydration of cementitious material.

Scanning electron microscope (SEM) technique has been successfully used in studying the pore structure in cementitious material for decades. The most common mode used to obtain images with SEM is the backscattered electron (BSE), which can provide information on the distribution and composition of hydration products. SEM technique requires several regards, including proper sample preparation, image acquisition and data interpretation. Sample preparation and image acquisition are referred to the literatures [14, 15, 19, 20] and will be described in the next section.

Images on representative areas of polished surface are obtained by backscatter SEM. Pixels corresponding to each of the components are separated by binary segmentation (threshold) based on the grey level [21, 22]. The unhydrated phases of cement particles appear brightest; the calcium hydroxide (CH) shows light grey and the other hydration products (C–S–H) look as various shades of darker grey. The pores, usually filled with epoxy resin in specimen preparation, appear uniformly black and can be accurately segmented. A picture of cement paste with 20 % replacement of M6 (coarse) at 7 days curing age is shown in Fig. 3. The area fraction can be calculated by applying image analysis software, i.e. Photoshop, imageJ and Optimas. Certain phase area fraction calculated from a 2-D image analysis then is assumed equal to the volume to the volume fraction in a simple way:

$$\textit{phase volume fraction}\,(\mu m^3) = \textit{phase area fraction}\,(\mu m^2) \tag{5}$$

In the blended specimen, MS shows the similar grey color as CH and C–S–H. However, it is easy to determine the volume fraction of MS from the beginning of the mixture design since it is inert filler and not change in the system.

If the amount of images is sufficient, this method can give a fair representation of reality. Based on the analysis of statistic theories, more than 15 pictures were taken for each sample [22].

A Philips-XL30-ESEM was used in this research. The emitted electrons are accelerated by an electric field up to energy of 1–30 kV. By means of electromagnetic lenses, the beam is focused at the surface of the specimen. The minimum spot size is 1 nm.

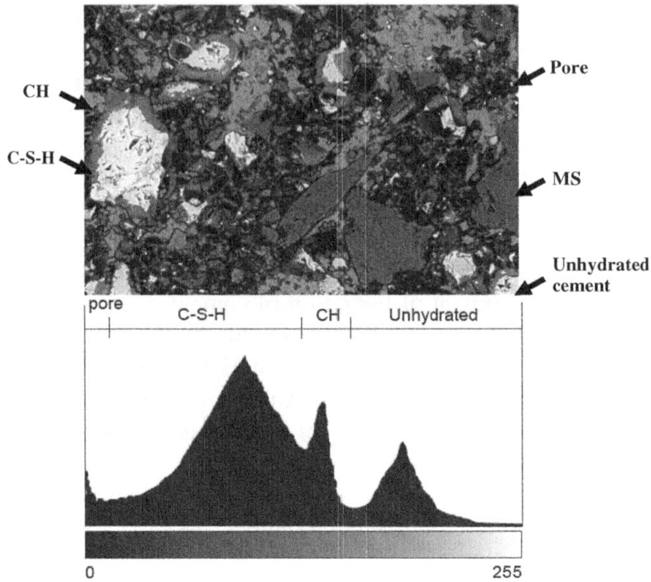

Fig. 3 SEM image of cement paste with 20 % replacement of M6 (coarse) at 7 days curing and the typical grey level schematic histogram [22]

2.2.3 Non-Evaporable Water Test

The degree of hydration of cement in paste sample was obtained by determining the non-evaporable water content, Wn/c. Wn/c is defined as the mass loss per gram of original cement, measured between the temperatures of 105 and 1,000 °C [23]. The degree of hydration was then determined as the ratio of the measured non-evaporable water content per gram of cement to the amount at full hydration. The Wn/c content of hydrated pastes on the ignited sample was calculated by Eq. (6):

$$\frac{W_n}{c}(\%) = 100 \cdot \frac{dried\,weight\,of\,paset - ignited\,weight\,of\,paste}{ignited\,weight\,of\,paste} \tag{6}$$

For the degree of hydration, α(t), at time t, it holds [24]:

$$\alpha(t) = \left[\frac{W_n(t)}{c}\right] / \left[\left(\frac{W_n}{c}\right)_{complete}\right] \tag{7}$$

2.3 Sample Preparation

All cement paste specimens with or without MS were mixed with tap water. A HOBART mixer was used for mixing. After 1 min of mixing at low speed and two minutes mixing at high speed, the pastes were poured into a plastic bottle (Fig. 4), of which the volume is approximately 70 ml. The bottles were shaken continuously in order to remove air bubbles and then sealed with a plastic tape. The specimens are stored in a climate oven of 20 °C until the age of testing. Curing age of 1, 3, 7, 28 and 90 days were taken into account. After achieving the required age, the plastic bottle was broken and the sample was removed and split into small pieces, about 1 cm^3 (Fig. 4), using hammer. These pieces of samples are separated into two groups. One is prepared for MIP and SEM tests; the other is prepared for non-evaporable water test.

2.3.1 Samples for MIP

Freeze-drying method was applied to prepare the samples. Several pieces of samples were first immersed into liquid nitrogen (−195 °C) for 5 min (Fig. 4). After freezing, the samples were placed in a freeze-dryer (Fig. 4) with temperature of −24 °C and vacuum of 10–1 Pa. Water loss was recorded each 24 h until a stable mass loss of 0.01 %/day was reached [22, 25]. Around 5–6 g sample is used

Fig. 4 **a** Small pieces of sample. **b** Emerge samples into liquid nitrogen. **c** Freeze-dryer. **d** Plastic bottle

for each MIP measurement. Mercury intrusion starts from a manual low-pressure test and then the high-pressure test.

2.3.2 Samples for SEM Tests

A high-quality original image is the prerequisite for accurate segmentation of features and subsequent quantification steps [17]. Proper sample preparation is critical to the successful images of microstructure [19]. In order to view in the SEM, around 30 g of paste samples are selected after stable mass is reached. Epoxy impregnation is applied after the samples are placed in the chamber for evacuation at 0 torr for 3 h. The impregnated samples are cured at atmospheric pressure at 35 °C for 24 h. Then the samples are ready for the cutting, grinding and polishing [19, 22]. The rough surface will be removed by a diamond saw, followed by a delicate grinding procedure. The sample will be ground on a rotating plate with p320, p500, p1200, p2000 and p4000 sand papers. Final polishing is done on a lap wheel with 6, 3, 1 and 0.25 μm diamond paste for 2–3 min each [19, 22]. Then the samples are ready to be observed in SEM. The relevant apparatuses are demonstrated in Fig. 5.

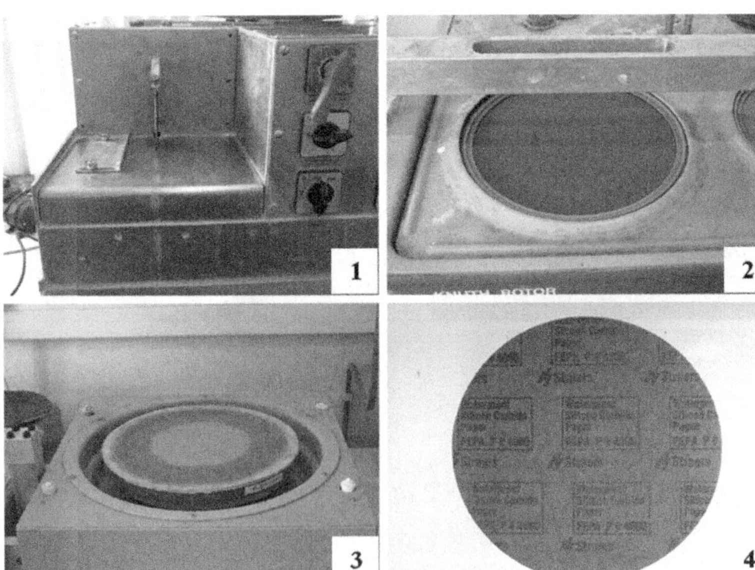

Fig. 5 **a** Sawing machine. **b** Grinding machine. **c** Polishing machine. **d** Sand paper

2.3.3 Samples for Non-Evaporable Water Tests

Samples for the non-evaporable water content (W_n/c) determination were ground to powder using a mortar and pestle. Around 4 g powder was prepared and then placed in 3 crucibles separately. To determine W_n/c content, 3 crucibles with around 1.1 g hydrated sample were dried in an oven at 105 °C for at least 12 h, and were then ignited at 1,000 °C in a furnace for 3 h. The mass of the crucibles and samples were measured before and after placing them in the furnace, respectively.

3 Results and Discussions

3.1 Effect of Micronized Sand on the Hydration Process

The degree of hydration of cement determined by the non-evaporable water tests is presented in Fig. 6. With the w/b of 0.40, the degree of hydration of cement in the blended samples is higher in comparison with the reference OPC paste sample. At the age of 28 days, around one third of cement material is still unhydrated in the reference cement paste. To some extent, at that moment the unhydrated cement particles acts as expensive filler material. In the modified cement pastes, a higher effective w/c ratio, because of the replacement of micronized sand, results in a higher rate of hydration reaction. In these blended mixtures the cement is used more effective.

The results also show that the degree of hydration of cement in the blended cement pastes increases with increasing fineness of the micronized sand. With the

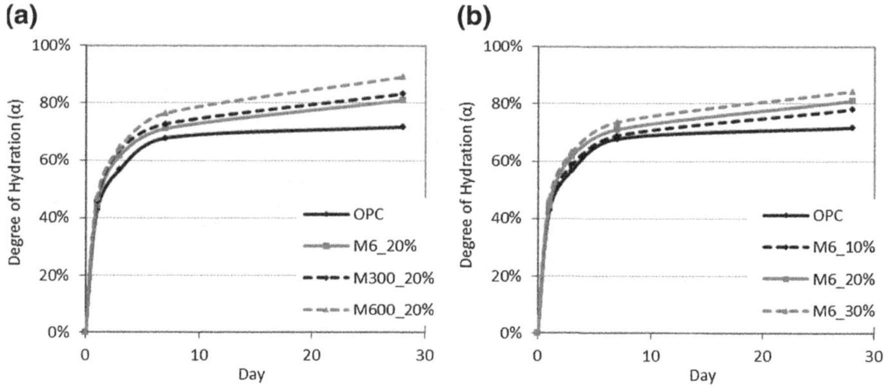

Fig. 6 Degree of hydration of cement in pastes containing micronized sand with **a** 20 % of M6, M300 and M600, **b** 10, 20 and 30 % of M6

same replacement percentage, i.e. 20 %, samples with M300 or M600 have higher degree of hydration in comparison with the sample with M6.

M300 and M600 have higher fineness than that of M6, which can be seen in Table 5 and Fig. 1. This indicates that mixtures with M300 and M600 have a lot small sand particles that are easy to be the nucleation sites during the hydration reaction. Therefore, the heterogeneous nucleation takes place easily in the samples with M300 and M600. This effect will influence the speed and degree of hydration. Although micronized sand is considered as inert filler, because of its potential to act as nucleation sites it can still indirectly modify the hydration kinetics.

3.2 *Effects of Micronized Sand on the Pore Structure of Cement Paste*

3.2.1 Effect of Micronized Sand Fineness

The total porosity of paste samples containing 20 % micronized sand with different fineness is presented in Fig. 7. Compared to the reference sample, i.e. OPC paste, the total porosity is higher in the pastes containing micronized sand. That is because the effective w/c is increased because of the addition of micronized sand. The total porosity of the blended pastes decreases with the fineness of the micronized sand in the corresponding samples before the age of 28 days. It is noticed that after 90 days the effect of the fineness of the micronized sand does not result in much difference in the porosity of the pastes. Obviously, the effect of micronized sand fineness on cement hydration at later stages is limited.

The cumulative and differential pore size distribution *curves* of cement paste containing 20 % micronized sand are shown in Figs. 8 and 9. The peak in Fig. 9, which is called "critical pore size", shifts to the smaller pores size at the age of

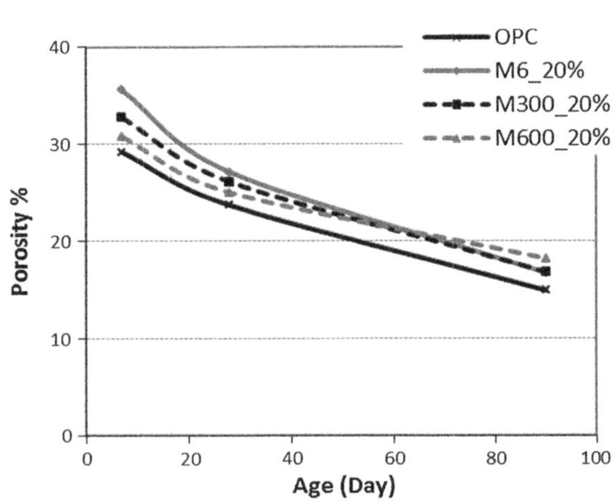

Fig. 7 Total porosity of paste samples with w/b = 0.40 as a function of the ages (20 % replacement)

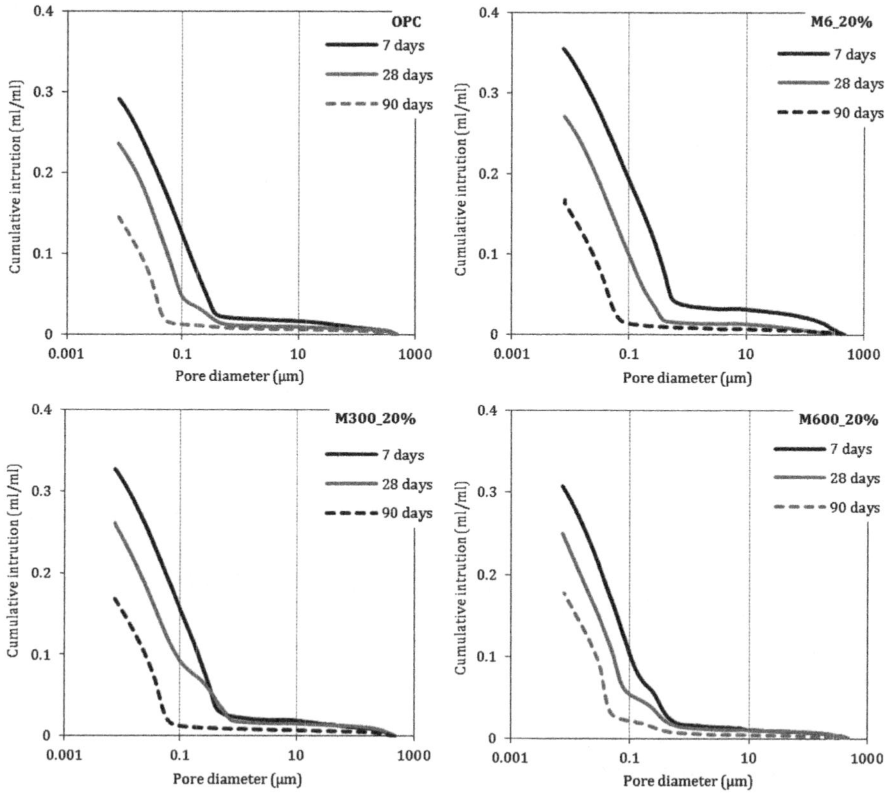

Fig. 8 The cumulative pore size distribution *curves* of cement blended with 20 % M6, M300 and M600 at the ages of 7, 28 and 90 days

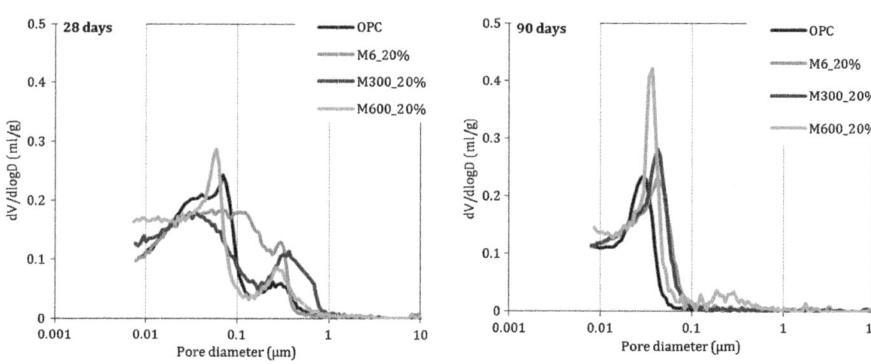

Fig. 9 The differential intrusion *curves* of cement blended with 20 % M6, M300 and M600 at the ages of 28 days (*left*) and 90 days (*right*)

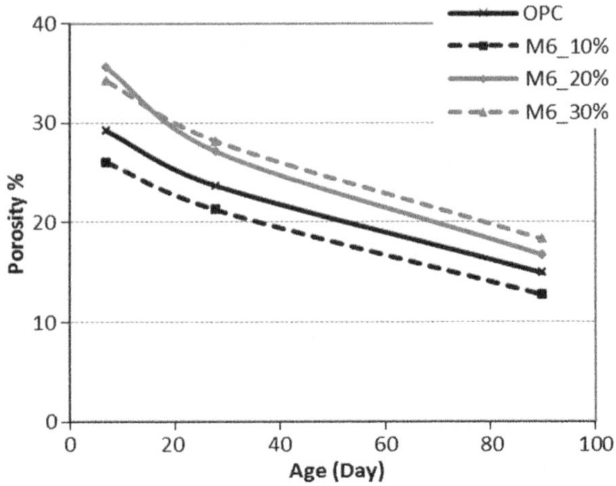

Fig. 10 Total porosity of paste samples with w/b = 0.40 as a function of the ages (different replacement of M6)

90 days, which means that the pore structure of the paste is further refined with progress of the hydration process. At the age of 90 days, the pores larger than 0.1 μm are vanished in all the samples. The differential intrusion curves of the blended pastes show a similar trend, but with higher differential pore volume in the range of the pore diameters between 0.01 and 0.05 μm compared with the reference sample.

3.2.2 Effect of Micronized Sand Content

The total porosity of paste samples containing M6 (coarse sand) with different replacement percentages is presented as a function of time in Fig. 10. The total porosity decreases along with the progress of the hydration process. The porosity in the blended samples is higher with increasing sand content. The "dilution effect" is becoming stronger with increasing sand content. M6_10 % shows lower porosity than the reference (OPC) paste. The microstructure can be refined by the presence of the micronized sand. In addition, the higher degree of hydration in M6_10 %, discussed in Sect. 3.1, contributes to the refinement of the microstructure of the paste. The dilution effect of micronized sand is more important and the packing effect of micronized sand is limited when the content of micronized sand is higher than 20 %.

The cumulative and differential pore size distribution curves of cement paste containing 20 % micronized sand are shown in Figs. 11 and 12. It can be seen that at the age of 28 days, 20 and 30 % coarse micronized sand (M6) results in a higher differential pore volume in the range of pore diameters between 0.1 and 0.5 μm compared with the reference and M6_10 % samples. After 90 days the pores larger than 0.1 μm are almost vanished in all the samples. The curve of differential intruded pore volume of sample M6_10 % has the same trend as the one of the reference OPC.

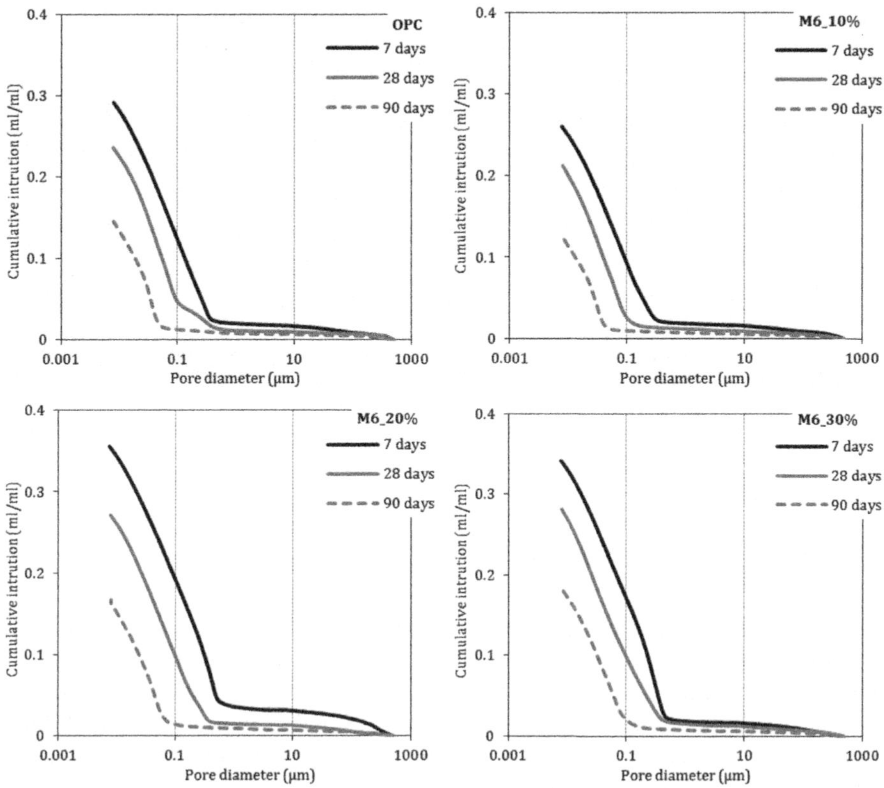

Fig. 11 The cumulative pore size distribution *curves* of cement blended with 10, 20 and 30 % M6 at the ages of 28 days (*left*) and 90 days (*right*)

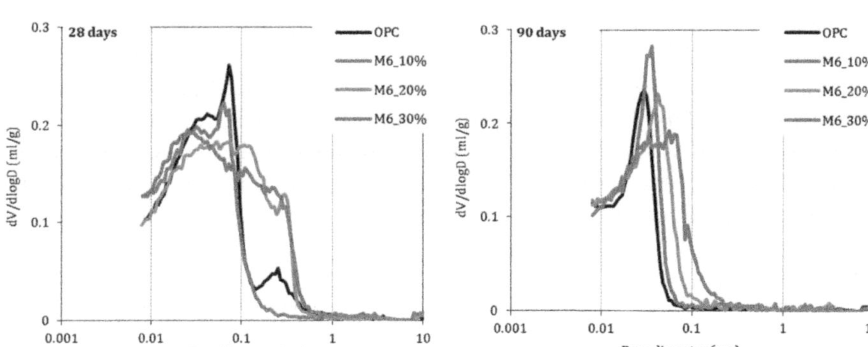

Fig. 12 The differential intrusion *curves* of cement blended with 10, 20 and 30 % M6 at the ages of 28 (*left*) and 90 days (*right*)

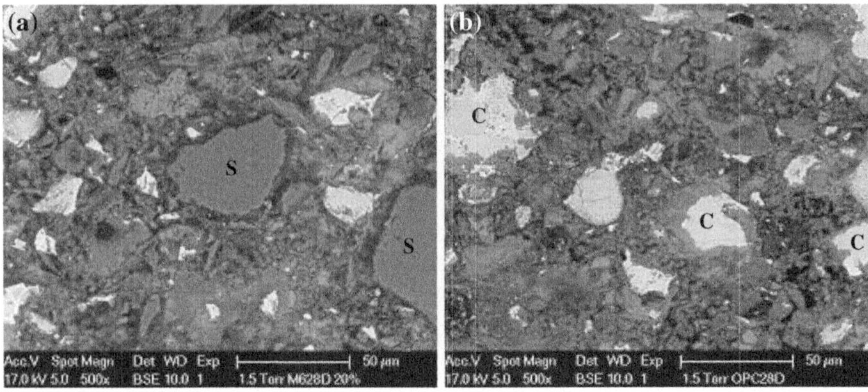

Fig. 13 BSE images of M6_20 % (**a**), and OPC (**b**) at the age of 28 days. (*C* unhydrated cement particle; *S* micronized sand particle)

3.3 Observation of Microstructure

A clear picture of the microstructure of cement paste, including the distribution of hydration products and the shape of pores, can be obtained from SEM observations. In Fig. 13, the BSE image of the sample M6_20 % is shown and compared with the reference paste at the age of 28 days (S and C indicate sand and unhydrated cement). It can be seen that in the blended pastes the fraction of unhydrated cement grains is lower and the unhydrated cement cores are smaller (Fig. 13a) than in the OPC paste (Fig. 13b).

The porosity of the paste samples was calculated based on 15 images. The thus obtained porosity is compared with the results from MIP measurements, presented in Table 7.

The data in Table 7 show that the porosity measured by MIP is much larger than that obtained from image analysis. The difference between these two measurements is in good agreement with the results found by Lange et al. [14] and Diamond et al. [26]. They have explained the reason for the difference: MIP technique measures the larger pores only by intruding mercury through smaller bottlenecks and thus the technique systematically misinterprets their size. Pore maybe considered as smaller than it should be by this so-called "ink-bottle" effect. On the other hand, the resolution in the images here is 0.18 μm/pixel, indicating that only pores larger than 0.18 μm can be detected. The limitation of the SEM

Table 7 Comparison of porosities determined from BSE image analysis and MIP measurement of paste samples at the age of 28 days

Sample	MIP (%)	SEM (%)
OPC	23.68	9.90
M6_20 %	27.11	15.60
M300_20 %	26.12	13.90
M600_20 %	25.00	11.70

image analysis technique is that only 2D information is available and the accuracy of the results is limited by the resolution of the images.

The combination of MIP measurement data and SEM image analysis allows to determine the development of microstructure and to get a better insight, not only in the pore size distribution, but also in the pore geometry and connectivity. MIP measurement can give accurate total porosity, whereas SEM images can visualize the pores. Furthermore, the combination of the microstructural information generated by SEM and MIP is also fundamental for validating the microstructure of cement paste generated from the numerical simulation.

3.4 The Thickness of the Interfacial Zone Between Micronized Sand and Hydration Products

In Fig. 13, it can be seen that the morphology of the reaction products around the micronized sand particles is different from that around cement particles. Figure 14 shows more details of the microstructure under the magnification of 2,000. Less hydration products precipitate on the surface of the micronized sand particles forming a clear interfacial zone between sand surface and hydration products. This new interfacial zone is distinguishing from the one between aggregate and paste, since it is in a lower level. The microstructure in this interfacial zone is more porous in comparison with the bulk matrix. It is expectable that this porous interfacial zone caused by the presence of micronized sand will affect the transport properties of the material. In this section, the interfacial zone around micronized sand will be studied by quantitative analysis of the SEM images.

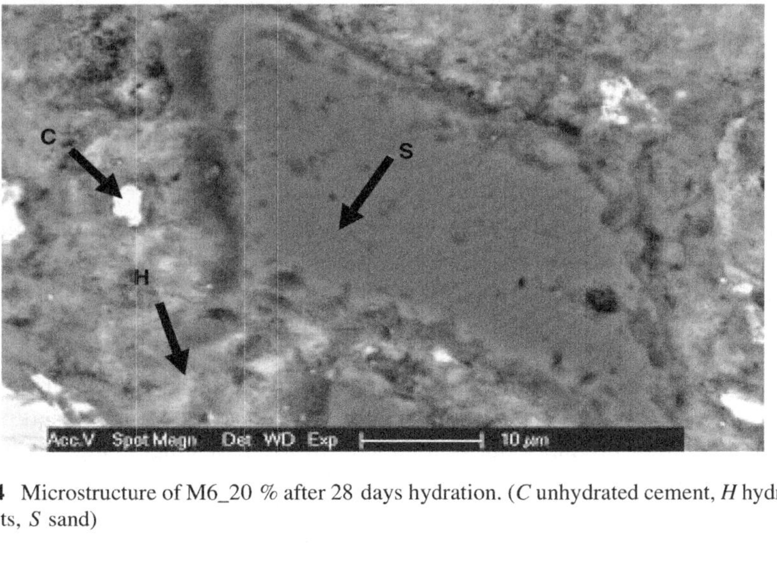

Fig. 14 Microstructure of M6_20 % after 28 days hydration. (*C* unhydrated cement, *H* hydration products, *S* sand)

Fig. 15 ESEM image with selected micronized sand particle

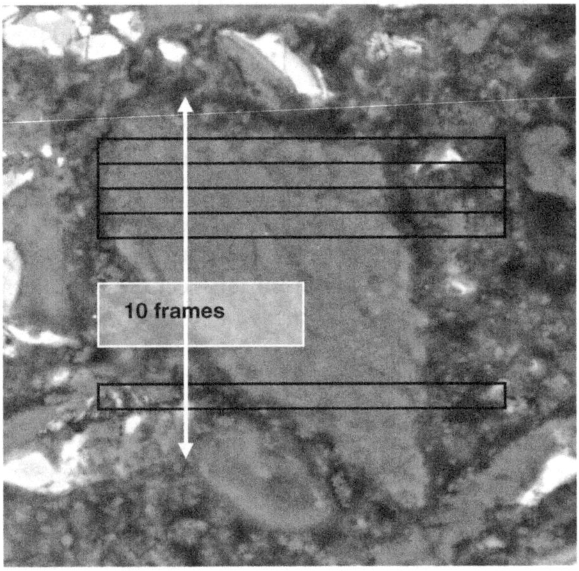

Fig. 16 Selected micronized sand particle with frames

Around 100 sand particles were captured from the images with a magnification of 1,000. The micronized sand particles were selected randomly and varied in size from 8 to 70 μm, for instance the one in Fig. 15. Each particle was subdivided into 10 frames (Fig. 16) and the left side of each frame was put along the surface of the micronized sand particles. The grey level of each frame was obtained by using image read software. The grey level is plotted over the distance from the sand particle (Fig. 17). The thickness of the interfacial zone between the sand particle and the hydration products can be obtained from the average value of the grey level of the frames.

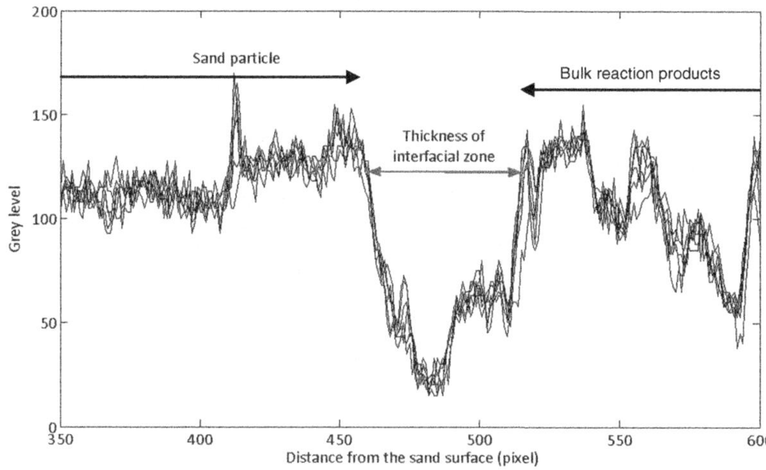

Fig. 17 Grey levels of selected frames

Fig. 18 Interfacial distance of particles in different mixtures (28 days)

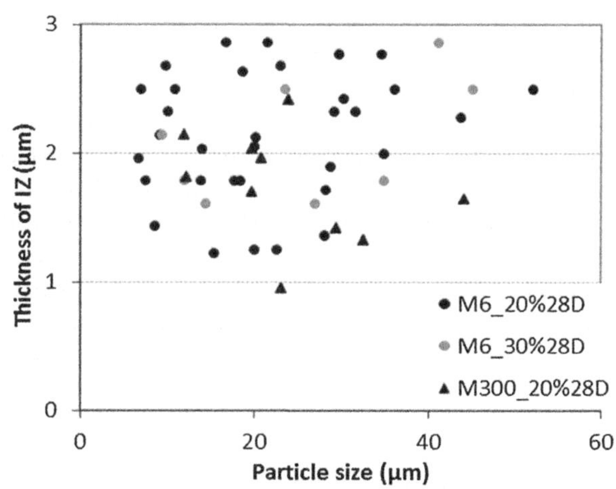

Parameters, including replacement percentage and fineness of micronized sand and ages of the samples were considered. The thickness of the interfacial zone plotted against the particle size is presented in Figs. 18 and 19. The thickness of interfacial zone varies from 1 to 3 μm and mainly fluctuates around 2 μm at the age of 28 days. The thickness of the ITZ is reduced to 1–1.5 μm at the age of 90 days. During the development of the hydration reaction, the amount of hydration products increases; more and more hydration products fill the pores and the microstructure becomes denser. There is no clear relation between the particle size of micronized sand and thickness of interfacial zone. It seems that the thickness is independent on the fineness or the percentage of micronized sand in the mixtures considered in this study. This value is determined by the interfacial

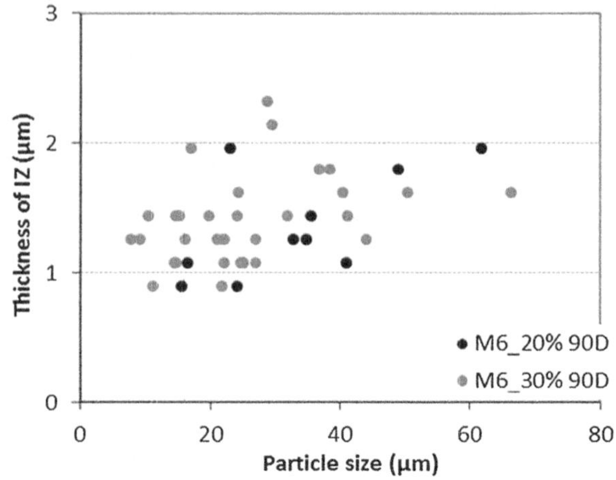

Fig. 19 Interfacial distance of particles in different mixture (90 days)

energy of the individual elements, as explained by [27]. The creation of a nucleus is accompanied by the formation of an interface at the boundaries of a new phase. An interfacial area appears when the hydration products precipitate on the surface of micronized sand. That is the important area that influences the transport properties.

It is notice that there are rarely sand particles detected under the size of 8 μm in Figs. 18 and 19. Heterogeneous nucleation takes place when the interfacial energy between the micronized sand particle and CH/C–S–H is smaller than the interfacial energy between CH/C–S–H and the pore solution. The interfacial energy of the element is mainly depended on the chemical composition of the material and inversely proportional to the surface area [27]. Small particles are easy to become the nucleation sites and embedded in the hydration products. The sand particles under the size of 8 μm are considered as the sites for simplicity in this thesis.

4 Summary Remarks

In this study, the degree of hydration of cement, porosity and pore size distribution of cement paste incorporating micronized sand were studied and compared with the reference OPC paste. The following conclusion can be drawn:

- Almost one third of the cement grains remains unhydrated at the age of 28 days (CEM I 42.5 w/c = 0.40), acting as an expensive filler material. The degree of hydration of cement in the blended samples is higher in comparison with the reference one. From this point of view, the use of micronized sand as a replacement material of cement is a potential solution.
- Although micronized sand is considered as an inert filler, it can indirectly modify the hydration kinetics and influence the microstructure of the paste. The

effect of fineness of micronized sand plays an important role in the hydration reaction before 28 days.

- The effects of micronized sand on the microstructure of the blended cement paste are the dilution effect, the packing effect and the heterogeneous nucleation effect. It is determined by the replacement percentage and fineness of micronized sand.
- The pore structure of cement paste is influenced by the presence of micronized sand. A different pore size distribution in the range from 0.1 to 0.5 μm appears in the blended samples in comparison with the reference paste. Pores in this range are defined as large capillary pores, which influences the permeability.
- The combination of the microstructural information from SEM and MIP measurements is more suitable to understand the microstructure other than only one of them.
- There is a clear interfacial zone between micronized sand and hydration products. This thickness is independent of the fineness or replacement level (up to 30 %) of micronized sand. This interfacial zone caused by the micronized sand is considered as the decisive factor which will influence the transport properties of the material.

References

1. Aligizaki, K.K.: Pore structure of cement-based materials. Taylor & Francis, London (2006)
2. Neville, A.M.: Properties of concrete. Harlow Longman, UK (1995)
3. Richardson, M.G.: Fundamentals of durable reinforced concrete. Modern Concrete Technology, New York (2002)
4. Garboczi, E.J.: Permeability, diffusivity, and microstructural parameters: A critical review. Cem. Concr. Res. **20**(4), 591–601 (1990)
5. Midess, S., Young, J., Darwin, D.: Concrete, 2nd edn. Prentice Hall, Englewood Cliffs (2002)
6. Ye, G., Liu, X., De Schutter, G., Poppe, A.M., Taerwe, L.: Influence of limestone powder used as filler in SCC on hydration and microstructure of cement pastes. Cem. Concr. Compos. **29**(2), 94–102 (2007)
7. De Vries, W.: The use of micronized sand as cement replacement. Delft University of Technology. Master, Delft
8. Moro, F., Bohni, H.: Ink-bottle effect in mercury intrusion porosimetry of cement-based materials. J. Colloid Interface Sci. **246**(1), 135–149 (2002)
9. Cook, R.A., Hover, K.C.: Mercury porosimetry of hardened cement pastes. Cem. Concr. Res. **29**(6), 933–943 (1999)
10. Diamond, S.: Mercury porosimetry: An inappropriate method for the measurement of pore size distributions in cement-based materials. Cem. Concr. Res. **30**(10), 1517–1525 (2000)
11. Zhou, J., Ye, G., van Breugel, K.: Characterization of pore structure in cement-based materials using pressurization-depressurization cycling mercury intrusion porosimetry (PDC-MIP). Cem. Concr. Res. **40**(7), 1120–1128 (2010)
12. Cook, R.A., Hover, K.C.: Experiments on the contact angle between mercury and hardened cement paste. Cem. Concr. Res. **21**(6), 1165–1175 (1991)
13. Léon, Y., León, C.A.: New perspectives in mercury porosimetry. Adv. Colloid Interface Sci. **76–77**, 341–372 (1998)

14. Lange, D.A., Jennings, H.M., Shah, S.P.: Image analysis techniques for characterization of pore structure of cement-based materials. Cem. Concr. Res. **24**(5), 841–853 (1994)
15. Diamond, S.: Considerations in image analysis as applied to investigations of the ITZ in concrete. Cem. Concr. Compos. **23**(2–3), 171–178 (2001)
16. Scrivener, K.L.: Backscattered electron imaging of cementitious microstructures: Understanding and quantification. Cem. Concr. Compos. **26**(8), 935–945 (2004)
17. Wong, H.S., Head, M.K., Buenfeld, N.R.: Pore segmentation of cement-based materials from backscattered electron images. Cem. Concr. Res. **36**(6), 1083–1090 (2006)
18. Wang, Y., Ye, G., Guo, Z., van Breugel, K., Phung, T.Q.: Microstructure characteristics in the cement paste containing micronized sand filler. International conference on microstructure related durability of cementitious composites, Nanjing, China (2008)
19. Bentz, D.P., Stutzman, P.E.: SEM analysis and computer modelling of hydration of Portland cement particles. Philadelphia **1215**, 60 (1994)
20. Stutzman, P.E., Clifton, J.R.: Specimen preparation for scanning electron microscopy. Twenty-first international conference on cement microscopy. Las vegas, Nevada (1999)
21. Diamond, S., Huang, J.: The ITZ in concrete—a different view based on image analysis and SEM observations. Cem. Concr. Compos. **23**(2–3), 179–188 (2001)
22. Ye, G.: Experimental study and numerical simulation of the development of the microstructure and permeability of cementitious materials. Delft University of Technology, Delft (2003)
23. Copeland, L.E., Hayes, J.C.: The determination of non-evaporable water in hardened Portland cement pastes. ASTM Bull. **194**, 70–74 (1953)
24. Molina, L.: On predicting the influence of curing conditions on the degree of hydration. CBI report Stockholm, vol. 5. Swedish Cement and Concrete Research Institute, Stockholm (1992)
25. Gallé, C.: Effect of drying on cement-based materials pore structure as identified by mercury intrusion porosimetry—A comparative study between oven-, vacuum-, and freeze-drying. Cem. Concr. Res. **31**(10), 1467–1477 (2001)
26. Diamond, S., Leeman, M.E.: Pore Size Distributions in Hardened Cement Paste by Sem Image Analysis. in Microstructure of Cement-Based Systems/Bonding and Interfaces in Cement-Based Materials. (1994)
27. Stumm, W.: Chemistry of the solid-water interface: Processes at the mineral-water and particle-water interface in natural systems. Wiley, New York (1992)

Durability of Interior Renderings in Schools: Optimization of Envelope Insulation for Mitigation of Mould Growth Risk

N. M. M. Ramos, I. M. Ribeiro, J. M. P. Q. Delgado and V. P. de Freitas

Abstract A high standard education is, nowadays, a clear objective for any modern society. In Portugal, the retrofit of educational buildings has an enormous relevance as demonstrated by the "Program for Retrofit and Upgrade of High Schools". Educational buildings specific characteristics such as high occupancy rates and high amount of time children spend in classrooms make them a special case compared to other buildings. Mould growth on envelope surfaces is frequently observed in these buildings, reducing the expected durability of renderings and coatings. The present work consisted of two main lines of research: one of simulation and other of optimization. The simulation part included modelling of Interior Surfaces Temperature, Indoor Air Temperature and Relative Humidity variation, before and after a hypothetical retrofit process. Optimization techniques were applied in the definition of insulation thickness to be numerically tested in the "after retrofit" scenario. The objective was to minimise insulation thickness in each building element of the envelope. The constraints were derived from the control of surface temperature values that would result in mould growth risk.

Keywords Building retrofitment · Hygrothermal behaviour · Mould · Optimization

---

N. M. M. Ramos (✉) · I. M. Ribeiro · J. M. P. Q. Delgado · V. P. de Freitas
LFC-Building Physics Laboratory, Civil Engineering Department Faculty of Engineering, University of Porto, 4200–465 Porto, Portugal
e-mail: nuno.ramos@fe.up.pt

I. M. Ribeiro
e-mail: iribeiro@fe.up.pt

J. M. P. Q. Delgado
e-mail: jdelgado@fe.up.pt

V. P. de Freitas
e-mail: vpfreita@fe.up.pt

V. P. de Freitas and J. M. P. Q. Delgado (eds.), *Durability of Building Materials and Components*, Building Pathology and Rehabilitation 3,
DOI: 10.1007/978-3-642-37475-3_4,

1 Introduction

A high standard education is, nowadays, a clear objective for any modern society. In Portugal, the retrofit of educational buildings has an enormous relevance as demonstrated by the "Program for Retrofit and Upgrade of High Schools" [1]. The retrofit of the first 205 schools represents an investment of 2,500 million Euros, intending for them to respond to present demands namely energy efficiency, comfort, indoor air quality (IAQ) and functional organization [2, 3].

In Portugal, as in European countries, educational buildings share many similar design, operation and maintenance features. Educational buildings specific characteristics such as high occupancy rates and high amount of time children spend in classrooms make them a special case compared to other buildings [4]. Therefore, the retrofit process should take into account the high internal heat gains due to high occupancy rates, the high solar gains due to glazing and the demanding needs for ventilation and IAQ [5, 6]. The regulations adopted in recent years were thought for new buildings and not for the retrofit of existing ones. Therefore their implementation may not lead to intended objectives. On the other hand, the costs for satisfying the demands are so high and imply such intrusive interventions that justify a careful technical reflexion.

Durability issues are also of main importance when looking at intensively used buildings. Considerable work has been internationally carried out in the area of durability and service life prediction as requisite tools for helping assess long-term effects for maintenance of building envelope systems, envelope components and related materials. Increasingly, building material and component manufacturers are seeking systematic methods to assess the likely risk to premature deterioration of existing products given specific climatic effects, or the most vulnerable exposure conditions of new products in specified systems. The importance of these aspects is reflected in several initiatives and activities at international level. For example, the joint CIB W080/RILEM Committees (71-PSL, 100-TSL and 140-TSL) have been responsible to help develop the necessary guides, methods, and techniques that will enable practitioners to select the appropriate tools to predict service life [7].

In this work, the focus was put on degradation of inside surfaces due to mould growth. This is a common problem in some existing Portuguese buildings due to the combination of a mild humid climate, discontinuous heating, adventitious ventilation and low quality envelope.

The objective of this work is to find the optimal insulation thickness to be added to envelope components regarding mould growth mitigation. Two major constraints were elected to conduct the optimization process. The resulting thermal resistance of walls and ceiling should prevent mould growth by avoiding surface condensation and contribute to the increase of minimum indoor temperature admitting no additional heating is available. It should be taken into account that due to high internal gains in classrooms and mild climate Portuguese conditions it may be possible, in some cases, to avoid additional heating. Also, due to economic restraints, adequate heating regarding comfort may not be implemented. The result of this optimization process should therefore be regarded as a safety net for the envelope, rather than the ideal solution for a properly operated building.

The demand for energy efficiency and IAQ has lead to solutions where the passive behaviour of the building becomes of minor importance when compared to systems. But for technical or economic reasons the full operation of the systems may be impossible and therefore the evaluation of the building performance for that scenario must be investigated.

The present work consisted of two main lines of research: one of simulation and other of optimization. The simulation was focused on the hygrothermal behaviour of a classroom using a simplified model developed by the authors. The Energy-Plus software [8, 9] was used to simulate the same scenarios, providing results for comparison. The optimization process was adapted to the simplified model since it uses linear functions.

2 Design Problem

Figure 1 presents a 3D model of the studied building, which is part of a school built on the late 1970s. This specific building holds two classrooms that have simultaneous occupation and can therefore be approximated with a one zone model.

The building envelope (Figs. 2 and 3) is composed of 15 cm concrete walls (Fig. 4), single glazed aluminium frame windows (Fig. 5) and a flat roof composed by a 25 cm concrete slab protected by cement-fibre plates. No thermal insulation

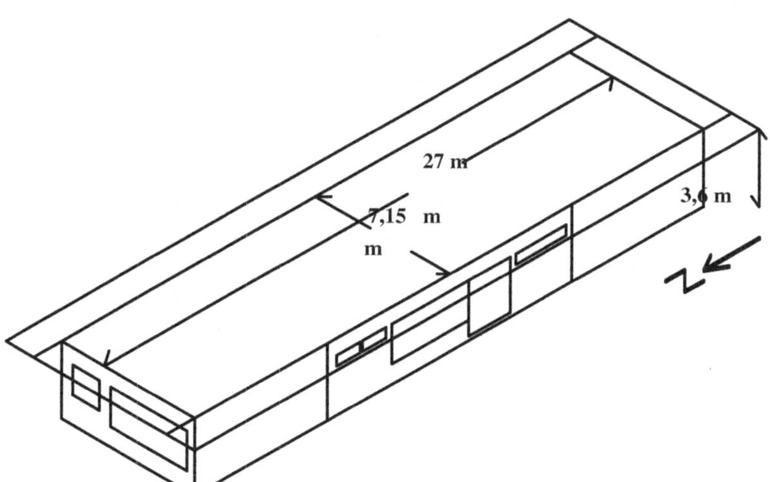

Fig. 1 Sketch of the building analyzed

Fig. 2 School building's East façade

Fig. 3 School building's north façade

was originally applied in the envelope opaque elements. The interior surfaces of those elements were often affected by mould growth (Fig. 6).

The building is located in Porto, in the so-called Warm-Summer–Mild-Winter zone. The exterior temperature will not drop below 0 °C and may reach 30 °C in the summer. The average temperature in January is 7.5 °C. For simulation purposes, the TMY format weather file for Porto, available on the Energy Plus web page, was used (Fig. 7).

The occupation of the two classrooms corresponds to a total of 50 persons with a metabolic rate of 70 W/m^2 (ISO 7730, [10]), in a schedule 8.30–13.00 and 14.00–18.30. No mechanical ventilation was applied in the original classrooms, therefore relying on adventitious ventilation and window opening by users. But

Fig. 4 Building façade

since the new regulations [3] demand 30 m^3/h per person, it was admitted in all calculations that, when occupied, an ach = 2.2 h^{-1} would prevail and, in the remaining periods, ach would drop to 0.6 h^{-1}.

The difference between the scenarios of pre or post retrofit was the insulation thickness added to the envelope elements. In this exercise, XPS [0.035 W/(m^2°C)] was considered as insulation material, applied on interior surfaces.

3 Numerical and Optimization Model

For these calculations an energy simulation needs to be performed to solve the design optimization problem described above. Energy-Plus, a building energy analysis tool was used to simulate the hygrothermal behaviour of the building. The differential equation that define the energy balance of a room is [9]

$$
\begin{aligned}
C_z \frac{dT_z}{dt} &= \sum_{l=1}^{sl} Q_l + \sum_{k=1}^{Ns} h_k A_k (T_{sk} - T_z) \\
&+ \sum_{i=1}^{Nzones} m_i C_p (T_{zi} - T_z) + m_{\text{inf}} C_p (T_\infty - T_z) + Q_{sys}
\end{aligned}
\tag{1}
$$

where Q is the internal gains by convection; $C_z(dT_z/dt)$ is the total energy stored in the air in each thermal zone; $h_k A_k (T_{sk} - T_z)$ is the heat transfer by convection to the surface of the element k; $m_i C_P (T_{zi} - T_z)$ is the heat transfer due to outside air infiltration; $m_{\text{inf}} C_p (T_\infty - T_z)$ is the heat transfer between different thermal zones and Q_{sys} is the heat transfer by the building outlet air.

Fig. 5 Windows

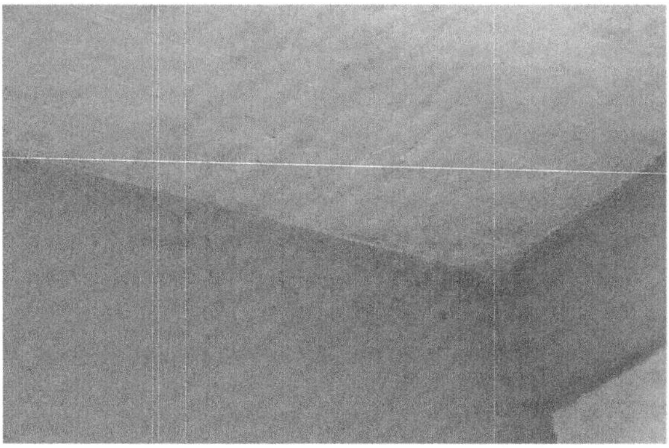

Fig. 6 Mould growth in interior surfaces of the envelope

The above differential equation is solved numerically in Energy-Plus using an explicit finite difference scheme, resulting in Eq. (2) for the energy balance.

$$T_z^t = \frac{\sum_{l=1}^{sl} Q_l^t + \sum_{k=1}^{Ns} h_k A_k T_{sk}^t + \sum_{i=1}^{Nzones} m_i C_p T_{zi}^t + m_{\text{inf}} C_p T_\infty^t + m_{sys} C_p T_{\text{sup}}^t}{\left(\frac{11}{6}\right)\frac{C_z}{\delta t} + \sum_{i=1}^{Ns} h_i A_i + \sum_{i=1}^{Nzones} m_i C_p + m_{\text{inf}} C_p + m_{sys} C_p} - \frac{\left(\frac{C_z}{\delta t}\right)\left(-3T_z^{t-\delta t} + \frac{3}{2}T_z^{t-2\delta t} - \frac{1}{3}T_z^{t-3\delta t}\right)}{\left(\frac{11}{6}\right)\frac{C_z}{\delta t} + \sum_{i=1}^{Ns} h_i A_i + \sum_{i=1}^{Nzones} m_i C_p + m_{\text{inf}} C_p + m_{sys} C_p} \tag{2}$$

The inside superficial temperature is given by

$$\frac{\rho C_p \Delta x\left(T_k^t - T_k^{t-\delta t}\right)}{\delta t} = \frac{\lambda\left(T_{k-1}^t - T_k^t\right)}{\Delta x} + \frac{\lambda\left(T_{k+1}^t - T_k^t\right)}{\Delta x} \tag{3}$$

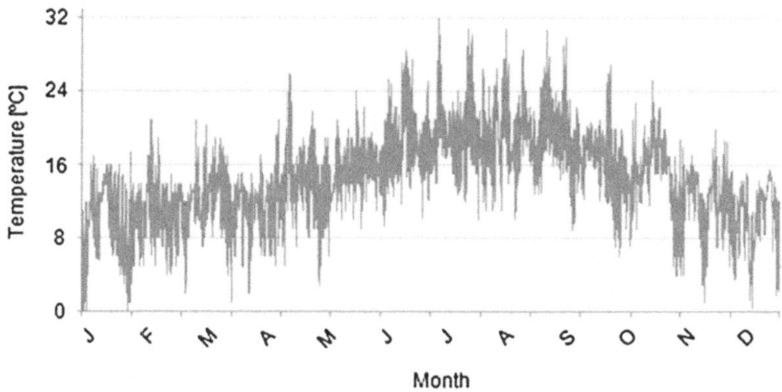

Fig. 7 Outdoor temperature

The aim of this study is to optimize the thickness of thermal insulation to be added in the retrofit process of the building envelope. The model seeks an optimal solution corresponding to the minimum use of the thickness of the different elements (concrete, glass and insulation) so that, without additional climatization:

- indoor air temperature throughout the whole occupation period is between the recommended values of comfort

$$T_{com} \leq T_z^t \leq 25^{\circ}C, \text{ if } \sum_{l=1}^{sl} Q_l^t > 0 \text{ for all } t \in \{1, \ldots Nt\};$$

- occurrence of surface condensation is not allowed

$$T_{si}^t \geq T_{dp}^t \text{ for all } k \in \{1, \ldots Np\}; \text{ and } t \in \{1, \ldots Nt\};$$

Since each interface between materials (concrete, glass and insulation) is a node with a predictable heat flow, it is necessary to include equations for its calculation. These equations result from Eq. (3) with the necessary modifications to the conditions of the school built. In this model, some simplifications have been taken into account. One such simplification is related to the thickness of the elements. In this study, the thickness of concrete and glass surfaces are given, while for each opaque surface the optimized thickness of thermal insulation takes the same value.

3.1 Case I: Model Without Solar Gains

The gains associated with solar radiation, an important factor of energy balance, were not included in the model. Thus, the energy balance obtained during the optimization process will only be adequate for colder days, usually associated with cloudy sky and rain, in which solar gains are negligible. The last simplification is the consideration of the internal gains by convection as an average value throughout the year, i.e., the values of the gains due to occupancy and lighting are considered as an average value during periods with occupation. All these considerations lead to the following nonlinear program: (P1) Minimize,

$$\sum_{i=1}^{Np} A_i\left(x_i^{ins} + x_i^{con}\right) + \sum_{j=N_p+1}^{Ns} A_j x_j^{glas} \tag{4}$$

subject to

$$T_z^t = \frac{\sum_{l=1}^{sl} Q_l^t + \sum_{k=1}^{Ns} h_k A_k T_{sk}^t + m_{\text{inf}} C_p T_\infty^t - \left(\frac{C_z}{\delta t}\right)\left(-3T_z^{t-\delta t} + \frac{3}{2}T_z^{t-2\delta t} - \frac{1}{3}T_z^{t-3\delta t}\right)}{\left(\frac{11}{6}\right)\frac{C_z}{\delta t} + \sum_{k=1}^{Ns} h_k A_k + m_{\text{inf}} C_p} \tag{5}$$

$$\rho^{con} C_p^{con}\left(x_i^{con}\right)^2 x_i^{ins}\left(T_{coni}^t - T_{coni}^{t-\delta t}\right) = \left[\lambda^{con} x_i^{ins}\left(T_\infty^t - T_{coni}^t\right) + \lambda^{ins} x_i^{con}\left(T_{si}^t - T_{coni}^t\right)\right]\delta t \tag{6}$$

$$\rho^{ins} C_p^{ins}\left(x_i^{ins}\right)^2\left(T_{si}^t - T_{si}^{t-\delta t}\right) = \left[\lambda^{ins}\left(T_{coni}^t - T_{si}^t\right) + h_i x_i^{ins}\left(T_z^t - T_{si}^t\right)\right]\delta t \tag{7}$$

$$\rho^{glas} C_p^{glas}\left(x_j^{glas}\right)^2\left(T_{sj}^t - T_{sj}^{t-\delta t}\right) = \left[\lambda^{glas}\left(T_\infty^t - T_{sj}^t\right) + h_j\left(T_z^t - T_{sj}^t\right)\right]\delta t \tag{8}$$

$$T_{si}^t \geq T_{dp}^t;\ T_{com} \leq T_z^t \leq 25^{\circ}C,\ \text{if} \sum_{l=1}^{sl} Q_l^t > 0 \tag{9}$$

$$x_i^{ins} \geq 0, \quad x_i^{con} \geq 0, \quad x_j^{glas} \geq 0 \tag{10}$$

with $i \in \{1, \ldots, Np\}$, $j \in \{Np+1, \ldots, Ns\}$, $k \in \{1, \ldots, Ns\}$, $l \in \{1, \ldots, sl\}$ and $t \in \{1, \ldots, Nt\}$.

The meaning of all parameters in this problem is presented below:

- N_t is the number of time intervals analysed;
- N_p is the number of opaque elements;
- N_s is the number of surface;
- $(Ns-Np)$ is the number of glass elements;
- δt is the time interval considered;
- Q_l^t is the internal gains at time t;
- h_k is the transfer coefficient of the surface k;
- A_k is the area of surface k;
- m_{inf} is the infiltrated air mass;
- $T_\infty{}^t$ is the outside air temperature at time t;
- $C/\delta t$ is the total amount of energy in the air during δt;
- $T_z{}^0$ is the average air temperature inside the 1st instant before the instant 1;

- T_z^{-1} is the average air temperature inside the 2nd instant before the instant 1;
- T_z^{-2} is the average air temperature inside the 3rd instant before the instant 1;
- ρ^m is the density of material m;
- $C_P{}^m$ is the specific heat capacity of material m;
- T_{dp} is the dew point temperature;
- T_{com} is the comfort temperature;
- x_i^m is the thickness of material m (con or glas) of surface i;
- λ^m is the thermal conductibility of material m.

The variables have the following meaning:

- x_i^{ins} is the thickness of thermal insulation of opaque element i
- T_z^t is the indoor air temperature at time instant t;
- $T_{sk}{}^t$ is the indoor superficial temperature of surface k at time instant t.

Note that $T_z^{t-k\delta t}$ is the average air temperature inside the k^{st} instant before the instant t.

3.2 *Case II: Model with Solar Gains*

The gains associated with solar radiation were included in a second version of the optimization model. The possibility of an optimization of the window thermal resistance was also considered in this model. The aim of this study was therefore to optimize the thickness of thermal insulation to be added in the retrofit process and an eventual replacement of windows of the building envelope. The model seeks an optimal solution corresponding to the minimum cost of the different elements (windows and insulation). The model underlying can be formulated as the following nonlinear program:

$$\text{(P2) Minimize: } \sum_{i=1}^{Np} A_i\left(x_i^{ins} V^{ins}\right) + \sum_{j=N_p+1}^{Ns} A_j V_U^{win} \tag{11}$$

subject to

$$T_z^t = \frac{\sum_{l=1}^{sl} Q_l^t + \sum_{k=1}^{Ns} h_k A_k T_{sk}^t + m_{\text{inf}} C_p T_\infty^t - \left(\frac{C_z}{\delta t}\right)\left(-3T_z^{t-\delta t} + \frac{3}{2}T_z^{t-2\delta t} - \frac{1}{3}T_z^{t-3\delta t}\right)}{\left(\frac{11}{6}\right)\frac{C_z}{\delta t} + \sum_{k=1}^{Ns} h_k A_k + m_{\text{inf}} C_p} \tag{12}$$

$$\rho^{con} C_p^{con}\left(x_i^{con}\right)^2 x_i^{ins}\left(T_{coni}^t - T_{coni}^{t-\delta t}\right) = \left[\lambda^{con} x_i^{ins}\left(T_\infty^t - T_{coni}^t\right) + \lambda^{ins} x_i^{con}\left(T_{si}^t - T_{coni}^t\right)\right]\delta t \tag{13}$$

$$\rho^{ins} C_p^{ins} \left(x_i^{ins}\right)^2 \left(T_{si}^t - T_{si}^{t-\delta t}\right) = \left[\lambda^{ins} \left(T_{coni}^t - T_{si}^t\right) + h_i x_i^{ins} \left(T_z^t - T_{si}^t\right) + x_i^{ins} \frac{RD^t}{\sum_{i=1}^{Np} A_i} \right] \delta t \tag{14}$$

$$\rho^{glas} C_p^{glas} \left(x_j^{glas}\right)^2 \left(T_{sj}^t - T_{sj}^{t-\delta t}\right) = \left[\frac{x_j^{glass}}{\frac{1}{U} - 0.17} \left(T_\infty^t - T_{sj}^t\right) + h_j \left(T_z^t - T_{sj}^t\right) \right] \delta t \tag{15}$$

$$T_{si}^t \geq T_{dp}^t;\; T_{com} \leq T_z^t \leq 25^\circ C, \text{ if } \sum_{l=1}^{sl} Q_l^t > 0 \tag{16}$$

$$x_i^{ins} \geq 0, \quad x_i^{con} \geq 0, \quad x_j^{glas} \geq 0 \tag{17}$$

$$1 \leq \mathrm{U} \leq 6 \tag{18}$$

with $i \in \{1,\ldots, Np\}$, $j \in \{Np +1,\ldots, Ns\}$, $k \in \{1,\ldots, Ns\}$, $l \in \{1,\ldots, sl\}$ and $t \in \{1,\ldots, Nt\}$.

The parameters incorporated in this new problem are:

- V^k is the cost of material k;
- $V_U{}^{win}$ is the cost of window which is a function of U;
- RD^t is the solar heat gains distributed to wall surface at time instant t.

While the only new variable introduced is:

- U the window heat transfer coefficient.

4 Results and Discussion

4.1 Simulation

The simulation of the building in the pre-retrofit scenario, using Energy Plus, provided a portrait of its hygrothermal behaviour. Energy Plus (Webpage: www.energy-plus.org) is the U.S. DOE building energy simulation program for modelling and optimizing building heating, cooling, lighting, ventilating, and other energy flows. It builds on the most popular features and capabilities of BLAST and DOE-2 but also includes many innovative simulation capabilities such as time steps of less than an hour, modular systems and plant integrated with heat balance-based zone simulation, multizone air flow, thermal comfort, and photovoltaic systems.

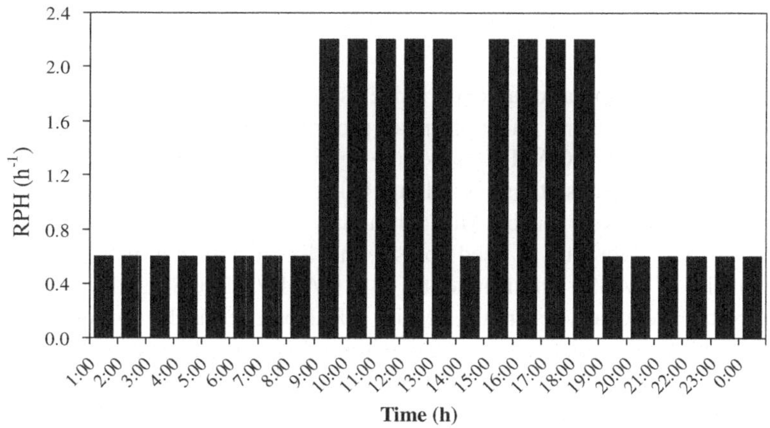

Fig. 8 Adopted schedule for ventilation rates

Its approach is highly detailed in aspects that affect hygrothermal behaviour such as solar gains calculation. The detailed introduction of relevant data is also of importance. Figure 8 presents the adopted schedule for ventilation rates, in the numerical simulations.

Figure 9 presents the indoor temperature variation in January. The existence of heat gains due to users and lighting as well as solar gains, provide a rather small difference between indoor and outdoor temperature. This is caused by the very high ventilation rate considered.

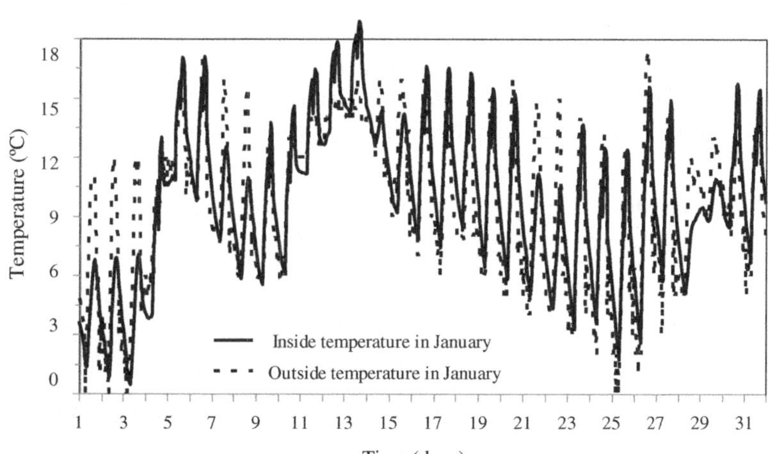

Fig. 9 Temperature variation during January in the pre-retrofit scenario

4.2 Optimization

In this section computational experience is reported on the solution of hydrothermal models by using the nonlinear programs P1 and P2. The commercial program Minos of the GAMS (General Algebraic Modelling System) collection has been used to process these problems [11].

This program developed by Murtagh and Saunders [12] is a specially adapted version of the solver that is used for solving linear and nonlinear programming problems in a GAMS environment [13]. GAMS/MINOS is designed to find solutions that are locally optimal (see Fig. 10). However, if the nonlinear objective and constraint functions are convex within this region, any optimal solution obtained will be a global optimum. Otherwise there may be several local optima, and some of these may not be global. In such cases the chances of finding a global optimum are usually increased by choosing a starting point that is sufficiently close, but there is no general procedure for determining what close means, or for verifying that a given local optimum is indeed global.

4.2.1 Case I: Model Without Solar Gains

The application of the Optimization Program (O.P.) to this specific case provided results at different levels. Not only the optimal (local) insulation thickness was spotted but also the evolution of indoor temperature and inside surface temperature was defined by hourly values. As it was possible to have the same temperature values simulated in Energy Plus, using the spotted insulation thickness, the temperature variations obtained in both models could be compared.

One of the days targeted for analysis was January 4, since it was one of the coldest days with occupation. The Optimization Program was run using the two criteria separately. Applying the first criterion, corresponding to the increasing of the inside temperature during winter, O.P. determined an insulation thickness of 4.5 cm. But with this insulation level the increase of indoor temperature for that

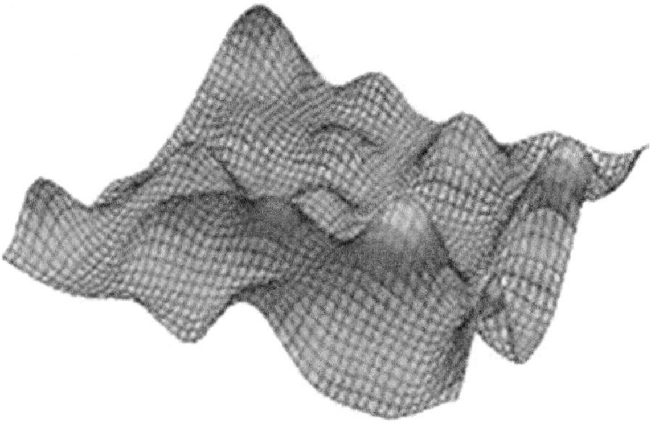

Fig. 10 Function with several local optima

day was less than 2 °C. For higher increments of temperature the problem is infeasible. Using the second criterion for optimization, corresponding to the elimination of surface condensation, the O.P. returned an insulation thickness of 2.4 cm.

The overall conclusion would be that relying only on indoor heat gains and passive behaviour, reaching hygrothermal comfort would be impossible. But the control of surface condensation, however, would be possible, using less than 3 cm insulation.

The results of the hourly values of temperature obtained in O.P. and Energy Plus were also compared, as presented in Fig. 11. The modelling of January 4 using O.P. provided excellent results, when comparing with the more complete modelling using Energy Plus. The quality of the simulation using O.P. cannot of

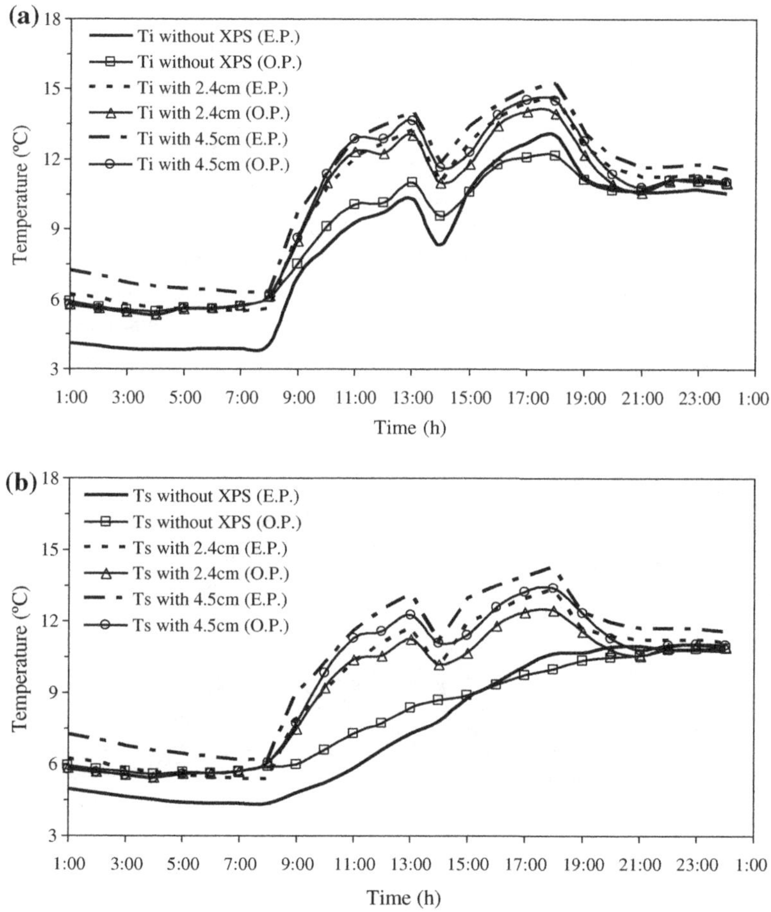

Fig. 11 Evolution of the (**a**) indoor temperature and (**b**) superficial indoor temperature on north façade, obtained by the optimization program (*O.P.*) and energy-plus (*E.P.*) with different insulation thicknesses, on January 4

course be achieved for days where high solar radiation prevails. That was the case for January 13 and, as it is presented in Fig. 12, the difference between O.P. and E.P. results is rather large. But winter days with high solar gains present a much lower risk of condensation.

To understand how far the results of O.P. could be interesting for finding an optimal insulation thickness regarding surface condensation control, the building was simulated in E.P. for several thickness values until condensation was eliminated. The results of that simulation are presented in Table 1. The results can be considered in agreement with the ones returned by O.P. if December 29 is disregarded. The comparison for that day is presented in Fig. 13. The graph shows that certain days can lead to an exaggerated value of insulation thickness due to specific hygrothermal behaviour with low overall significance.

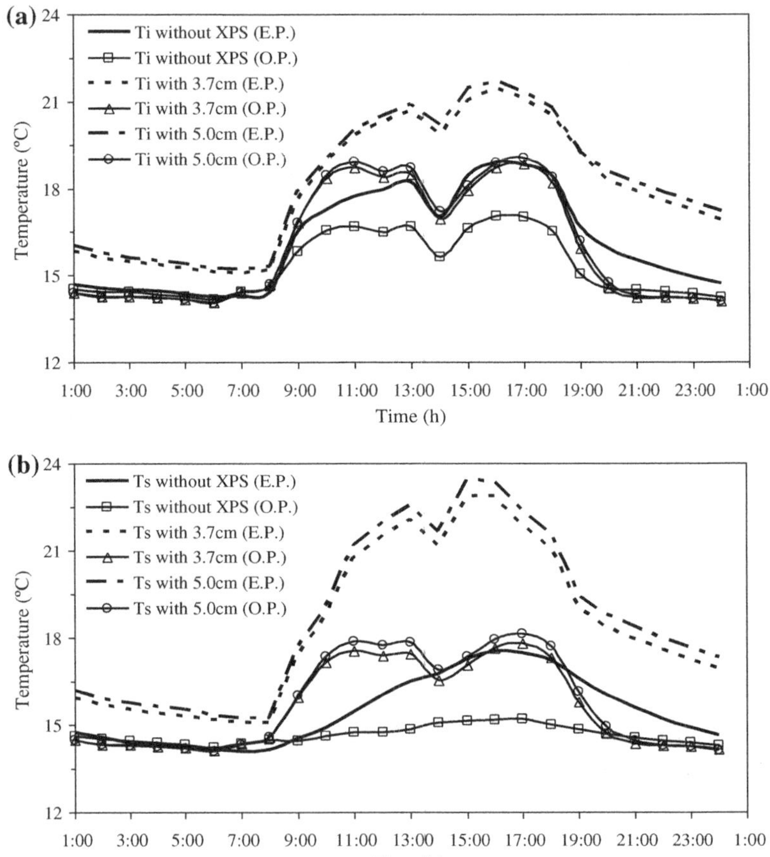

Fig. 12 Evolution of the (**a**) indoor temperature and (**b**) superficial indoor temperature on north façade, obtained by the optimization program (*O.P.*) and energy-plus (*E.P.*) with different insulation thicknesses, on January 13

Table 1 Results of energy-plus simulations with different thicknesses of thermal insulation

Insulation thickness (cm)		0	1	2	3	4	5	6	7	8	9
T_{min}		0.4	1.6	2.3	2.9	3.3	3.7	4.0	4.2	4.4	4.6
T_{max}		30.7	33.1	34.6	35.5	36.3	36.8	37.3	37.6	37.9	38.1
Results for an annual simulation	$P_{5\ \%}$	7.5	8.8	9.7	10.2	10.6	10.9	11.2	11.4	11.6	11.7
	T_{ave}	15.8	17.5	18.5	19.2	19.7	20.1	20.4	20.6	20.8	21.0
	$P_{95\ \%}$	23.6	26.0	27.4	28.3	29.0	29.5	29.9	30.3	306	30.8
No of days with superficial condensations on the walls		33	12	4	2	1	1	1	1	1	0
No of days with superficial condensations on the ceiling		126	22	4	4	1	1	1	1	1	0

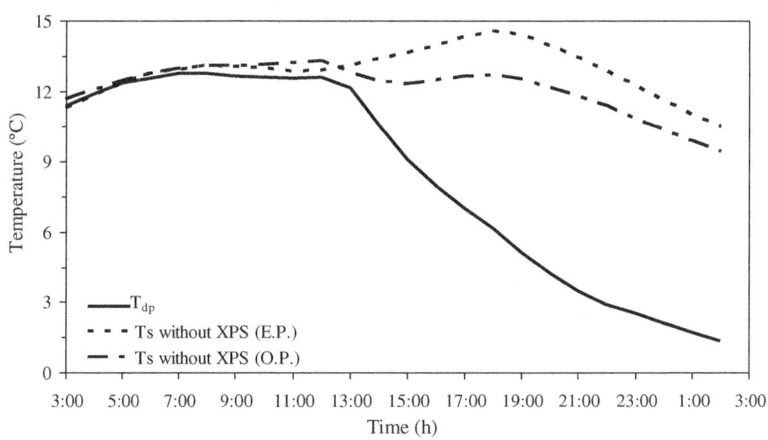

Fig. 13 Evolution of the superficial indoor temperature on north façade, obtained by the optimization program (*O.P.*) and energy-plus (*E.P.*), on December 29

4.2.2 Case II: Model with Solar Gains

January 4 and 25 were targeted for analysis as were the days with minimum and maximum radiation in that month (Fig. 14).

The Optimization Program, including solar gains and admitting the optimization of the window thermal resistance determined an insulation thickness of 4.5 cm and a U value of the window of 6.0 W/m^{2}C. The temperatures assessed for that day, presented in Fig. 15 show a close course to the obtained in Energy plus. It's also possible to see that the period more severe for the determination of insulation thickness was the end of the day, after the students leave the room. Changing the windows demonstrated not to be relevant for the specific objective of this optimization case.

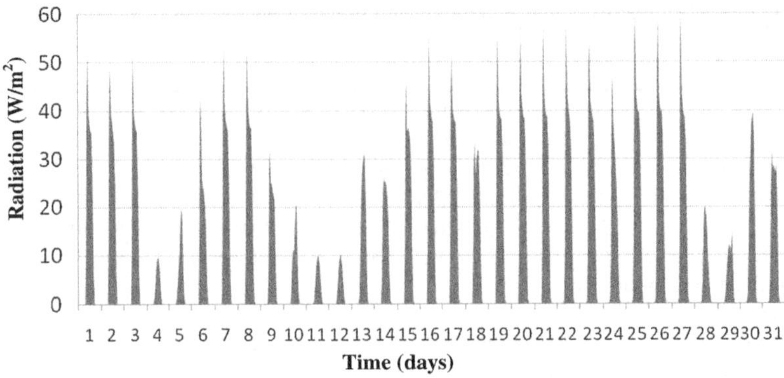

Fig. 14 Radiation in January

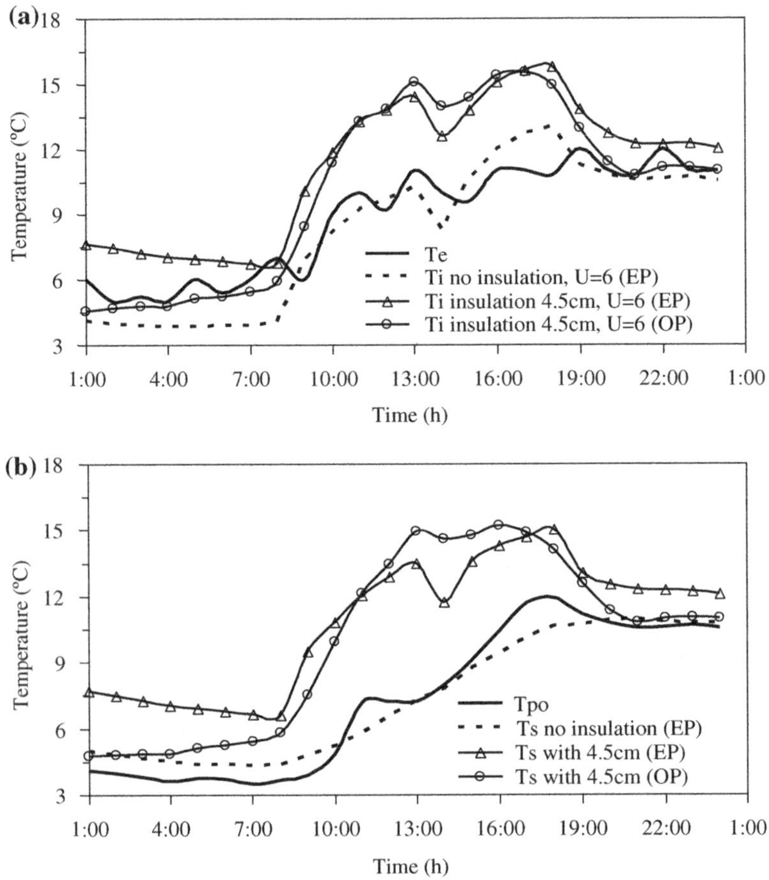

Fig. 15 Evolution of the (**a**) indoor temperature and (**b**) surface temperature, obtained by the optimization program (*O.P.*) and energy-plus (E.P.), on January 4

The results of the hourly values of temperature obtained in O.P. and Energy Plus were also compared for January 25, as presented in Fig. 16. The modelling of that specific day didn't provide an agreement as good as before. In this specific day, solar radiation was important, leading to a not so good performance of O.P. simpler model. Nevertheless, solar gains were included, leading to the possibility of a first assessment of this scenario.

To understand how far the results of O.P. could be interesting for the determination of the optimal insulation thickness regarding surface condensation control, the building was simulated in E.P. for several envelope configurations. The results of those simulations are presented in Fig. 17. The results show that a small thickness of insulation added to the wall can make a huge difference in terms of

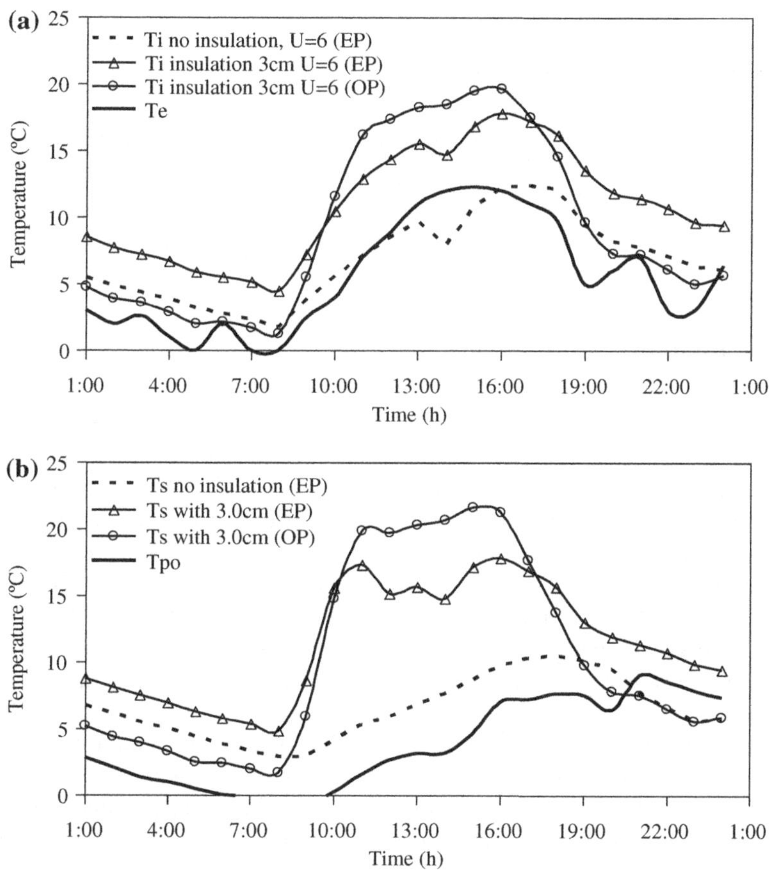

Fig. 16 Evolution of the (**a**) indoor temperature and (**b**) surface temperature, obtained by the optimization program (*O.P.*) and energy-plus (E.P.), on January 25

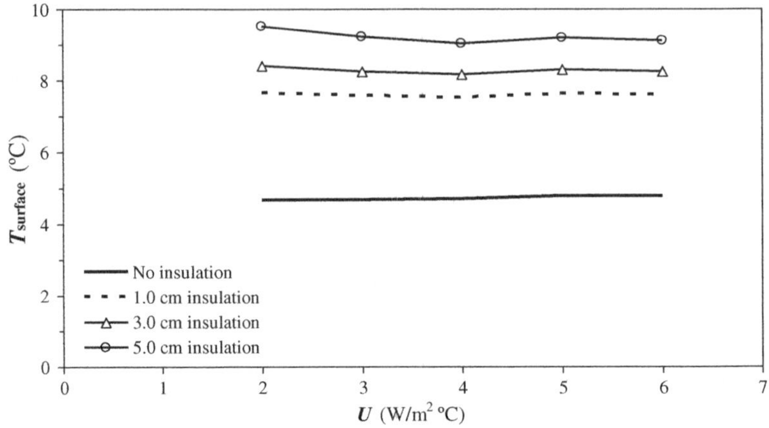

Fig. 17 Evolution of the superficial indoor temperature on north façade, obtained by the optimization program (*O.P.*) and energy-plus (*E.P.*), on December 29

resulting minimum surface temperature while the window resistance will have a very small influence. Hence, the O.P. pointed directly to a result that's feasible while considering, in a simplified model, the more relevant parameters to this problem.

5 Conclusions

Thermal comfort, energy efficiency, indoor air quality and durability must be analysed in depth for sensitive buildings of which schools are an example. A simple model, combined with an optimisation tool was able to provide a local optimal solution to the design of an envelope intended to prevent surface condensation. The scenario analysed in this example puts in evidence that a detailed analysis of the hourly hygrothermal behaviour can be very important. If a detailed analysis is available, the decision on using additional insulation will be based not only on energy efficiency versus comfort but also on durability aspects that can depend, for instance, on avoiding defacement caused by mould due to surface condensation.

In this specific case, it was shown how less than 3 cm additional insulation would be sufficient to reduce the risk of severe mould growth.

The optimization problem, consisting of minimizing a linear function on a set of linear and nonlinear constraints, using the solver MINOS of the commercial software GAMS proved to be an interesting aid for retrofit processes analysis. The limitations of the O.P. are linked to a not so perfect calculation if solar gains are included.

A set of aspects must be added, however, so that a complete design methodology can be of use, providing a complete risk analysis scheme. Especially the stochastic nature of the hygrothermal processes must be included, leading to a non deterministic approach. The authors believe that the developed optimization program will be useful due to the linear functions adopted on simulation.

Acknowledgment J.M.P.Q. Delgado would like to thank Fundação para a Ciência e a Tecnologia (FCT) for financial support through the grant SFRH/BPD/84377/2012.

References

1. PMEES: Programa de Modernização das Escolas do Ensino Secundário, Resolução do Conselho de Ministros n°1/2007, Lisboa, Portugal (in Portuguese) (2007)
2. SCE: Sistema nacional de certificação energética e da qualidade do ar interior nos edifícios, DL no 78/2006, Diário da República, I Série-A, Abril, Lisboa (in Portuguese) (2006)
3. RSECE: Regulamento dos sistemas energéticos de climatização dos edifícios, DL no 79/200 Diário da República, I Série-A, Abril, Lisboa (in Portuguese) (2006)
4. Grimsrud, D., Bridges, B., Schlte, R.: Continuous measurements of air quality parameters in schools. Build. Res. Inf. **34**(5), 447–458 (2006)
5. Freitas, V.P.: Educational building retrofitting. In: Proceedings of the 3rd International Symposium on Building Pathology, Durability and Rehabilitation-PATORREB, Porto, Portugal (2009)
6. Holm, A., Hellwig, R., Sedlbauer, K.: Retrofitting of a school with an integral aspect. In: Proceedings of the 8th Nordic symposium of Building Physics, Copenhagen, Denmark (2008)
7. Jernberg, P., Sjöström, C., Lacasse, M.A.: TC 140-TSL: Prediction of service life of building materials and components (joint with CIB W80): State-of-the-art report. Mater. Struct. **30**(1), 22–25 (1997)
8. Crawley, D.B., Lawrie, L.K., Winkelmann, F.C., Pedersen, C.O.: Energy-plus: energy simulation program. ASHRAE J. **42**(4), 49–56 (2000)
9. Energy-Plus: Engineering reference. In Energy-Plus Documentation Main Menu (2009)
10. ISO 7730: Ergonomics of the thermal environment analytical determination and interpretation of thermal comfort using calculation of the PMV and PPD indices and local thermal comfort criteria. Switzerland (2005)
11. Robinson, S.M.: A quadratically-convergent algorithm for general nonlinear programming problems. Math. Program. **3**(1), 145–156 (1972)
12. Murtagh, B.A., Saunders, M.A.: MINOS 5.1 User's Guide. Report SOL 83-20R (Revised), Department of Operations Research, Stanford University (1987)
13. Broke, A., Kendrick, D., Meeraus, A., Raman, R.: GAMS a User's Guide. GAMS Development Corporation, New York (1998)

Influence of Air Lime type and Curing Conditions on Lime and Lime-Metakaolin Mortars

Paulina Faria and Ana Martins

Abstract Air-lime mortars with or without pozzolanic components were largely used in the past. Due to natural or accidental degradation the application of repair mortars it is often necessary. Repair mortars have to be compatible with the masonries of historic buildings and should be as durable as possible (without compromising the previous compatibility condition). Within this context and associating the improvement of mortar characteristics with the necessity of sustainable construction practices, mortars formulated with limes and the addition of pozzolans have been studied. Each type of mortar presents its specificities. In pure lime mortars the setting occurs by carbonation and in lime-metakaolin mortars it occurs both by carbonation and hydration. A crucial question in order to optimize the characteristics of the mortars (and its applicability) is related to the curing conditions, which potentiate differently the reaction and consumption of the calcium hydroxide. This article describes an experimental campaign with different pure air lime mortars and lime-metakaolin mortars, cured under different conditions of relative humidity and CO_2 content. Properties of the mortars, mainly in terms of mechanical behaviour and open porosity, capillary water absorption, drying capacity and resistance to chloride contamination, are obtained, compared and discussed. The benefits in some properties revealed by the different mortars are correlated with the laboratorial curing conditions and with in situ application possibilities.

Keywords Air lime · Metakaolin · Mortar · Curing · Characterization

---

P. Faria (✉)
Nova University of Lisbon, 2829-516 Caparica, Portugal
e-mail: paulina.faria@fct.unl.pt

A. Martins
Polytechnic Institute of Setubal, 2839-001 Lavradio, Portugal
e-mail: ana.martins@estbarreiro.ips.pt

V. P. de Freitas and J. M. P. Q. Delgado (eds.), *Durability of Building Materials and Components*, Building Pathology and Rehabilitation 3,
DOI: 10.1007/978-3-642-37475-3_5,

1 Introduction

Air lime-based mortars are present in all Portuguese ancient buildings, in different types of application. The most common types of application in ancient buildings are as renders, plasters, ceramic glazed tiles adherence layers and masonry joint mortars.

These mortars are composed by air lime, as unique or at least main binder, sand and sometimes pozzolans.

An important role is played by the aggregates, as their mineralogical type, shape maximum size and particle size distribution, influence the structure and the behaviour of the mortars [1–4].

Lime-based mortars are ecological mortars, in comparison with mortars with cement, because air lime is obtained by calcination at low temperature, approximately half the temperature needed for cement production. They are also compatible with historic masonries, what does not happened with cement-based mortars [5].

Air lime can be purchased and used as a hydrated powder, after hydration of quicklime with a minimum of water, or as putty, obtained by hydration of quicklime with excess of water [6, 7]. The hydration of the quicklime occurs with rising temperature and traditionally can be held together with the addition of vegetal or animal fat, for water repellence of the air lime.

The pozzolans, fine materials rich in silicate and aluminate in amorphous form [8], although not being a binder because pozzolans do not react alone with water, can partially substitute the air lime. In the presence of water, the pozzolans react and combine with the calcium hydroxide of the air lime, developing calcium silicate and calcium aluminate hydrates which confer hydraulic properties to the mortars [9] and can also increment its durability, while generally maintaining its compatibility with old masonry materials [10, 11].

An available and promising pozzolanic material is metakaolin, obtained by dehydrating kaolinitic clays at around 600 °C, bellow temperatures that cause the formation of a vitreous phase and crystallization of other phases such as mullite [12]. Kaolinitic clays are available in Portugal, although many quarries are no longer active due to lack of demand. Kaolin for metakaolin production can also be obtained from some industrial by-products or from kaolin rejected from other industries (as the ones rejected for fine ceramic production or from kaolinitic sand washing). Metakaolin is an amorphous material, with high specific surface and also high content of acidic oxide ($SiO_2 + Al_2O_3 > 90$ %) [13]. Due to lack of other traditional pozzolanic materials, such as natural pozzolans or fly ash from thermoelectric power plants, and the abundance of kaolins, the Portuguese industrial and research sectors are working towards metakaolin production [14] and optimization of application [15–17].

Pure air lime mortars harden by carbonation, while air lime-pozzolan mortars harden by carbonation but also cure by hydration. The carbonation process occurs by combination of $Ca(OH)_2$ with CO_2 from the environment and depends on the presence and transport of CO_2 throughout the mortar. The carbonation of air lime mortars affect the pore structure of the mortars and, in consequence, its properties. There is a change in the volume of pores associated with the transition of the binder from calcium hydroxide to calcium carbonate. It seems that pores below 0.1 μ m are not involved in the carbonation process [18], what might explain why the carbonation of air lime mortars can continue for many years.

In lime-metakaolin mortars the amorphous silicates and aluminates react with CO_2, producing CSH gel and several calcium silicate and aluminate hydrates (as C_2ASH_8 and C_4AH_{13}) [13]. This pozzolanic reaction is a slow process as well; depends on the presence of uncarbonated lime ($Ca(OH)_2$), the reactivity of the pozzolan, which also depends on its specific surface, and the presence of water.

The presence of water, as moisture, is therefore important for the CO_2 transport for carbonation and for the hydration of compounds by pozzolanic reaction.

The microstructure that is established and, consequently, the characteristics of lime-metakaolin mortars depend on which of the two reactions prevails [13] and that then are affected by on the reactivity of the constituents, their proportion in the mortars and the curing conditions.

Further advantages of air lime-metakaolin mortars are their lower environmental impact, when compared to cement mortars. This is due to lower energy consumption during the air lime and the metakaolin production, the possibility of using kaolinitic by-products or kaolin rejected by other industries partially substituting binders [19, 20] and the absorption of CO_2 by carbonation. Also an advantage can be the light colour of the mortars, that can be changed a little by the chosen sands or other aggregates; indeed the colour of the mortars may be important for joint repointing and to unpainted renders.

Since the beginning of the 20th century and until nowadays air lime mortars have been replaced in ancient buildings, mainly in plasters and renders, by cement mortars and, due to this reason, the thousands of years knowledge of lime mortars craftsmen abruptly decreased. In the last decades, the origin of many defects that appeared in ancient buildings was correctly attributed to the cement mortars that have been applied. Many researchers, all over the world, have been trying to fundament the advantages of air lime-based mortars when compared with cement-based mortars. Fortunately also the knowledge of lime mortars craftsmen tends to be regained [21, 22].

The main problems of cement mortars when applied as substitution renders in ancient buildings are their mechanical, chemical and physical incompatibility with the masonries and with other old mortars. In fact, cement mortars are much stiffer and stronger than the old masonry walls, and cement mortar renders induce stresses at the interface with those walls. Later on it tends to break by the wall that the render was supposed to protect.

Frequently cement mortars release salts, namely sulphates, which also contribute to the contamination of those walls.

Water from different sources is often present in old walls such as by capillary rising from the ground, problems with roofs, migration of the rain water through the porous structure of exterior layers of walls and water vapour generated inside the building, which migrate through the thickness of walls, and its protective layers, towards the exterior. The water can transport salts from the outside and also salts that were already inside the walls. When the water front faces a layer that is much less permeable to water vapour in comparison to the wall materials—some paint layers [23] or some substitution mortar layers, for instance—, the water, eventually transporting dissolute salts, concentrate in the previous layers, often the exterior surface of the original walls, which are weaker than the impermeable rendering layers.

In cold climates the water in the wall can originate problems of freeze/thaw, generating stresses and weakening the surrounding material. When salts like chlorides are involved, they can easily go through cyclic crystallization/dissolution processes, involving stresses that also weaken the old materials [24].

Even if the exterior rendering seems in good conditions, behind its thickness often there are voids, due to material that lost cohesion. Later on the apparently good substitution render detaches, showing a huge degradation in the wall itself.

Nevertheless air lime mortars also have disadvantages mainly regarding actual construction constrains. In fact, in construction sites rapid construction schedules and fast resistant gains are often pursuit and these are not easily achieved with pure air lime mortars.

Frequently no one cares if cement renders will behave properly and if they will really protect the walls; the short term apparent resolution of the problem is generally the only constrain and that is why in some countries, cement mortar plasters and renders continue to be used in interventions on ancient buildings.

But it should also be remarked the fact that lime renders need different application procedures. Sometimes lime renders are applied with the same techniques as if they were cement renders, and that may lead to unsuccessful new uses of lime mortars. Some of those different procedures are: the air lime-based mortars should be applied with low consistency compared with cement mortars because their workability is excellent; if the water is added in order to achieve similar consistency to the one needed for cement mortars, it will be too much and the mortar layer will be to porous; the lime render have to be applied in separate thin layers, with about a week between them to achieve some carbonation; the lime mortar rendering layer should be re-tight over the base after suffering initial shrinkage to achieve better compactness and nullify that shrinkage.

The characteristics of the walls where the mortars are applied alter significantly the properties of the mortars and that should also be considered when formulating a mortar. But in the study that is presented here, the main focus is on the different properties obtained with air lime-based mortars cured under different relative humidity conditions and CO_2 content.

In face of the results, several conclusions can be taken in order to optimize in situ curing conditions of pure air lime mortars and of air lime-metakaolin mortars, and to define possibilities to accelerate laboratorial curing of specimens.

2 Experimental Campaign

2.1 Preparation of the Material and of the Mortar Samples

For the preparation of mortars two commercial "washed" sands were used as aggregates: a 0/4 sand with coarser particles and a 0/2 sand only with finer particles; they were used in a mixture of one volume of finer sand and two volumes of coarser sand (Fig. 1). The mixture of sands intended to enlarge the grading curve of the mortar aggregate and to minimize the volume of voids (which was 35 % for the 0/4 sand, 40 % for the 0/2 sand and 32 % for the mixture) while maximizing the loose bulk density.

Two calcium limes were used as binder in different mortars: a powder hydrated lime EN 459-1 CL90-S commercialized by Lusical (designated in the following text as air-lime AL); a water repellent lime putty EN 459-1 CL 90-S PL commercialized by Fradical (designated as PL) [25]. From what is known, an olive oil by-product is incorporated in the lime putty production, as water repellent natural product.

While a lime putty stay uncarbonated since it is always covered by a water layer, care should be taken to ensure that a powder hydrated lime also stays uncarbonated [26]. In fact, it is common that powder hydrated lime is commercialized in paper bags, not completely air and water vapour tight. Some samples were collected from closed bags directly received by the factory while others were collected from similar bags, after being stored at interior laboratory conditions for some time. A higher content of carbonated lime was found in samples that have been stored. The hydrated lime that was used was recently produced and was stored carefully; the paper bags were kept inside other plastic bags.

A metakaolin was used as a pozzolan (Mk) produced from thermal and granular treatment of a Portuguese kaolin from Grupo Lagoa (Table 1). This company was,

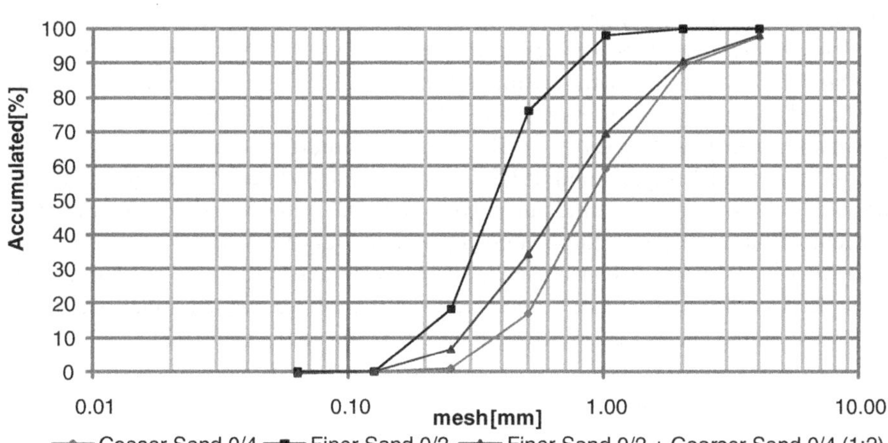

Fig. 1 Particle size distribution *curves* of the sands and their volumetric mixture

Table 1 Metakaolin characterization [19, 45]

Mk	SiO_2	Al_2O_3	Fe_2O_3	K_2O	TiO_2	P_2O_5	CaO	LOI
(%)	52.17	44.5	0.45	0.15	1.42	0.12	0.01	1.42

by that time, optimizing the product [14]. When the pozzolanic reactivity was tested by the Chapelle test determined by NF P 18-513 [27] it did not tested "reactive" (a value of 265 mg $Ca(OH)_2$/g metakaolin was registered while the reference value would be 700 mg Ca(OH)2/g metakaolin). Its specific surface by the Blaine method was 9,310 cm^2/g. Other metakaolins can be found with much higher specific surface and reactivity by the same tests, such as 33,760 cm^2/g and 860–1,320 mg $Ca(OH)_2$/g metakaolin for commercial metakaolin Argical M 1200 S [14, 19].

All mortar volumetric compositions were 1:3—1 volume of binder (calcium lime (or calcium lime + metakaolin) and 3 volumes of sand (1 volume of finer sand and 2 volumes of coarser sand). The volumetric composition was chosen in order that the volume of binder would optimize the filling of the voids, left by the volume of sand. The volumetric compositions of the constituents were transformed in precise weight values by the loose bulk density [28], presented in Table 2 although PL is not a granular material.

The dried weight of a specimen of lime putty registered that the lime putty had 59 % of water content. The lime content of the lime putty PL multiplied by its bulk loose density, compared with the loose bulk density of the hydrated lime AL, showed that the lime content in the lime putty mortars was 30 % higher than the one in mortars with powder hydrated lime.

In lime-metakaolin mortars 20 % of the mass of the volume of lime was replaced by identical mass of metakaolin. In terms of weight composition, it represented 25 % of the mass of the remaining lime. In mortars with lime putty, once the volume of lime was heavier, compared with the same volume of powder hydrated lime, the 20 % mass content in metakaolin was higher but was still 25 % of the mass of the remaining lime.

Four different mortars were prepared: mortar AL with powder hydrated lime; mortar AL+Mk with the same hydrated lime and 20 % of metakaolin substitution; mortar PL with lime putty; mortar PL+Mk with the same lime putty and 20 % of metakaolin substitution (Table 3). By weight the mortar compositions were 1:12 for mortars with lime AL and 1:3.5 for mortars with lime PL; in this last case, including the water of the lime putty. By weight but without considering the water content of the lime putty, the composition of mortars PL was 1:9 and of mortars PL+Mk was 1:11. As it can be seen, similar volumetric ratios of mortar constituents lead to different weight compositions and the lime putty mortar constitution (even excluding the water) is stronger in terms of binder (Table 3).

Table 2 Loose bulk density of mortar materials

Material	AL	PL	Mk	Sand 0/2	Sand 0/4
Loose bulk density (g/cm^3)	0.38	1.26	0.71	1.50	1.55

Table 3 Mortar and curing designation, volumetric and weight composition, water/binder ratio and flow table consistency

Mortar	Mortar/curing designation	Volumetr. comp. (binder: sand)	Weight comp.		Water/binder ratio		Flow
			(Binder: sand)	(Lime: Mk: sand)	Added	Existent	(mm)
AL	AL_D	1:3	1:12	1:0:12	2.4		155
	AL_S						
	AL_C						
	AL_H						
AL+MK	AL+Mk_D	1:3	1:12	1:0.25:15	2.4		154
	AL+Mk_S						
	AL+Mk_C						
	AL+Mk_H						
PL	PL_D	1:3	1:3.5	1:0:3.5		0.6	151
	PL_S						
	PL_C						
	PL_H						
PL+MK	PL+Mk_D	1:3	1:3.5	1:0.25:4.6	0.2	0.7	152
	PL+Mk_S						
	PL+Mk_C						
	PL+Mk_H						

The mixture of the mortar components was mechanical and always identical: the water was added in the mechanical mixer tank, followed by the air lime and the sand (previously hand homogenized); 30 s of mechanical mixture at low speed; another 30 s to scrape the material inside the tank and mechanical mixture for three more minutes at high speed. The procedure was based on EN 196-1 [29] and EN 1015-2 [30] but the period of mixture was enlarged because the one defined in the standard was considered inadequate for air lime-based mortars. In lime putty mortars PL no water was added and only a little amount was added for PL+Mk mortars; the other constituents were mixed in a similar way.

The water/binder ratio is registered in Table 3. The existent water was determined by the water content of the lime putty of mortars with PL.

The mortar samples were mechanically compacted in two layers inside prismatic metallic moulds 40 × 40 × 160 mm. The general samples of each mortar were subjected to four types of curing conditions until the age of test—at 7 days for shrinkage evaluation, from 30 to 120 days for carbonation tests, at 60 days for mechanical tests, at 120 days for capillary and drying tests—, at 20 °C temperature, inside conditioned chambers: 50 % relative humidity (RH)—curing identified by D; 65 % RH—standard curing identified by S; 65 % RH and 5 % carbon dioxide—curing identified by C; 95 % RH—curing identified by H (Table 3). Six samples of each mortar were subjected to each curing conditions. After tested at 60 days for mechanical properties, one half of each sample was kept in interior summer conditions and tested for compressive strength and open porosity at

120 days; after tested at 120 days for physical properties, one half of each sample was kept in interior conditions and tested for resistance to chlorides contamination at 17 months.

2.2 Testing Program and Preliminary Results

For each type of mortar multiple mixings were made, due to the mechanical mixer capacity and the number of mortar samples needed for the experimental campaign. For each type of mortar, when needed, the same quantity of water was always added. The quantity of water was added so that all the mortars seemed to provide good workability for application in real conditions. The influence of the amount of water in the fresh mortars was evaluated by the consistency flow table test [31].

Mortar's shrinkage inside the moulds was evaluated, with six samples of each mortar/curing condition, before demoulding, at the age of 7 days—except mortars cured inside the carbonation chamber (cure C) that could only be demoulded (without registering any visual shrinkage) at the age of 21 days.

At the age of 7 days those C samples were almost as soft as at the moment of moulding; at the age of 14 days the problem persisted and only at the age of 21 days, with particular care, they could be demoulded. A possible justification for this occurrence was a possible saturation of carbonate ions at the only exterior surface of the samples (still inside the moulds), forming a solution rich in hydrogen carbonates, from the reaction of carbon dioxide with water, that strongly diminished the carbonation velocity or even stopped the carbonation front in the mortar sample exterior face. It showed that confined rich CO_2 environments are not adequate for laboratory initial curing of lime-based mortars.

Nevertheless, and except for mortars C, shrinkage inside the moulds of the different mortar submitted to diverse curing conditions was registered, showing that shrinkage evaluation since moulding—and not only after demoulding—is important to lime-based mortars.

The carbonation velocity intended to be evaluated by the phenolphthalein method. A phenolphthalein solution at 0.5 % in alcohol was applied in freshly cut surfaces 2 cm thickness of three samples of each type at the ages of 30, 60, 90 and 120 days. It was obvious that mortar C achieved complete carbonation during the test; for the other curing conditions the test colour change (and the carbonation) seemed to be very slow, generally a little faster in mortars D and S, and a little slower in those in cure H. For lime-metakaolin mortars, a trend could not be seen using this test and other method should be pursuit [32], especially taking into account the influence of the pozzolanic reaction on PH.

At 60 days of age, three samples of each mortar and curing were dried in an oven at 60 °C until constant mass, with weight variation in 24 h not higher than 0.1 %. The mentioned drying of the samples intended to stop (or at least minimize) the curing at the age of test and to homogenize the moisture content of samples.

The mortar samples were used for dynamic modulus of elasticity determination using fundamental resonance frequency [33] and three points bending flexural strength determination [34]. One half of each sample from the flexural test was used to compressive strength determination at 60 days [34]. As mentioned before, from 60 to 120 days the other half of each sample was kept in interior summer environment at 30 ± 3 °C temperature and 50 ± 5 % RH. Those conditions were not particularly beneficial for the lime-based mortars curing, due to the lack of moisture for carbon dioxide transport and for pozzolanic reaction. At 120 days those half samples were used to compressive strength determination and afterwards the tops of those half samples, which were perfectly undamaged, were used for open porosity determination by vacuum and hydrostatic weighing [35, 36].

At 120 days, the half of three samples of each mortar and curing, resulting from the ones used before for the carbonation determination, were dried in an oven at 60 °C until constant mass. After cooling in dry environment, they were used for capillary water absorption determination (Capillary Coefficient in terms of initial capillary absorption velocity and Capillary Absorption in terms of total adsorbed water) [37, 38]. The lateral faces of the samples were not watertight and the test was held inside a box with saturated environment; the samples were placed over a geotextil with 2–5 mm water high.

When completely saturated by capillary water, the samples were directly used for the drying index determination [23, 39], also without watertight faces. This situation allowed drying to occur over a large surface and without being unidirectional. During drying the mortar samples were kept in environmental conditions of 20 ± 3 °C temperature and 50 ± 5 % RH.

After this test the samples were kept in interior environment at medium temperature of 25 ± 3 °C and 57 ± 5 % RH. At the age of 17 months the half samples of each mortar and curing that have been used for capillary and drying tests were dried in an oven at 60 °C until constant mass and submitted to a resistance to chlorides contamination test [40]. After cooling in a dry environment, they were immerged in a sodium chloride solution with 1,000 g NaCl in 3.4 L of water for 24 hours and dried again until constant mass. By the difference between the dry masses of each sample after and before immersion, the percentage of retained chlorides was determined. The samples were then placed inside a climatic chamber where they were exposed to repeated cycles of 12 h at 90 % RH and 12 h at 40 % RH, with a constant temperature of 20 °C. During those cycles the samples were weekly weighed to determine the mass variation that occurred and the type of degradation.

3 Results

Results of flow table consistency are presented in Table 3. For all mortars preparation a comparable consistency flow of 153 ± 3 mm was always reached. For mortars with lime AL the water/binder ratio and the consistency did not change with the metakaolin partial substitution; for mortars with lime PL the partial substitution of lime putty PL by powder metakaolin implied an increment on the total water/binder ratio (considering the existent plus the added water) for a similar consistency. That can be justified by the fact that the partial weight substitution of lime putty by powder metakaolin is indeed a big volume of powder instead of putty (composed by calcium hydroxide plus water).

Test results of mortars for mechanical characteristics (dynamic modulus of elasticity, flexural and compressive strength) and internal structure (open porosity) are presented in Table 4; total capillary absorption, capillary coefficient, drying index, retained chlorides and weight variation at 42 cycles after chlorides contamination are presented in Table 5.

It is expected that mortars cured with high CO_2 optimize carbonation and cured with high RH optimize hydration. For that reason in pure air lime mortars curing with high CO_2 is expected to be the most favorable; in air lime-metakaolin mortars the fact that curing with high CO_2 content optimize carbonation, with calcium hydroxide consumption, can diminish the possibility of hydration because there is a lack of $Ca(OH)_2$. That is why with air lime-metakaolin mortars the balance is unknown.

3.1 Mechanical Characteristics

3.1.1 Dynamic Modulus of Elasticity

The dynamic modulus of elasticity E_d is associated to the deformability of the mortars; this means that mortars with low E_d would to be more deformable than mortars with higher E_d.

Table 4 Test results (average values and standard deviation) of dynamic modulus of elasticity, flexural and compressive strength and open porosity of mortars and curing

Mortar/curing	$E_{d\ (60d)}$		$R_{f\ (60d)}$		$R_{c\ (60d)}$		$R_{c\ (120d)}$		O.P.	
(ID)	(MPa)	StDv	(MPa)	StDv	(MPa)	StDv	(MPa)	StDv	(%)	StDv
AL_D	2,671	10	0.2	0.1	0.4	0.1	1.0	0.0	30	0.2
AL_S	2,627	42	0.2	0.0	0.5	0.0	0.9	0.0	30	0.3
AL_C	5,028	227	0.8	0.1	1.4	1.0	1.2	0.2	29	0.4
AL_H	2,412	98	0.2	0.0	0.4	0.1	0.9	0.1	31	0.3
AL+Mk_D	3,023	123	0.3	0.0	0.5	0.2	1.1	0.1	30	0.1
AL+Mk_S	2,822	71	0.3	0.1	0.4	0.2	1.0	0.0	30	0.2
AL+Mk_C	3,691	504	0.7	0.1	1.2	0.3	1.2	0.2	29	0.3
AL+Mk_H	3,194	76	0.3	0.0	0.7	0.3	1.3	0.0	29	0.2
PL_D	1,529	29	0.2	0.0	0.3	0.1	0.6	0.0	35	0.4
PL_S	1,455	4	0.2	0.0	0.3	0.0	0.6	0.0	35	0.1
PL_C	4,587	179	0.7	0.1	1.3	0.4	1.7	0.2	35	0.3
PL_H	1,232	43	0.2	0.0	–	–	0.5	0.0	35	1.0
PL+Mk_D	2,153	52	0.3	0.1	0.5	0.0	0.9	0.0	34	0.1
PL+Mk_S	2,132	61	0.2	0.0	0.5	0.1	0.9	0.1	35	0.1
PL+Mk_C	4,518	147	0.8	0.0	1.6	0.1	1.9	0.2	35	0.4
PL+Mk_H	2,167	86	0.4	0.1	0.7	0.0	1.0	0.0	35	0.1

Table 5 Average values of total capillary absorption, capillary coefficient, drying index, retained chlorides and weight variation at 42 cycles after chlorides contamination of mortar/curing

Mortar/curing (ID)	Capillary absorp. (kg/m^2)	Capillary coef. ($kg/m^2 \cdot min^{0.5}$)	Drying index (–)	Ret.chlor. (%)	Weight var. cycle 42 (%)
AL_S	13.21	1.14	0.25	2.6	−20.0
AL	13.90	1.20	0.23	2.6	−11.0
AL_C	12.42	1.17	0.24	2.2	2.8
AL_H	14.40	1.32	0.23	2.6	2.6
AL+Mk_S	13.85	1.07	0.15	2.5	−10.6
AL+Mk	14.47	1.12	0.21	2.6	−7.9
AL+Mk_C	16.91	0.96	0.13	2.6	4.1
AL+Mk_H	13.33	0.92	0.24	2.3	4.8
PL_S	4.08	0.09	0.35	1.4	−54.7
PL	3.27	0.03	0.33	1.4	−44.9
PL_C	1.54	0.01	0.27	0.2	−1.4
PL_H	3.28	0.05	0.42	1.8	−48.7
PL+Mk_S	4.11	0.13	0.56	1.3	1.7
PL+Mk	2.74	0.13	0.47	1.3	0.7
PL+Mk_C	6.49	0.03	0.25	1.7	−46.1
PL+Mk_H	4.40	0.07	0.51	1.7	0.3

As shown in Table 4 and Fig. 2, mortars with powder hydrated lime AL present higher dynamic modulus of elasticity than mortars with lime putty PL, what may induce a higher deformability of lime putty mortars due to the decrease of portlandite crystal dimensions of the putty when compared to powder lime [41]. Mortars with metakaolin AL+Mk and PL+Mk present higher E_d than similar mortars without metakaolin AL and PL—except on the case of mortars with powder hydrated lime cured with high CO_2 content AL+Mk_C, with a very high standard deviation. With regard to mortars of each type (mortar and curing), the

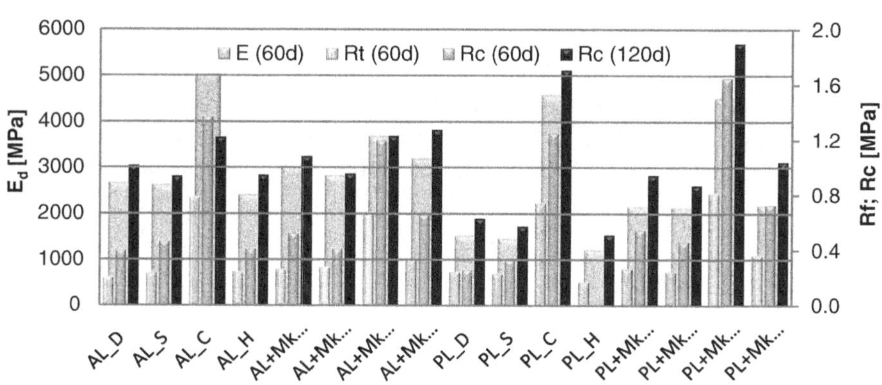

Fig. 2 Dynamic elasticity modulus versus flexural and compressive strength of mortars

higher E_d is always registered by samples C cured with high CO_2 content; the following values of E_d are registered by samples cured at 50 or 65 % RH (cure D or S) for mortars without metakaolin and by samples cured at 95 % RH (cure H) for mortars with metakaolin.

3.1.2 Flexural Strength

In what concerns flexural strength of mortars, higher values induce better resistance to cracking; but compressive strength values of rendering mortars should not be too high and do not overpass those of the wall where the mortars are to be applied.

Regarding the flexural strength at 60 days of age (Table 4 and Fig. 2), mortars with lime putty generally register a slight increase comparatively with mortars with powder hydrated lime. Except for mortar AL+Mk_C, mortars with metakaolin present slightly higher values of flexural strength than similar pure lime mortars. In what concerns each type of mortar with different type of cure also mortars cured with high content of CO_2 register the higher results of R_f.

3.1.3 Compressive Strength

Respecting the compressive strength, an increase of the results generally occurs from 60 to 120 days of age of the mortars (Table 4 and Fig. 2), although the alteration of environmental condition where the samples were kept meanwhile. In terms of percentage, the increase was lower with cure C because the acceleration of carbonation curing.

At 120 days mortars with powder hydrated lime AL register higher values than those with PL—except for mortar with lime putty PL cured with a high content of CO_2.

Mortars with metakaolin generally register an increase of compressive strength compared to the similar ones without this pozzolan. In what concerns each type of mortar with different types of curing, as for the case of E_d and R_f, also mortars cured with high content of CO_2 register the highest results of R_c, except for AL+Mk_C; as happened before for E_d, the following values of R_c are registered by mortars cured at 50 or 65 % RH (cure D or S) among mortars without metakaolin and by mortars cured at 95 % RH (cure H) among those with metakaolin.

3.2 Internal Structure

In terms of internal structure, mortars were tested for open porosity determination. It is higher for mortars with lime putty PL compared with mortars with powder hydrated lime AL; those last mortars are then denser than the previous. Results of similar mortars with or without metakaolin are almost the same. Regarding each type of mortar with different types of curing, only mortars with powder lime AL cured with high content of CO_2 present a lower open porosity and a higher compactness; that can be related to a rapid carbonation evolution of this mortar.

Results of open porosity can justify some of the mechanical characteristics obtained although they cannot justify the higher mechanical characteristics of mortars cured with high CO_2 content; but results underline the particularly different internal structure that may occur in mortars with lime AL compared with lime PL. One of the reasons can be due to the fact that the lime putty PL was water repellent. Further studies about the microstructure of mortars need to be carried on.

3.3 Physical Characteristics

3.3.1 Capillary Absorption

As expected, there is a strong difference of capillary coefficient and total capillary absorption between mortars with powder air lime AL and with lime putty PL; the last mentioned mortars are much less absorbent than the others due to the water repellent natural product incorporated in the lime putty production, which, from what is known, is an olive oil by-product (Table 5 and Fig. 3).

Capillary coefficient of mortars with hydrated lime AL is lower for mortars with metakaolin; for mortars with lime putty PL, capillary coefficient is a little higher

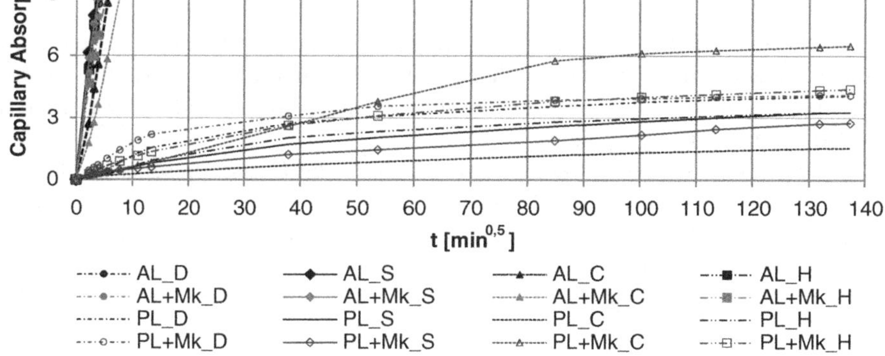

Fig. 3 Capillary absorption *curves* of mortars

when metakaolin partially substitutes the lime. Regarding each type of mortar with different types of curing, generally mortars cured with high content of CO_2 register the lower capillary coefficient results; only mortar with powder lime with metakaolin cured in humid conditions AL+Mk_H present a slightly lower value compared to AL+Mk_C. But these results should be analyzed together with the total capillary water absorption.

In what concerns the total capillary water absorption, mortars with lime putty PL also have a very low absorption compared with those with AL. Comparing mortars with lime AL, there is an increase of total capillary water absorption when metakaolin is used, except mortar with humid curing AL+Mk_H.

These last mentioned mortar present the best behaviour in terms of capillary absorption test, among the ones with AL+Mk and particularly in terms of capillary coefficient among mortars with lime AL; mortar AL_C present the best behaviour among the ones with AL without Mk, in terms of capillary absorption, and one of the best concerning capillary coefficient. Comparing mortars with lime PL, the best behaviour in terms of capillary absorption test is registered by mortar PL_C. Among mortars with PL+Mk, the mortar with lower capillary coefficient register the highest values of total absorption and the mortar with lower total capillary absorption register the higher capillary coefficient.

3.3.2 Drying

Results of capillary absorption must be analyzed together with the drying capacity of mortars, fundamental for the elimination of water, once absorbed (Table 5 and Fig. 4).

Mortars with powder hydrated lime AL register lower values of drying index compared to mortars with water repellent lime putty PL, what means the moisture can be easily and faster eliminated from AL mortars than from PL mortars.

Among mortars with lime AL, mortars with metakaolin generally present lower values of drying index; between mortars with lime PL, mortars without metakaolin present lower values—except for mortar C cured with high CO_2 content, with similar values.

All mortars AL without metakaolin cured in different conditions present very similar values; among mortars with the same lime but with metakaolin, mortar AL+Mk_D and AL+Mk_C register the lowest values, and humid cured mortar H present the highest drying index.

Among mortars with lime PL, mortar with curing C also presents the lowest value and mortar with curing H register the highest value; among mortars with metakaolin PL+Mk, a low drying index is register by mortar C while the other mortars present the highest values.

Comparing the capillary coefficient with the drying index of mortars (Fig. 5) it can be noticed that, in general terms, there is an inverse correlation between the capillary coefficient and the drying index, showing that with this type of mortars, faster capillary absorption is correlated with easier drying and vice versa.

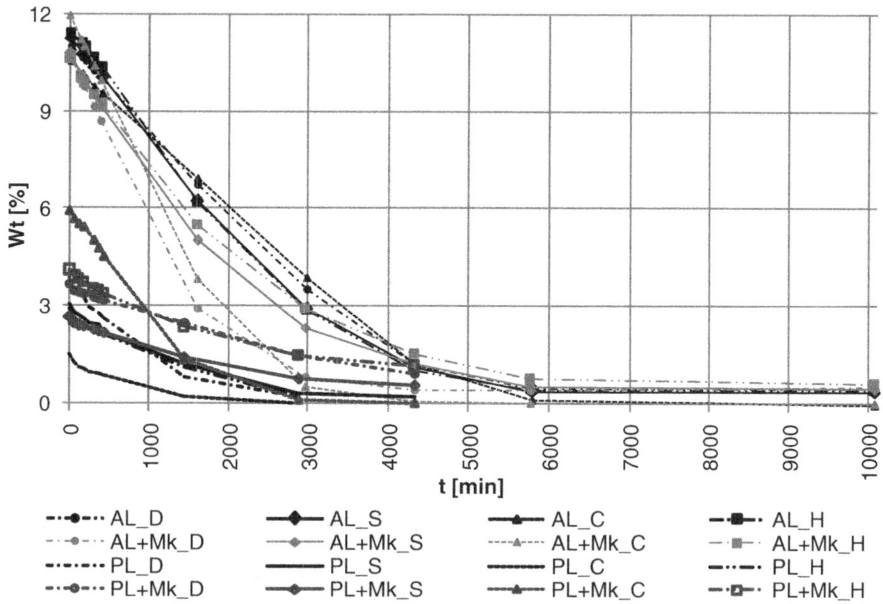

Fig. 4 Evaporation *curves* of mortars

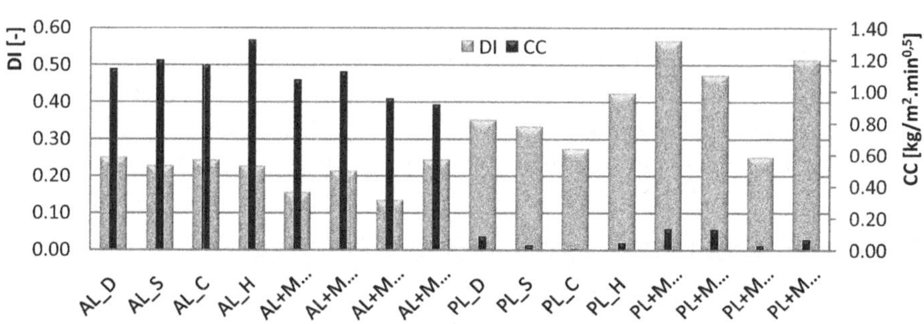

Fig. 5 Capillary coefficient versus drying index of mortars

3.4 Resistance to Chlorides Contamination

Table 5 presents the retained chlorides and the weight variation of mortars at 42 cycles after chlorides contamination. Weight variations due to humid and drying cycles after chloride contamination are presented in Fig. 6.

The deterioration of the samples generally occurred by superficial disaggregation, as can be seen in Fig. 7. But the degradation of lime putty mortars was much more visible and intense than the one that occurred in powder hydrated lime mortars. In Fig. 7 some similarity can be observed between the PL mortar sample

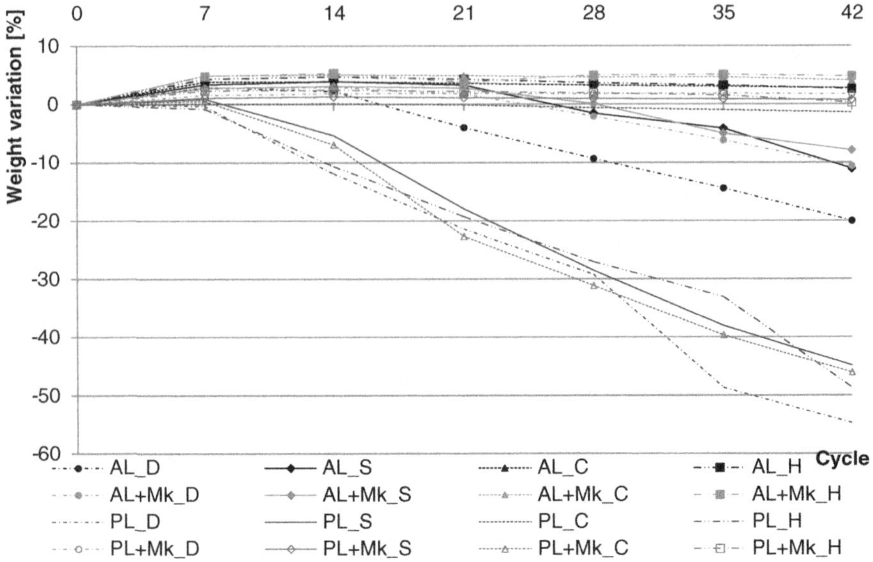

Fig. 6 Weight variation as a function of the number of cycles after chlorides contamination

degradation and degradation that occurs in plasters and renders of buildings with problems of capillary rising and salts contamination when a vapour barrier is formed.

At Table 5 it can be noticed that the percentage of retained chlorides is high for all the mortars with powder lime AL; only a little lower for AL_C and AL+Mk_H. It is much lower for all the mortars with water repellent lime putty PL. But the percentage of retained chlorides is particularly low for PL_C mortar, showing that

Fig. 7 General view of mortar samples at 42 cycles after chlorides contamination and zoom view of an AL mortar sample only with superficial disaggregation and a PL mortar sample with great mass loss

something enables the mortar with pure lime putty with a water repellent agent cured with high level of CO_2 to retain chlorides and that cannot be justify by the measured open porosity.

Observing Fig. 6, presenting the percentage of weight variation of the mortar/curing over the resistance to chlorides contamination test, the mortars behaviour can be divided in three groups: the mortars that suffers a huge degradation almost since the first cycles, losing a high percentage of mass—as the cases of PL_D, PL_H, PL_Mk_C and PL_S; the ones that resist the first cycles but then begin loosing mass between the 14th and the 21st cycles—as the AL_D, AL_S, AL+Mk_D and AL+Mk_S; the mortars that maintain their integrity and mass over the multiple cycles—all the other mortars. Among those last mentioned mortars there is the PL_C mortar with very low percentage of chloride retained. In general terms and except the PL_C mortar, all the mortars with pure lime putty with water repellent agent presented low resistance to chloride contamination; the substitution by metakaolin seemed to improve PL mortar resistance to chloride, except, in this case, for the PL+Mk_C mortar. In terms of AL mortars, with or without metakaolin, the humid curing and curing with high CO_2 content (H and C) seemed to be most favorable in what concerns the chlorides attack.

4 Discussion

Firstly, a remark about the reactivity of the air lime itself. Knowing that the lime putty particles diminish their size and increment their reactivity as long as they are covered by a film of water and, therefore, without carbonating [6], care should be taken with the powder hydrated lime, as mentioned at 2.1. In fact, these powder hydrated limes are sometimes commercialized in paper bags and the contact with humidity and atmospheric CO_2 is not totally prevented. It must be ensured that the powder hydrated lime that is going to be used was not stored since production in humid and cold environments without the bags being completely covered by plastics, because if that does not happened, someone can be using powder hydrated lime partially carbonated and acting as a filler instead of a real binder and without noticing the fact.

In what concerns the metakaolin, natural clays—other than pure kaolinite—or even industrial by-products can be interesting pozzolanic materials when thermally activated and the substitution of air lime by this type of material can result in energetic and environmental gains [19, 20].

The experimental campaign highlighted several aspects: the good workability of air lime-based mortars, even with relatively low flow table consistencies; the difficulty but the need of evaluating shrinkage since moulding in lime-based mortars; the phenolphthalein test inadequacy to evaluate lime-metakaolin carbonation due to changes registered on PH by pozzolanic reaction; the increase of mechanical characteristics when lime was partially substituted by metakaolin, although the metakaolin that was used was not chosen by its reactivity, but by the fact of being an available Portuguese metakaolin, and the lime-metakaolin proportion was not optimized; the good deformability, expressed in terms of the low dynamic modulus of elasticity, evidenced by all mortars but particularly by mortars with lime putty, with similar flexural strength results; the higher open porosity of mortars with lime putty compared to powder hydrated lime mortars, but with similar mechanical resistances when metakaolin replacement occurred; the low capillary water absorption of analyzed lime putty mortars, produced with a water repellent agent, but also the greater difficulty to dry of these mortars (while mortars with lime without water repellent agent absorbed rapidly and more quantity of capillary water but could release that moisture more easily); an improvement of the behaviour of powder hydrated lime mortars to initial water capillary absorption and drying capacity when metakaolin was added and the inverse situation for lime putty mortars with water repellent agent; the improvement of the resistance to chlorides of powder hydrated lime mortars with and without metakaolin with humid and carbonated curing and the generally week resistance to chlorides attack of lime putty mortars without metakaolin—for curing other than C or with metakaolin with cure C.

Comparing the obtained values with the general requirements concerning some characteristics for rendering, plastering and repointing substitution mortars for ancient buildings [42] it can be seen that all the mortars with powder hydrated lime AL satisfy those requirements in terms of mechanical characteristics, while the mortars with lime putty PL can only satisfy all those requirements when metakaolin is added. In fact the lime substitution for metakaolin and the optimization of curing conditions may increment the mechanical strength but mortars do not strengthen enough to generate stresses that might lead to failure of the ancient walls; thus the mechanical compatibility is ensured.

Concerning the behaviour in the presence of water, the comparison with mentioned general requirements can only be done in terms of the capillary coefficient. While all the mortars with AL lime satisfy the requirements, these may not be satisfied by the mortars with the PL lime because of the too low values that are presented by these mortars.

Regarding the influence of different curing conditions, the most important aspects detected were: the initial difficulty of hardening of lime mortars inside the moulds when exposed to high levels of CO_2; the increase of carbonation evolution and of mechanical characteristics, after the initial hardening, when cured with high level of CO_2; also a different behaviour in terms of resistance to chlorides was highlighted especially with lime putty mortar with curing C. Although that cure situation is not reproducible in situ, the acceleration of cure of pure air lime mortars that, after initial hardening and demoulding, are submitted to higher CO_2 environments during a defined period of time, can help in the preparation of lime mortar samples to be tested and characterized, but must be further studied to assure there are no important changes at microstructural level.

Accordingly this study is being extended to further characterization of mortar samples submitted to carbonation curing and other curing conditions, combining some of the previous and different ones. The aim is to define cure conditions that potentiate pure lime mortars characteristics, in order to be able to prepare reproducible laboratory specimen that can be used as aged substrate to the application of other products to be tested (as the cases of paint systems and consolidants), but also trying to optimize curing conditions that can be reproduced in situ. The optimization of curing conditions, that can be reproduced in situ for lime-metakaolin mortars, will also continue to be tested, namely taking into direct consideration the substrate influence.

For the time being and from the obtained results of conditions reproducible on site, curing at 65 % RH seemed to be the most appropriate for pure air lime mortars, but the resistance to chlorides attack—as well as to other salts, like sulphates—have to be deeply studied.

For lime-metakaolin mortars, and although the general improvement registered in the mortar's characteristics with partial substitution of lime by metakaolin, the authors think that a substantial improvement can yet be achieved with the use of a more reactive metakaolin—especially with a higher specific surface that should lead to a higher reactivity—and an optimized proportion between each type of lime and the metakaolin [43].

In lime-metakaolin mortars the amount of calcium hydroxide must be able to react with the silicates and aluminates of the metakaolin but also to carbonate. It is important to be aware of the kinetic of both the pozzolanic reaction and the carbonation process, in order to potentiate the best conditions during mortar formulation and curing. It is expected that a higher proportion of binder, for instance a volumetric proportion 1:2 of binder:aggregate, may potentiate the pozzolanic effect, because although some calcium hydroxide became carbonated, it is more likely there would be some left (uncarbonated) to hydrate.

From the analyzed results, and although not reproducible in situ, curing with high level of CO_2 generally potentiated the lime-metakaolin mortar characteristics—except for resistance to chlorides attack with the analyzed water repellent lime putty. Among the curing conditions that can be closer to in situ situations, humid curing can potentiate lime-metakaolin mortars characteristics. Humid curing seems fundamental both for the continuity over time of the hydration—the pozzolanic reaction—and the CO_2 transport—for the carbonation process.

In most Portuguese exterior environmental conditions, and even in summer time, cycles occur, between night and day, ranging from very humid to dryer conditions. That situation provides some level of humidity to the renders but may not be enough; nevertheless a geotextile covering, frequently wetted, could be recommended to be applied in situ over lime-metakaolin renders during the first ages, let say the two first weeks [44].

The addition of fine sepiolite to air lime-metakaolin mortars, acting as a water reservoir in pozzolanic systems, can also be a possibility for low-humidity applications [15].

Non water repellent lime-metakaolin mortars may also be appropriate to walls with capillary rise problems.

For interior plastering of old walls in very humid environments, the application of lime-metakaolin mortars should be advantageous.

Lime-metakaolin mortars should also be advantageous for repointing masonries and substitution layers supporting ceramic glazed tiles in ancient buildings, as an alternative to old air lime pure mortars, as these are applications where the contact with CO_2 is minimized.

5 Conclusions

In general terms it can be highlighted that:

- a partial substitution of air lime by metakaolin can be advantageous in terms of the characteristics of the mortars;
- water repellent putty lime should be used with care because its behaviour in terms of moisture and chloride attack is not always the most appropriate for rendering mortars;
- according to, the type of air lime mortar, with or without metakaolin, the choice of the binder, should take into consideration the environmental conditions where the mortars will be applied;
- great care should be taken with the curing conditions in order to optimize and potentiate the mortar characteristics and their applications in ancient buildings.

Acknowledgments The authors acknowledge the support of the Polytechnic Institute of Setubal where the experimental work was conducted, the availability of limes Lusical and Fradical and of metakaolin from Grupo Lagoa, and the financial support from the Portuguese Science and Technology Foundation to projects LIMECONTECH (PTDC/ECM/100234/2008) and METACAL (PTDC/ECM/100431/2008).

References

1. Faria-Rodrigues, P., Henriques, F., Rato, V.: Current mortars: Influence of the type of binder and aggregate (in Portuguese). In: 2° Congresso Nacional de Argamassas de Construção, APFAC. Lisboa, LNEC (2007)
2. Konow, T.: Aggregates grain size distribution. A major influence on many properties of lime mortars for restoration. In: EUROMAT 2003—European Congress on Advanced Materials and Processes, EPFL, Lausanne (2003)
3. Rato, V.: The influence of morphologic microstructure on the performance of mortars (in Portuguese). PhD thesis Civil Eng., Nova University of Lisbon (2006)
4. Stefanidou, M., Papayianni, I.: The role of aggregates on the structure and properties of lime mortars. Cem. Concr. Compos. **27**, 914–919 (2005)
5. Faria-Rodrigues, P., Henriques, F.: Current mortars in conservation: an overview. Restor. Build. Monum. **10**(6), 609–622 (2004)
6. Faria, P., Henriques, F., Rato, V.: Comparative evaluation of aerial lime mortars for architectural conservation. Cult. Herit. **9**(3), 338–346 (2008)

7. Margalha, G., Veiga, R., Santos Silva, A., Brito, J.: Traditional methods of mortar preparation: the hot lime mix method. Cem. Concr. Compos. **33**, 796–804 (2011)
8. Instituto Português da Qualidade: Pozzolans for concrete, mortars and grouts. Definitions, requirements and evaluation of conformity (in Portuguese). Caparica, NP 4220:2010 (2010)
9. Charola, E., Faria-Rodrigues, P., McGhie, A., Henriques, F.: Pozzolanic components in lime mortars: correlating behaviour, composition and microstructure. Restor. Build. Monum. **11**(2), 111–118 (2005)
10. Faria-Rodrigues, P.: Resistance to salts of lime and pozzolan mortars. In: RILEM Proceedings PRO 067—International Workshop on Repair Mortars for Historic Masonry, pp. 99–110 (2009)
11. Veiga, R., Velosa, A., Magalhães, A.: Experimental applications of mortars with pozzolanic additions: characterization and performance evaluation. Constr. Build. Mater. **23**, 318–327 (2009)
12. Velosa, A., Rocha, F., Veiga, R.: Influence of chemical and mineralogical composition of metakaolin on mortar characteristics. Acta Geodyn. Geomater. **6, 1**(153), 121–126 (2009)
13. Fortes-Revilla, C., Martínez-Ramírez, S., Blanco-Varela, M.T.: Modelling of slaked lime-metakaolin mortar engineering characteristics in terms of process variables. Cem. Concr. Compos. **28**, 458–467 (2006)
14. Ferraz, E., Andrejkovicová, S., Santos Silva, A., Rocha, F., Velosa, A.: Use of modified Chapelle test to evaluate pozzolanic reactivity of metakaolins (in Portuguese). In: 4° Congresso Português de Argamassas e ETICS. APFAC, Coimbra (2012)
15. Andrejkovicová, A., Ferraz, E., Velosa, A., Santos Silva, A., Rocha, F.: Fine sepiolite addition to air lime-metakaolin mortars. Clay Miner. **46**, 621–635 (2011)
16. Faria, P., Silva, V., Flores-Colen: Mortars of natural hydraulic lime and artificial pozzolans: laboratorial evaluation in Portuguese. In: 4° Congresso Português de Argamassas e ETICS. APFAC, Coimbra (2012)
17. Velosa, A., Veiga, R.: Lime-metakaolin mortars: properties and applications. SB07—Sustainable Construction, Materials and Practices. Challenges of the Industry for the New Millenium, Lisbon (2007)
18. Lawrence, R., Mays, T., Rigby, S., Walker, P., Ayala, D.: Effects of carbonation on the pore structures of non-hydraulic lime mortars. Cem. Concr. Res. **37**(7), 1059–1069 (2007)
19. Pontes, J.: Reactivity of artificial pozzolans for mortar and concrete (in Portuguese). MSc thesis Civil Eng., Nova University of Lisbon (2011)
20. Tironi, A., Trezza, M., Scian, A., Irassar, E.: Kaolinitic calcined clays: factors affecting its performance as pozzolans. Constr. Build. Mater. **28**, 276–281 (2012)
21. Faria, P., Tavares, M., Menezes, M., Veiga, R., Margalha, G.: Traditional Portuguese techniques for application and maintenance of historic renders. In: RILEM Proceeding PRO 078 HMC2010—2nd Historic Mortars Conference and RILEM TC 203-RHM, Repair Mortars for Historic Masonry Final Workshop, pp. 609–617 (2010)
22. Sandstrom Malinowski, E.: Historic mortars revived. Developing local materials and crafts for restoration. In: RILEM Proceedings of PRO 067—Repair Mortars for Historic Masonry, pp. 328–338 (2009)
23. Brito, V., Dias Gonçalves, T., Faria, P.: Coatings applied on damp building substrates: Performance and influence on moisture transport. J. Coat. Technol. Res. **8**(4), 513–525 (2011)
24. Gonçalves, T.: Salt crystallization in plastered or rendered walls. PhD thesis Civil Eng., LNEC and Technical University of Lisbon/IST (2007)
25. CEN: Building lime. Part 1: Definitions, specifications and conformity criteria. Brussels, EN 459–1:2010 (2010)
26. Dheilly, R., Tudo, J., Sebau Bi, Y., Quéneudec, M.: Influence of storage on the carbonation of powdered $Ca(OH)_2$. Constr. Build. Mater. **16**(3), 155–161 (2002)
27. Association Française de Normalisation—AFNOR: Métakaolin, addition pouzzolanique pour bétons. Définitions, spécifications, critéres de conformité. Paris, NF P 18-513:2010 (2010)

28. CEN: Tests for mechanical and physical properties of aggregates. Part 3: Determination of loose bulk density and voids. Brussels, EN 1097–3:1998 (1998a)
29. CEN: Methods of testing cement. Part 1: Determination of strength. Brussels, EN 196–1:2005 (2005)
30. CEN: Methods of test for mortars for masonry. Part 2: Bulk sampling of mortars and preparation of test mortars. Brussels, EN 1015–2: 1998/A1:2006 (1998b)
31. CEN: Methods of test for mortars for masonry. Part 3: Determination of consistency of fresh mortar (by flow table). Brussels, EN 1015–3: 1999/A1:2004/A2:2006 (1999a)
32. Lawrence, R., Mays, T., Walker, P., Ayala, D.: Determination of carbonation profiles in non-hydraulic lime mortars using thermogravimetric analysis. Thermochim. Acta **444**(2), 179–189 (2006)
33. CEN: Natural stone test methods. Determination of the dynamic modulus of elasticity (by measuring the fundamental resonance frequency). Brussels, EN 14146:2004 (2004)
34. CEN: Methods of test for mortars for masonry. Part 11: Determination of flexural and compressive strength of hardened mortar. Brussels, EN 1015–11:1999/A1:2006 (1999b)
35. CEN: Natural stone test methods. Determination of real density and apparent density, and of total and open porosity. Brussels, EN 1936:2006 (2006)
36. RILEM Commission 25 PEM: Essais recommendés pour mesurer l′alteration des pierres et évaluer l′efficacité des methods de traitment. Matér. Constr. 13(75) (1980)
37. CEN: Methods of test for mortars for masonry. Part 18: Determination of water absorption coefficient due to capillary action of hardened mortar. Brussels, EN 1015–18:2002 (2002)
38. CEN: Conservation of cultural property. Test methods. Determination of water absorption by capillarity, Brussels, EN 15801:2009 (2009)
39. Commissione Normal: Misura dell′indice di asciugamento (drying index), Roma, CNR/ICR, Doc. 29/88 (1991)
40. Faria-Rodrigues, P.: Mortars for ancient masonry renderings. The influence of binders (in Portuguese). PhD thesis Civil Eng., Nova University of Lisbon (2004)
41. Hansen, E., Tagle, A., Erder, E., Baron, S., Connell, S., Rodríguez-Navarro, C., Van Balen, K.: Effects of ageing on lime putty. In: RILEM Proceedings PRO 012—Historic Mortars. Characteristics and Tests, pp. 197–207 (1999)
42. Veiga, R., Fragata, A., Velosa, A., Magalhães, A., Margalha, G.: Lime-based mortars: viability for use as substitution renders in historical buildings. Archit. Herit. **4**(2), 177–195 (2010)
43. Gameiro, A., Santos Silva, A., Veiga, R., Velosa, A.: Hydration products of lime-metakaolin pastes at ambient temperature with ageing. Thermochim. Acta 535, 36–41 (2012)
44. El-Turki, A., Ball, R., Holmes, S., Allen, W., Allen, G.: Environmental cycling and laboratorial testing to evaluate the significance of moisture control for lime mortars. Constr. Build. Mater. **24**(8), 1392–1397 (2010)
45. Gomes, A.: Renderings with fibers for rehabilitation (in Portuguese). MSc Thesis Civil Eng., University of Aveiro (2010)

Effects of Ageing and Moisture on Thermal Performance of ETICS Cladding

Bruno Daniotti, Riccardo Paolini and Fulvio Re Cecconi

Abstract Herein we present the results of an experimental activity to assess the degradation evolution and the loss in hygrothermal performances of External Thermal Insulation Composite Systems with rendering (ETICS). We exposed to accelerated ageing a sample of ETICS (with vinyl resin added base coat and acrylic thick finishing coat onto expanded polystyrene) applied to a masonry wall. The sample was aged with a repetition of two macro-cycles consisting of 125 UV cycles, 125 summer thermal shock cycles, 50 winter thermal shock and freeze–thaw cycles. To assess the hygrothermal performances over time, we measured the thermal transmittance and the time shift when exposed to sinusoidal forcing, while degradation was surveyed with photographs at fixed positions. We note that, with ageing, the studied building component offered higher water absorption capability and, thus, the thermal resistance presented a decreasing trend, while an increasing trend was observed for heat capacity, and time shift. With regard to the evolution of degradation, we observed a very strong influence of the thermal shock and dilatation-contraction events, resulting in blistering and deformation of the top coat.

Keywords External thermal insulation composite systems (ETICS) · Thermal admittance · Time shift · Decrement factor · Moisture

B. Daniotti (✉) · R. Paolini · F. Re Cecconi
Architecture, Built Environment and Construction Engineeering Department, Politecnico di Milano, Via Ponzio 31, 20133 Milano, Italy
e-mail: bruno.daniotti@polimi.it

R. Paolini
e-mail: riccardo.paolini@polimi.it

F. Re Cecconi
e-mail: fulvio.rececconi@polimi.it

V. P. de Freitas and J. M. P. Q. Delgado (eds.), *Durability of Building Materials and Components*, Building Pathology and Rehabilitation 3,
DOI: 10.1007/978-3-642-37475-3_6,

1 Introduction

Exterior insulation cladding systems are a very common building technology—since the 1960s—both in Continental Europe (where they are usually applied to a masonry or concrete substrate and are called ETICS), and in North America (where they are usually applied on a wood frame and are called EIFS).

In different countries and regions the same kind of building component is realised in similar but often slightly different fashions, and so also different names may exist for the cladding system and its parts. Herein we refer to ETICS as they are described in the European Technical Approval Guideline for the External Thermal Insulation Composite Systems with rendering (ETAG 4) [1]. Thus, all the considerations and the results we present relate to cladding systems made of insulation boards mechanically fixed and bonded with a cement glue to a masonry or concrete substrate, and finished with a thin mortar—in ETAG's terminology defined as base coat—including a glass fibre mesh as reinforcement, covered by a paint or a thick, mineral or resin based, finishing coat, also known as top coat. Further in the text we will also refer to the thick finishing coat as RPAC. However, despite differences in terminology, and more relevantly, different building materials, exterior insulation cladding systems offer similar durability issues in different countries. The most relevant reported degradation modes are: mould growth, cracking over joints of insulation boards and blistering due to summer thermal shocks, water absorption [2–4], and air leakages combined with wind driven rain; the latter is true especially for EIFS [5]. In synthesis, the most relevant problems occur in presence of water (by itself or in synergy with other agents). Furthermore, an increase in the moisture content in the insulator affects all hygrothermal performances.

Given the importance of moisture influence on degradation and hygrothermal performances, several Heat and Moisture Transport (HMT) numerical studies have been performed to assess the response to moisture of exterior insulation cladding systems [6, 7]. However, software models are normally not able to cope with some phenomena (e.g. gravity), or they adopt simplifications, such as assuming a coefficient describing the percentage of water hitting the façade that does not bounce away from the surface (a detailed study about the error introduced considering a uniform rain distribution over the surface of a wall has been developed by Abuku et al. [8]). Sometimes, constraints are given by the lack of input data, for instance driving rain, especially with a sampling time shorter than one hour, or properties of interfaces between layers—relevant in heat and moisture transfer [9]—or of singular points. Actually, building details are the main points of penetration of rain in EIFS [10]. Especially for ETICS, the main moisture flow takes place, beyond connections with other building components (e.g. windows), through joints between insulation boards [11, 12].

Considered all these issues reported about exterior cladding systems, with accelerated ageing and testing in the laboratory, we assessed ETICS cladding durability, comprising the survey of the degradation of the finishing system and the monitoring of the evolution of the hygrothermal performances. Herein, we present results showing the evolution of the thermal resistance of the ETICS cladding, and of the masonry substrate, due to accelerated ageing and increase in moisture content. Moreover, we discuss changes in the unsteady state hygrothermal performance, measured with laboratory tests on the aged sample, and computed with computer modelling (assessing the influence on results of the stochastic variability of input parameters, influenced by water content). We also performed tests to characterize the capillary water absorption and the optical and radiative response of the finishing coats, which are useful to understand the performance loss and to explain the development of degradation, surveyed during ageing with photographs.

2 Description of the Experimental Programme

The experimental programme here presented has been developed to study the durability of ETICS, and includes accelerated laboratory ageing, long-term outdoor exposure, computer simulations, and comparison and analysis of results. Here we report about the results of laboratory activities.

2.1 Accelerated Ageing Procedure

One might argue that is not necessary to design new accelerated ageing procedures, since several standards detailing cycles to assess the durability of different materials already exists. However, the purpose of those standard accelerated ageing procedures is to provide an assessment of the suitability for use; they do not provide a service life prediction and they are not based on ISO 15686-2 [13] nor they make reference to it, as documented by Daniotti & Re Cecconi [14]. Also, in the EOTA guidance document to the assessment of the working life of products [15], some distinctions were already introduced between different ageing approaches: direct (one or more characteristics or performances directly affecting the service life are measured), indirect (one or more characteristics or performances that are indicators of performance loss are measured), and torture tests ("short term tests where the conditions of test are significantly more severe than the service conditions"). Introducing a similar classification, we can distinguish:

- Torture cycles, which reproduce extreme conditions, much harsher than those ever encountered in use;
- "Standard" cycles, whose aim is to allow the comparison of the durability of different materials or building components when exposed to a standard cycle. The output is not a predicted service life; and

- "Real" accelerated ageing cycles (ISO 15686 based), whose objective is to predict a Service Life in reference conditions, and to assess the overall performance decay and degradation evolution over time.

In a few words, existing standard procedures do assess the evolution of performances or degradation, but the time axis is missing (i.e. is not correlated with natural time, but it is just a sequence of ageing stages). The difference is the possible use of those data for life cycle assessments and costing, including estimation of the building energy demand over a period longer than just the first few years of life of a building, and, in general, about the sustainability.

2.1.1 Agents Included in the Accelerated Procedure

To decide how to produce accelerated ageing in the laboratory to predict the service life, first of all, we performed a wide analysis concerning degradation factors and mechanisms, and failure modes relating to ETICS (early results of this analysis are presented in [2], while other case studies are analyzed in [16]). More in detail, we focused on which agents need to be reproduced (i.e. which are relevant, and with which intensity), and which characteristics of the building component influence the response to those agents, thus, what to observe to identify the degradation and the performance decay evolution. Hence, we included in the ageing cycle:

- Summer and winter thermal shocks;
- Freeze–thaw cycles;
- Driving rain;
- UV radiation; and
- Cyclic variations in temperature and relative humidity.

Some agents that, actually, are relevant for the degradation and influence the performances of an ETICS, were not included into the ageing cycle:

- Environmental pollution and dirt deposition on the finishing surface—they were not included since at the time it was not possible to couple the dirt deposition on the surfaces. A protocol has been recently developed by Sleiman et al. [17], even though commercial weathering chambers do not allow spraying of "dirt & dust cocktails", and no apparatus has been developed to couple weathering (including freeze–thaw) and soiling in the same ageing protocol. (We also note that, when considering the effect of environmental pollution, also the characterization of the air mix within the chamber is essential to allow the reproducibility of the test);
- Mould growth—the biological growth processes are of a different order of duration compared with the physical ageing due to, for instance, thermal shocks. Moreover, we focused our attention on South and West facing walls, which are more relevant for Italian climates. Finally, also for mould growth, accelerated

protocols have been developed (e.g. [18]), but they are, often, still stand-alone tests (i.e. they do not include other agents);
- Vibrations—the effects of vibrations onto an ETICS largely depend on the constraints of the substrate—the masonry wall—to the structural frame);
- Impacts—often impacts cause immediate failures, and there is a lack of detailed information about frequency and intensity of impacts events onto a façade, to properly design an accelerated ageing phase, more than initial type testing.
- Wind uplift—similarly as for the above mentioned agents, there are technical constraints in reproducing wind uplift cycles (i.e. different machines and different sample shapes and dimensions).

2.1.2 Weather Data Analysis and Ageing Cycle Design

Once decided which agents to include in the protocol, to assess their intensity to be reproduced in the laboratory, we conducted a survey of existing standards (or harmonized guidelines) detailing accelerated ageing procedures, both regarding specifically ETICS [1, 19, 20] and procedures meant also for other materials, but considering one or more of the agents that we wanted to take into account. In addition, to consider both intensity and frequency, we analysed the climatic data of Milan's context: monthly averages to assess the long term behaviour and characterise the climate (30 years: 1961–1990, averaged according to EN ISO 15927-1 [21]) offered in the database given in the Italian standard UNI 10349 [22], and hourly data[1] from the Test Reference Year for Milano Linate (WMO Station 160800; data period 1961–1990). We also note that data series from a Reference Year describe the most probable weather conditions, excluding extreme stresses, which are indeed relevant for assessing ageing. Thus, we are conscious that, for this purpose, using complete data series for a long period is recommended.

In addition to these problems, intrinsic in the adoption of a reference year, almost all data (i.e. for almost all locations) are collected at airports, thus out of urban areas, where the population is concentrated, and, especially, where are concentrated the categories of buildings that consume more energy (e.g. office buildings, hotels, etc.), especially for cooling. On the other hand, climate data

[1] Hourly weather data sets are dealt according to ISO 15924-4 [23], so as to select a reference year from a data set of at least 10 years (recommended), normally longer. The reference year is actually a collection of real calendar months, each of those is the best fit with a computed year representative of the most frequent conditions in the long term for that data set (not simply computed as an average). Then, an adjustment, by interpolation, is applied to the last eight hours of each month and the first eight hours of the following calendar month, to "join" the best real months composing the reference year. A reference year for dynamic energy simulation must include at least: dry bulb air temperature; direct normal solar irradiance and diffuse solar irradiance on a horizontal surface; relative humidity, absolute humidity, water vapour pressure, or dew-point temperature; and wind speed at a height of 10 m above the ground.

collected within urban areas are seldom available over a long period of time, necessary to build historical series. Thus, the aim of using weather data is to pre-size the proportions between ageing cycles and to pre-size the intensity of the agents. The correct proportion between real years of exposure and accelerated ageing cycles will come from the comparison between laboratory and natural exposure (see ISO 15686-2 [13]), which is necessary to validate the ageing cycle for a given building material or component.

In other words, the aim of pre-designing—or pre-sizing—ageing cycles is to avoid a large number of trials (up to convergence of degradation got with long-term and with short-term exposure) and to get an ageing cycle suited for the climatic context. Then, according to the time re-scaling method (detailed in [24]; with an application of the method presented in [25]), when the same degradation level is got, a comparison is set between long-term ageing time (natural ageing) and the ageing cycles reproduced in accelerated laboratory tests, in order to gain the rate of ageing cycles per year. This kind of comparison is valid only if there is not a large initial error in designing the cycle (i.e. the ageing cycle reproduces the ageing events in a similar proportion) or the transport error may cause misleading results. Thus, the service life in very different contexts cannot be assessed with the same ageing cycle, and a comparison between natural and accelerated ageing would not be possible (there may be completely different failure modes).

Analysing hourly weather data, we counted how frequent critical ageing events are in Milano's context (the filtering procedure is detailed in [26]). Then, once we selected only the ageing parts of the year, we could "compress" them (see Fig. 1) into an ageing cycle, reproducing, hence, only critical events.

In particular, we considered as ageing events, in winter season, low (below 0 °C) temperatures and high relative humidity; in summer season, we computed how many times the surface temperature reaches high values together with high relative humidity (as indicator of rain). We also considered exposure to UV radiation. Thanks to this analysis, three ageing cycles (UV cycle, winter thermal shock plus freeze–thaw cycle, and summer cycle) have been designed and the

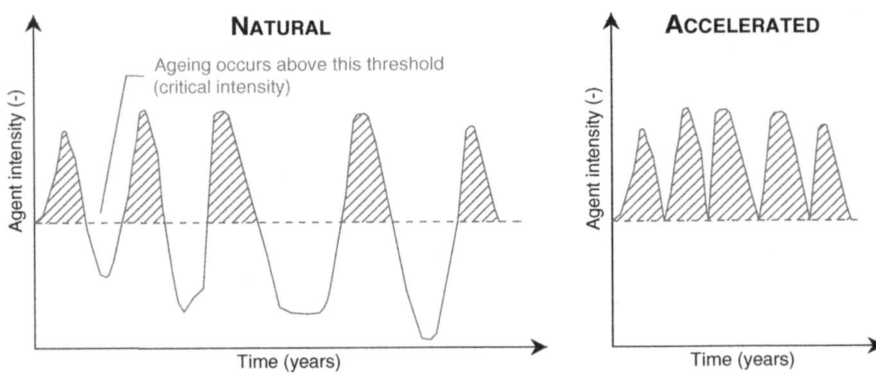

Fig. 1 Acceleration of ageing time is thought as excluding non-ageing parts of stress data series

Table 1 Ageing cycles. During freeze cycles relative humidity was not controlled. Rain was sprayed with 45° angle of incidence (1 Lt m^{-2})

Cycle	Times	Phase	$T_{air,chamber}$ (°C)	$RH_{chamber}$ (%)	$T_{air,lab}$ (°C)	RH_{lab} (%)	Duration (min)
UV	25	UV	35	15 ± 2	20 ± 2	50 ± 5	60
Winter	10	Rain	15 ± 2	100	20 ± 2	50 ± 5	60
		Freeze	−20 ± 2	–	20 ± 2	50 ± 5	180
		Winter heat	30 ± 2	50 ± 5	20 ± 2	50 ± 5	60
Summer	25	Dry heat	70 ± 2	15 ± 2	20 ± 2	50 ± 5	60
		Rain	20 ± 2	100	20 ± 2	50 ± 5	60

proportion between the summer cycles and the winter one set to 2.5 (see Table 1). Ageing sub-cycles are repeated in order to maintain the proportion 2.5. So in each complete cycle CX there are 25 UV cycles, 10 winter cycles and 25 summer cycles. Complete ageing cycles CX therefore are assembled in groups of five so to get a macro-cycle TX. The overall duration of a macro-cycle (i.e. 125 UV cycles, 125 summer thermal shock cycles, 50 winter thermal shock and freeze–thaw cycles) is more or less one month.

In synthesis, to design an ageing procedure, first, we isolated the agents, and their mechanisms and effects, identifying their critical intensities (or thresholds) beyond which ageing occurs. Then, we analysed existing standards and we performed a weather data analysis to assess the frequency of events trespassing the critical thresholds previously identified; hence, we pre-sized the proportions between ageing cycles. However, the true proportion between summer and winter cycles has to be determined with time re-scaling, comparing the degradation of specimens aged with natural exposure with that of samples aged in the laboratory.

2.2 Specimens Description

Four kinds of specimens have been built with masonry wall as substrate, EPS as insulator and polymeric additive in the base coat mortar: two with acrylic resin and two with vinyl resin, and for each type one with thick finishing coat and one with acrylic painting. For each of those, a sample is aged and one not, used as system characterization reference. For each type, there is a sample of a complete wall (measuring 1 m × 1 m) including masonry and ETICS to age, and a similar sample not to age, namely to test as reference for system characterization. Also four sets of small samples (only the ETICS) were prepared for measuring single hygrothermal and mechanical characteristics (also in this case there is a series to be aged, and one for reference). Here we discuss the results regarding a sample with vinyl resin added to the base coat, and thick acrylic top coat.

To study each considered combination of ETICS, small samples are stored inside a climatic chamber and a big sample is used as door of the chamber. The big samples (100 × 100 × 22.45 cm) are both mechanically fixed and bonded, and three insulation panels (two panels 50 × 50 cm in the lower part, and one 100 × 50 cm in the upper part) are used in order to reproduce one T joint in the middle of the specimen. Plastic anchors in polypropylene fix the insulation boards in five positions (in four middle points and in coincidence with the T joint) and aluminium profiles are used to reproduce boundary conditions and in coincidence with profiles there is a superposition for 10 cm of two sheets of glass fibre mesh. The substrate is a masonry wall, made of non load bearing aerated clay bricks 12 × 25 × 25 cm (holes percentage equal to 64 %), with, as bed mortar, a pure cement mortar with water/binder ratio equal to 0.56, and cement to sand ratio equal to 1/2.84. With regard to the plasters, on the two sides of the wall have been laid 1.5 cm thick lime cement plasters, with a water/binder ratio equal to 0.67, cement to sand ratio equal to 1/4, and cement to hydraulic lime to cement proportion equal to 0.5 (all ratios are expressed in volume). In both cases, were used cement CEM II/A-L 32.5, and EN 196-1 [27] normalized sand. Then, as adhesive for expanded polystyrene boards and as base coat reinforced with a glass fibre mesh (155 g m^{-2}), a pure cement mortar with polymeric additive (vinyl or acrylic) has been used with cement to sand ratio equal to 1/3, and additive to cement ratio equal to 0.05.

As insulation material, we employed panels of 6 cm thick expanded polystyrene (having volumic mass equal to 25 kg m^{-3}, and classified as EPS 150 according to EN 13163 [28]). Finally, two kinds of finishing systems characterized by Powder Volumic Concentration (PVC) are compared: an acrylic paint with PVC 40, and a 1.5 mm thick finishing coat with PVC 80, that consists of sand (0/1 mm) and an organic binder (acrylic resin) plus additives (see Fig. 2). Both finishes are red, with solar absorbance $\alpha = 0.75$, and the surface emissivity was measured as $\varepsilon = 0.90$. Despite the value of the solar absorbance might appear as very high, we note that most of top coats of ETICS' are rough, and rough surfaces offer higher solar absorbance than flat surfaces of the same colour [29]. The interior side has been finished with a gypsum plaster. An overview of physical and hygrothermal properties of the layers is offered in Table 2 (data from ISO 10456 [30]) for the tested sample, which, as a whole, is 22.45 cm thick, weighting 148 kg.

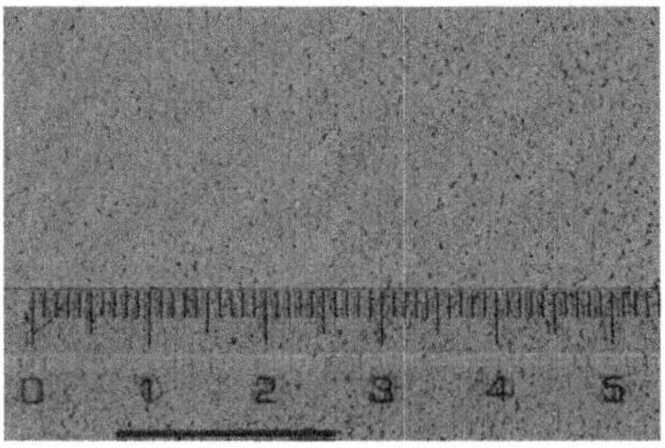

Fig. 2 Top coat with acrylic binder

Table 2 Physical and hygrothermal properties of the layers according to ISO 10456 [30]

n°	Layer	t (m)	ρ (kg/m^3)	λ_d (W m^{-1} K^{-1})	$c_{p,dry}$ (J kg^{-1} K^{-1})	μ (−)
1	Gypsum	0.005	900	0.300	1000	10
2	Plaster	0.015	1600	0.902	1000	10
3	Wall	0.12	650	0.385	840	7
4	Plaster	0.015	1600	0.902	1000	10
5	Adhesive	0.003	1800	1.001	840	15
6	EPS	0.06	25	0.034	1250	35
7	Base coat	0.005	1800	1.001	840	15
8	RPAC	0.0015	1100	1.000	840	7

Specimens have been built inside the laboratory spraying the plasters and the base coat to avoid shrinkage, and a curing time of three months was observed after the laying was completed (five months after the completion of the masonry wall) before ageing the first sample, in order to get the same initial water content for all specimens.

2.3 Weathering and Measurement Apparatus

For big samples, temperature and relative humidity sensors were placed to get three profiles (see Fig. 3): one in section, one in coincidence with EPS panels joints, and one in coincidence with bed mortar joint (see Fig. 4). In each profile, sensors are placed on the interior and the exterior surface of the sample, and between the insulator and the flattening plaster. A heat flux plate is placed on the interior surface in the main section (see Fig. 5).

A data-logger (BABUC A by LSI-LASTEM) recorded the data transmitted by the temperature sensors (Pt100 analogue sensors by LSI-LASTEM), the relative humidity sensors (HIH-3610 Series by Honeywell), and the heat flux plate (HFP01 by Hukseflux) and a relative humidity and dry bulb air temperature sensor measuring the conditions in the laboratory close to the sample.

The conditions inside the weathering cell are recorded by sensors of dry bulb and wet bulb air temperature and relative humidity inside the cell itself. The features of the weathering chamber (CH 1200 SP by Angelantoni Industrie S.p.A.) are detailed in Table 3, while the spectrum of the UV lamp is given in Fig. 6. Then, the specimen to age is positioned as a door of the climatic chamber (see Fig. 7) with a steel frame (separated from it by XPS panels) and it is insulated at boundaries (in order to get a monodimensional heat flow).

Fig. 3 **a** The aged sample during construction phase and instrumented with temperature and relative humidity sensors between masonry wall and EPS. **b** The sample finished with the top coat. The glass fibre mesh is turned round the edges of the specimen, to provide anchor

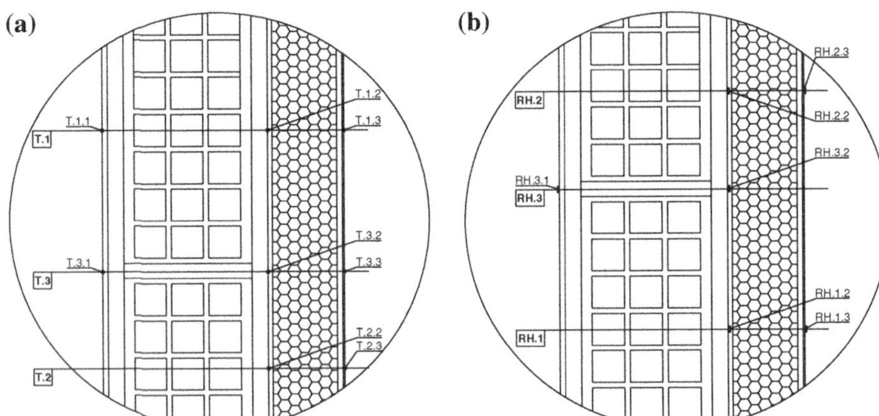

Fig. 4 Temperature sensors profiles in the vertical section (**a**) and relative humidity sensors profiles in the vertical section (**b**)

2.4 Survey of Degradation and Performances Over Time

To measure the performances over time and the evolution of degradation with ageing, we performed both non-destructive and destructive tests (also tests that do not physically destroy the samples, but alter their properties—thus the test

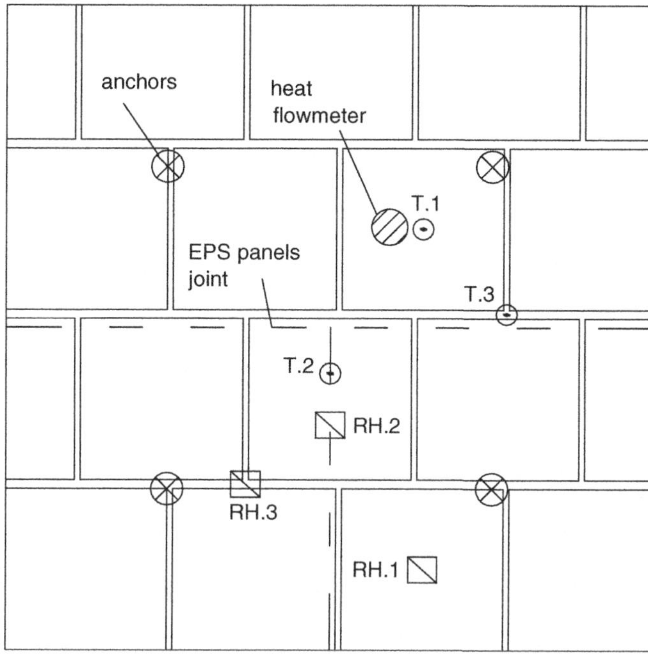

Fig. 5 Positions of the sensors profiles and the het flux meter

Table 3 Features of the weathering chamber

Feature	Value
Internal volume	1.152 m3
Precision in temperature field (−40/+180 °C)	±0.25 °C/±0.3 °C
Temperature variation speed	4 °C/min
Precision in relative humidity field (from 10 to 98 %) precision	±1 %/±3 %

repeated twice would provide qualitatively different results—are considered as destructive).

2.4.1 Destructive Tests

Destructive tests are carried out on the reference sample and on the aged one (at the end of service life) to observe variations in water absorption, water vapour permeability, tensile bond strength of the adhesive and of the base coat to the insulator, and tensile strength of the base coat (details about the samples and reference to test methods are given in Table 4). A set of small specimens is stored inside the climatic chamber during ageing.

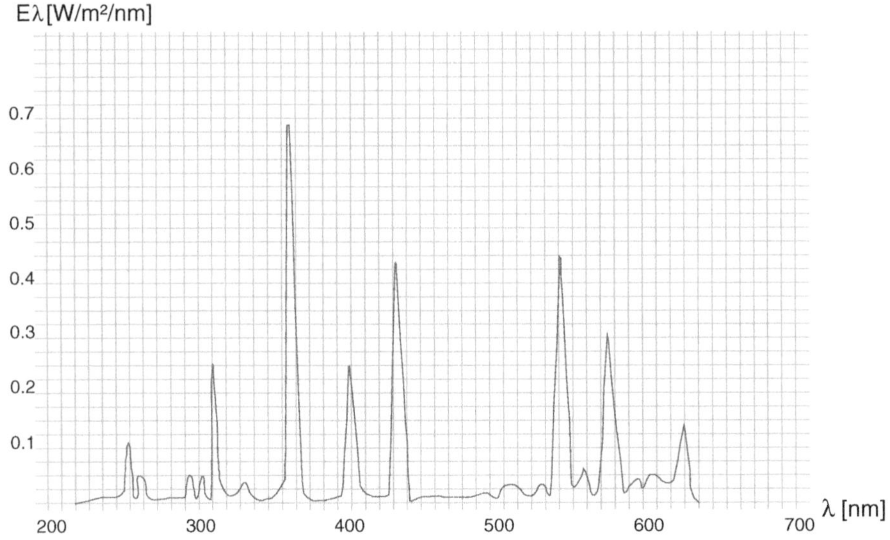

Fig. 6 Spectrum of the UV lamp (125 W)

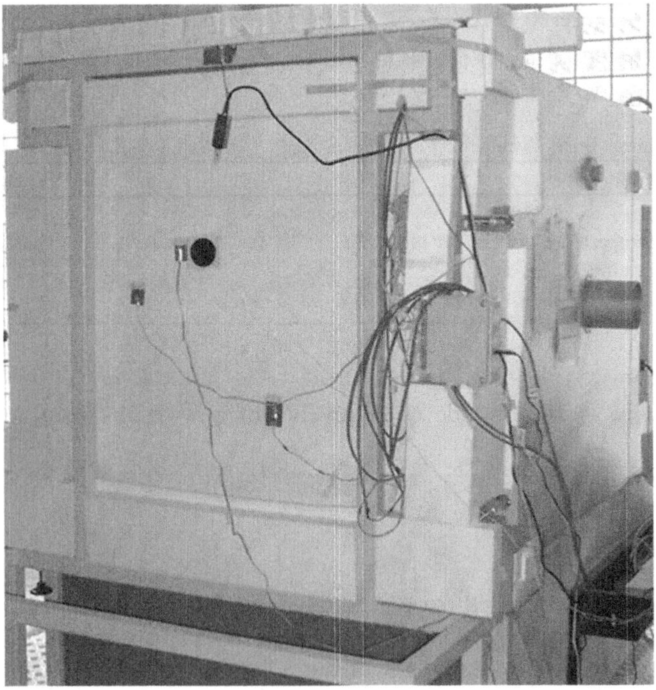

Fig. 7 The sample linked to the climatic chamber

Table 4 Destructive tests, dimension and number of samples and standard test method used

Test	Sample dimensions	n°	Standard
Water absorption (capillarity test)	200 × 200 mm	3	ISO 15148 [35]
Water vapour permeability	Round samples (surface > 5,000 mm^2)	5	ISO 12572 [50]
Render strip tensile test	600 × 100 mm. Mesh length: 800 mm, (100 mm leaning out at edges)	3	ETAG 004 § 5.5.4.1 [1]
Tensile bond strength of adhesive and base coat to insulator	200 × 200 mm	3	EN 13494 [51]

2.4.2 Non-destructive Tests

All throughout ageing time, temperatures, relative humidity, and heat flow across the sample are measured and, every macro-cycle, characterization tests are carried out in order to assess changes in thermal resistance, moisture transfer properties, dynamic thermal response, including the time shift. Infrared thermographies and capillarity non-destructive tests are also carried out and photos are taken in order to survey degradation evolution. Non-destructive tests are performed (in this order):

- Photographic degradation survey in six fixed positions (as indicated in Fig. 8) with illumination according to ASTM E 312-06 [31]. The sample was illuminated with two halogen lamps (500 W each), each of those pointed towards the centre of the sample with angle of incidence of the beams of 45°, while the camera was 5 cm distant from the sample. The evolution of degradation is checked with a method similar to that detailed in ISO 4628 [32] (in lack of a specific standard procedure for ETICS);
- Non-destructive capillary absorption test at low pressure (Karsten's method according to NORMal 44–93 [33]);
- SINa cycle (to measure the time shift φ and the decrement factor f_a). It is a sine wave (see Fig. 9) reproducing the outdoor air temperature for Milano (Italy) the 21st July (peak yearly condition) according to UNI 10349 [22]. Two sine curves, with period of 24 h, are repeated to get a stable harmonic regime;
- SINb cycle (to measure the time shift φ in high thermal stress conditions). It produces the surface temperature with absorbance of the finishing coat set as 0.75, for the 21st July in Milano's context (in clear sky conditions). Also in this case, two 24 h cycles are repeated (Fig. 10);
- Thermal Inertia cycle. Assessment of time constants and thermal capacities. The cycle consists of three steps: 24 h at 20 °C; 24 h at 70 °C; and 24 h again at 20 °C (while the relative humidity was set constant as 50 %);
- CON cycle—Measurement of the thermal resistance according to ISO 9869 [34]. The temperature is kept constant inside the chamber at −20 °C for 96 h;

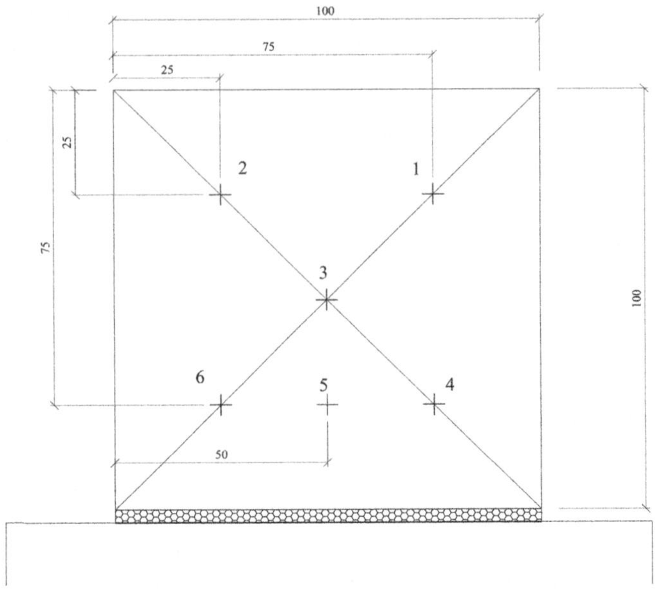

Fig. 8 Photographic survey points (P1 over a fixing; P2 over a fixing; P3 over T joint between insulation boards; P4 over fixing; P5 over vertical joint between insulation boards; P6 over fixing). The measures are expressed in centimetres

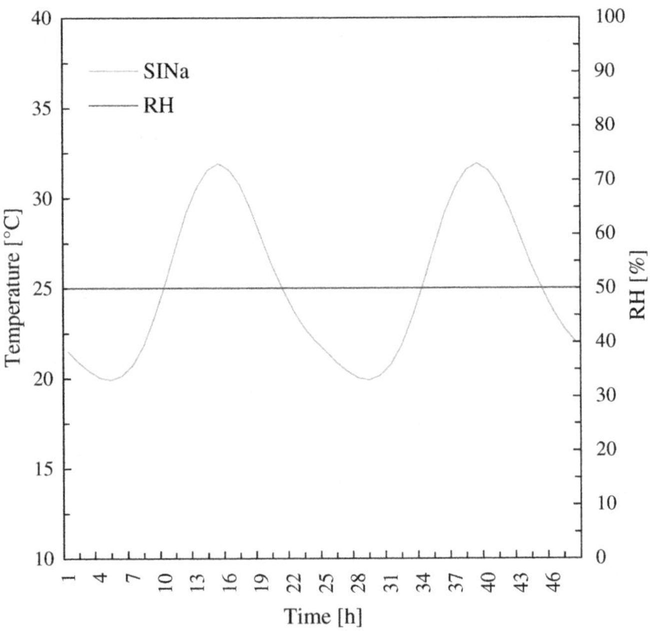

Fig. 9 Test forcing for SINa cycle (reproducing the outdoor air temperature in Milano, 21st July)

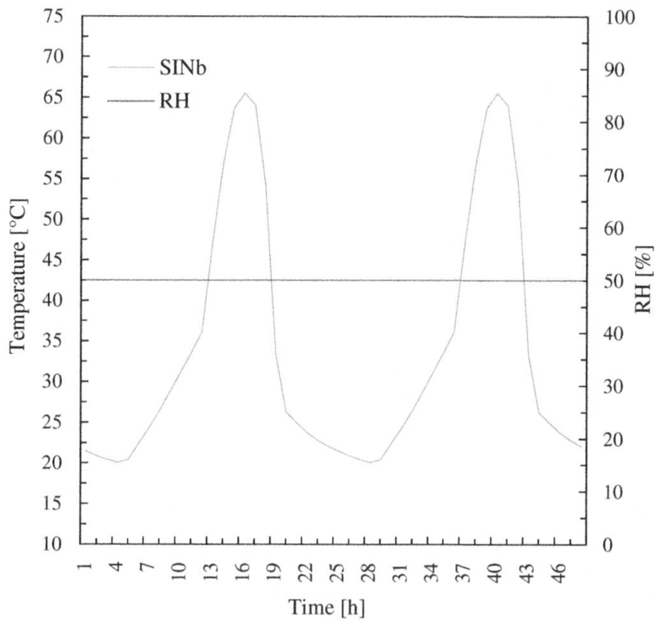

Fig. 10 Test forcing for SINb cycle

- Infrared thermografies—At the end of ageing time the specimen is turned round of 180° and a thermography is performed for 24 h (T = −20 °C inside the chamber) to assess the evolution of thermal bridges in coincidence with joints between insulation panels;
- RHst cycles—Relative Humidity stabilization. Assessment of stabilization time of relative humidity (and so water content) only for the exterior layers (base coat and finishing). Duration: 8 h (4 phases of 2 h each). Conditions: T[°C] = 35 [°C] constant. RH %: 2 h at 20 %, 2 h at 50 %, 2 h at 80 %, 2 h rain 1 Lt/m^2. A longer exposure (to assess moisture changes inside the whole sample) could influence the overall thermal capacity and alter ageing tests storing an excessive moisture amount.

3 Capillary Water Absorption

To understand how much water permeates the ETICS, we performed capillary water absorption tests of two kinds: at low pressure with the Karsten's method on large samples (1 m^2), and by partial immersion (according to ISO 15148 [35]), on small samples measuring 200 × 200 mm.

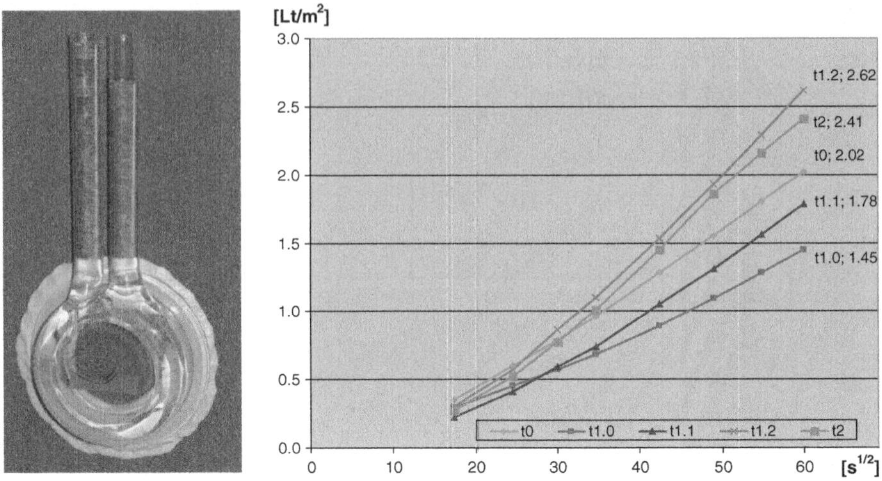

(a) (b)

Fig. 11 Measurement apparatus (**a**) and first results (water absorption in Lt m^{-2} as a function of the square root of absorption time expressed in seconds) (**b**)

3.1 Karsten's Method

To survey water absorption at different ageing times on the big sample, we used the Karsten's method (as detailed in NORMaL 44–93 [33]). During the test the volume of water (in mL) absorbed by the exterior surface of the complete specimen have been recorded at different times. Water absorption is then expressed as function of square root of time (Fig. 11).

At ageing time T1.0 a relevant absorption loss can be noticed, that could have two possible causes: water saturation conditions and increase in cross-linking of polymeric binder of finishing coat and curing of cement matrix of base coat. At ageing times T1.2 and T2 an increase in water absorption is measured. It is important to stress that this kind of test is useful because it is non-destructive, but provides a significant standard deviation in its results and it will need validation with other tests, especially capillary absorption destructive tests.

3.2 Water Absorption by Partial Immersion

Since we acknowledge that measurements with Karsten's method are strongly influenced by the operator, and boundary effects are often non-negligible, we consider those measurements as useful to identify a trend, while to achieve more detailed information we performed tests to measure the water capillary absorption by partial immersion (following the procedure described in ISO 15148 [35]).

Table 5 Water absorption coefficients of the tested samples

Sample	$A_{w,24}$ (kg m^{-2} $s^{-0.5}$) (average on three samples)
T0-EPS-VIN-RPAC	0.00549
T0-EPS-ACR-RPAC	0.00104
T0-MW-ACR-RPAC	0.00104
T0-EPS-ACR-PAINT	0.00062
T0-MW-ACR-PAINT	0.00116
T0-EPS-ACR (no top coat)	0.00128
AG-EPS-VIN-RPAC	0.00660

T0 stands for non-aged, while AG for aged. Then, the second part identifies the insulation material (EPS for the expanded polystyrene, and MW for the mineral wool), while the third part of the name of the sample identifies the base coat (with acrylic or vinyl resin added). Finally, RPAC or PAINT indicate that the sample is finished, respectively, with a thick top coat or a paint (acrylic in this case)

The tests were performed on non-aged samples (of several ETICS combinations) and at the end of ageing macro-cycle T2 (see Table 5).

We observe that the lowest water absorption coefficient is that of samples with a base coat with acrylic resin added and an acrylic paint as finishing. Then, more or less the same values are observed for samples with an acrylic resin added base coat and thick finishing coat (RPAC), regardless the insulation type. Interestingly, non-finished samples (T0-EPS-ACR with no top coat) offer a capillary absorption performance of the same order of magnitude of the samples finished with the thick top coat, which, indeed, transports all the water available at the surface.

Then, comparing the capillary water absorption of the aged samples with that of the same non-aged samples, we note that the water absorption coefficient increased by roughly 20 %.

In a second phase, we repeated the tests on a new set of samples, achieving consistent results (see Fig. 12 and Table 6). We note that already after 5 min the water content is already increased by about 0.25 kg m^{-2}, and about 0.5 kg m^{-2} after 20 min of partial immersion.

We performed also tests to assess the water saturation within the samples (see Fig. 13); we observe that a steady water content in the samples is reached only after a long period (about 6 days) of partial immersion, and that the final value of absorbed water (2.125 kg m^{-2}) is about once an a half the value achieved after 1 day of partial immersion (about 1.375 kg m^{-2}). Then, with a de-sorption test, we observed that the drying process—in laboratory conditions—could be considered completed after roughly nine days, when the samples regained their initial weight (see Fig. 14).

We tested also new samples (see Table 7) of just the base coats with vinyl or acrylic resin added, or with no resin added to the mix, and with different thickness, thus 5 or 10 mm, finished with acrylic paint or thick finishing coat (RPAC). For the base coat samples differently finished, the largest water absorption is that of samples with no resin added and finished with a thick top coat (RPAC), while the lowest absorption coefficient is shown by the samples with acrylic paint over a resin added base coat.

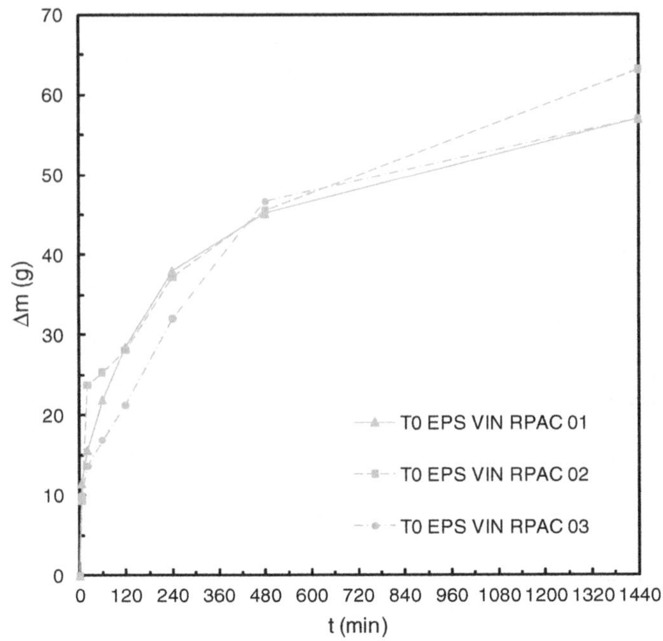

Fig. 12 Water absorption of T0-EPS-VIN-RPAC samples

Table 6 Water absorption coefficients of the second set of ETICS samples (each set is made of three samples, measuring 200 × 200 mm, according to ISO 15148)

Sample	$A_{w,24}$ (kg m^{-2} $s^{-0.5}$) (average on three samples)
T0-EPS-ACR-PAINT	0.0007
T0-EPS-ACR- NO PAINT	0.0011
T0-EPS-ACR-RPAC	0.0011
T0-EPS-VIN-RPAC	0.0050
T0-MW-ACR-PAINT	0.0007
T0-MW-ACR-RPAC	0.0015

4 Optical Properties

Measurements of the optical and radiative properties of the finishing coats used in the experiment have been performed on new samples: thick top coat over reinforced base coat (see Fig. 15 top) and acrylic paint over reinforced base coat (see Fig. 15 bottom). For each finishing we measured six samples, to assess the standard deviation. We measured the spectral reflectance with a spectrophotometer (Perkin-Elmer Lambda 900 with Labsphere integrating sphere) according to ASTM E 903 [36], and with a solar spectrum reflectometer (by Devices & Services) according to ASTM C 1549 [37], while the thermal emittance was measured

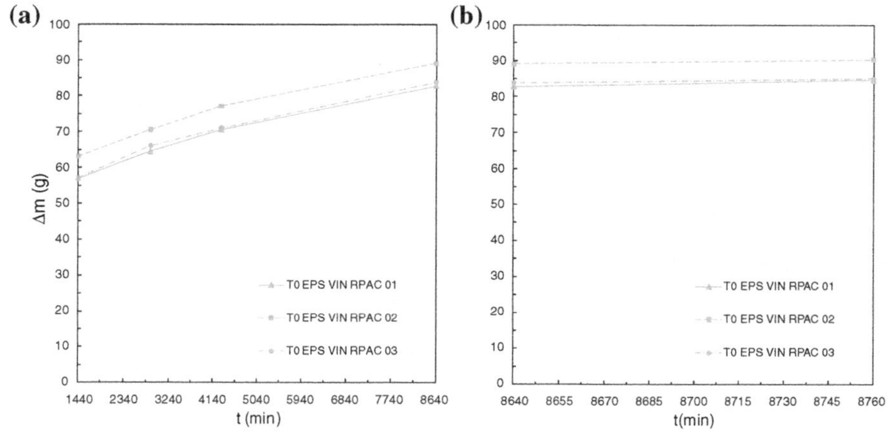

Fig. 13 Saturation curves

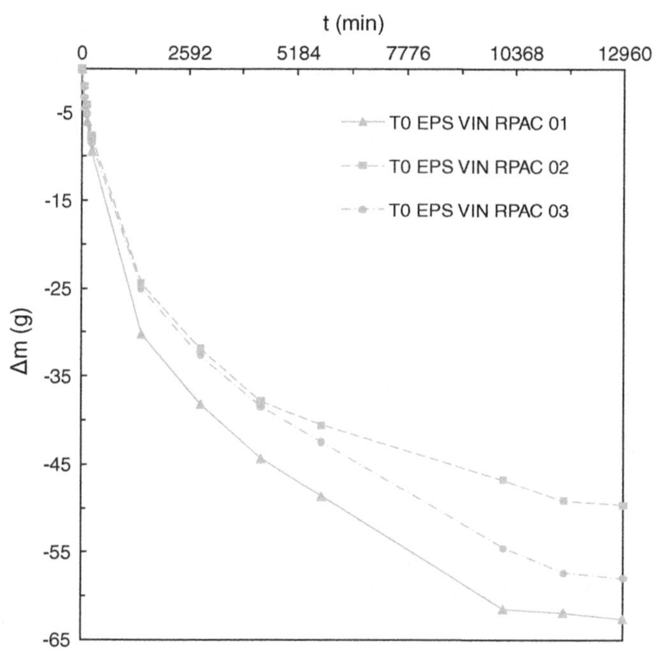

Fig. 14 De-sorption curves

by means of a portable emissometer (by Devices & Services) according to ASTM C 1371 [38]. The spectral reflectance (measured from 300 nm to 2500 nm each 5 nm) has been then weighted with the spectral irradiance distribution of the solar global radiation on the horizontal plane to compute the solar reflectance, the

Table 7 Water absorption coefficients of eight sets of samples (each set is made of three samples, measuring 200 × 200 mm, according to ISO 15148) of cement base coats with no resin added (CEM) or with vinyl (VIN) resin added, finished with paint or thick top coat (RPAC). The base coat are 5 or 10 mm thick (excluded the top coat)

Sample	Aw (kg m^{-2} $s^{-0.5}$) average (on three samples)
Base coat CEM 10 mm PAINT	0.0010
Base coat CEM 10 mm RPAC	0.0017
Base coat CEM 5 mm PAINT	0.0009
Base coat CEM 5 mm RPAC	0.0015
Base coat VIN 10 mm PAINT	0.0007
Base coat VIN 10 mm RPAC	0.0013
Base coat VIN 5 mm PAINT	0.0006
Base coat VIN 5 mm RPAC	0.0010

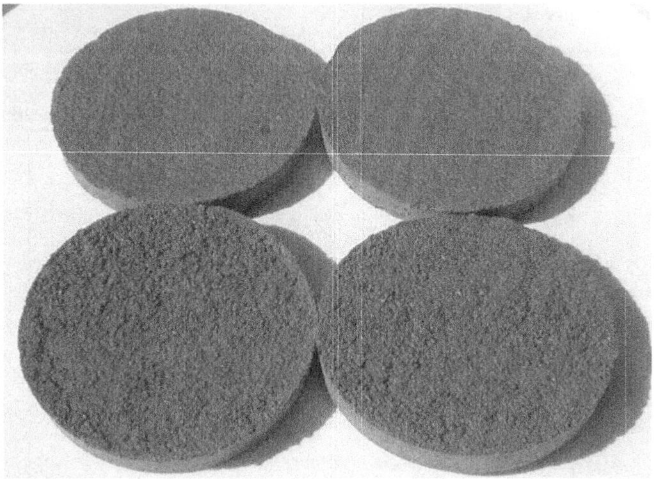

Fig. 15 Samples of the finishing coats used to characterize the optical and radiative properties

reflectance in the ultraviolet band (UV, 300–400 nm), in the visible (VIS, between 400 and 700 nm), and in the near infrared (NIR, between 700 and 2500 nm).

The thick (1.5 mm) top coat, resulted being a cool coating (i.e. having reflectance higher in the NIR than in the visible portion of the solar spectrum), showing an average solar reflectance equal to 0.230 (see Fig. 16). More in detail, while the reflectance in the UV and in the visible parts of the spectrum offer a quite homogeneous behaviour, in the NIR region, values are more spread, between, roughly, 1000 and 2000 nm.

The acrylic paint, over the reinforced base coat, showed as well higher reflectance in the NIR than in the visible band, with an average value of solar reflectance of 0.259 (see Fig. 17). Similarly to the thick top coat, the acrylic paint samples showed a very consistent response in the UV and in the visible region (the

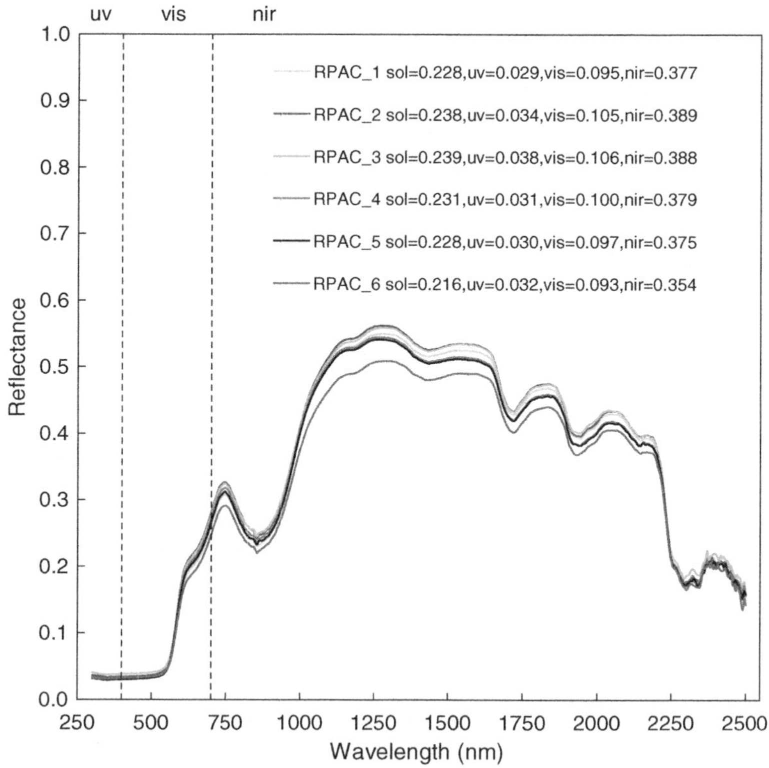

Fig. 16 Spectral reflectance of the top coat (finished coarse)

difference between the different samples with regard to the visible reflectance is negligible), while in the NIR the standard deviation is noticeably higher (values in the NIR range from 0.403 to 0.424).

The spectral measurements have been performed with direct illumination (the spectrophotometer shoots on the sample a beam 3 mm wide and 20 mm high, with angle of incidence of 8°), capturing the diffuse reflection of the samples, while other measurements on the same samples have been performed with a solar spectrum reflectometer (see Table 8) according to ASTM C 1549 [37], to characterize the diffuse reflection with diffuse illumination (for air mass 1.5, with the spectral distribution given in ASTM G 173 [39] for the global irradiance on a south facing surface tilted of 37°, referred as G 173), and the diffuse reflection when the object is hit by direct radiation (as above, but considering the spectral distribution of beam radiation, referred as b173). Moreover, measurements with a solar reflectometer are useful to characterize samples offering a rough surface, whose discontinuities might not be completely detected with a spectrophotometer. For all the samples of top coats (both the RPAC and the acrylic paint over the base coat) we note a slightly different response (roughly 0.01, namely about 4 %) between

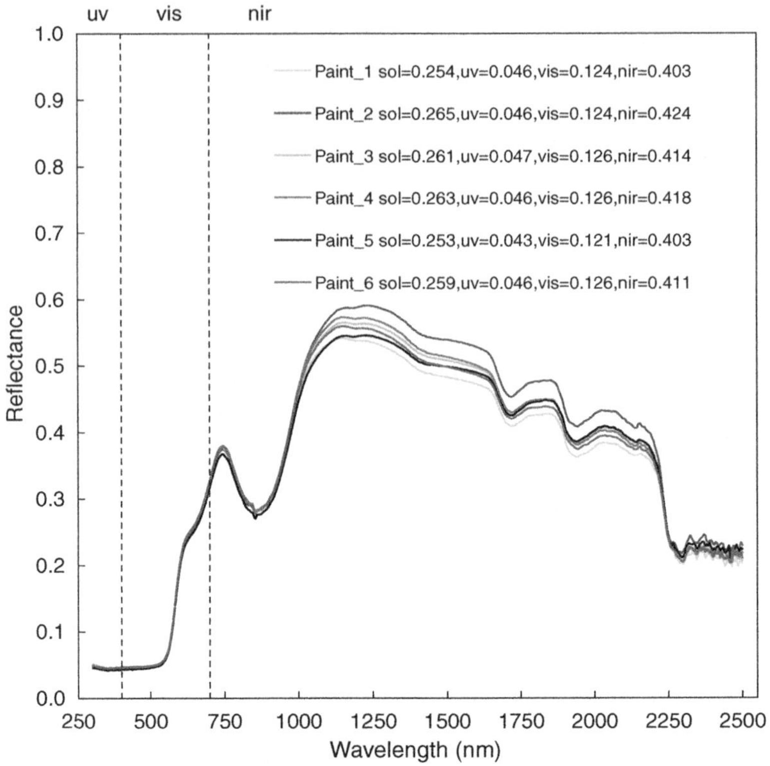

Fig. 17 Spectral reflectance of the acrylic paint over the reinforced base coat

Table 8 Solar reflectance of the samples measured with the SSR

Measurements with SSR/method	G173	b173
RPAC 1	0.253	0.264
RPAC 2	0.260	0.271
RPAC 3	0.265	0.275
RPAC 4	0.257	0.268
RPAC 5	0.255	0.266
RPAC 6	0.238	0.248
Paint 1	0.282	0.292
Paint 2	0.294	0.305
Paint 3	0.285	0.296
Paint 4	0.294	0.306
Paint 5	0.282	0.294
Paint 6	0.289	0.301

the solar reflectance of the sample when hit by beam radiation only, and when hit by global radiation. In general, we expect some differences in the directional response of the samples, since they present a rough surface; however, we do not expect large divergences for different angles of incidence since the aggregates in the top coat are covered by a matte finishing and the surface porosity is quite homogeneous. Thus we conclude that, modelling the hygrothermal behaviour of ETICS with this kind of finishing coat, using the bulk values of solar reflectance—namely neglecting the directional behaviour—and reproducing the surface temperature variations in the laboratory without a directional solar simulator, in this case, does not introduce large errors.

On the same non-aged samples we also measured the thermal emittance according to ASTM C 1371 by means of a portable emissometer. As for most of building materials, for the top coat samples we observed an average value equal to 0.88, while for the samples with the acrylic paint over the reinforced base coat we measured a thermal emittance of 0.90.

Once characterized the initial optical and radiative performance of finishing coats, we decided to measure their performances when wet. However, to perform measurements on wet samples, we had to apply a quartz glass window on the reflectance port of the spectrophotometer, to prevent damages to the integrating sphere. This alters results, thus we performed the measurement of the spectral reflectance of the dry sample with and without the glass window, so that we can compare the measurements of the wet samples to that of the dry sample with the glass window. We achieved an increase in water content of 106.10 g m^{-2} and 171.89 g m^{-2} after respectively 5 and 20 min of partial immersion in water of the samples (according to ISO 15148), and, after having removed the water adhered on the surface, we measured the spectral reflectance. We observe that the reflectance, for the considered sample, decreases with moisture content of about 5 % (Fig. 18). We also note that the largest dips, after water absorption, occur in the NIR region, while the visible spectral response is quite similar when the sample is dry and wet. We measured the wet samples also with the solar reflectometer (Table 9), and in this case we did not need a glass window to protect the instrument.

Also with this measurement technique we observe a decrease of the solar reflectance with moisture content. We note that with both measurement techniques, and after other measurements, that the decrease in solar reflectance is not proportional to the water content. More in detail, it is the water content in the first few millimitres below the surface that influences the radiative behaviour. We point out that in this case we measured samples offering a quite low initial solar reflectance (0.25 in average); thus, one might expect a more relevant loss in solar

Table 9 Solar reflectance of dry and wet samples measured with a solar reflectometer

SSR	G173	b173
RPAC dry	0.242	0.252
RPAC wet 5′	0.232	0.242
RPAC wet 20′	0.230	0.240

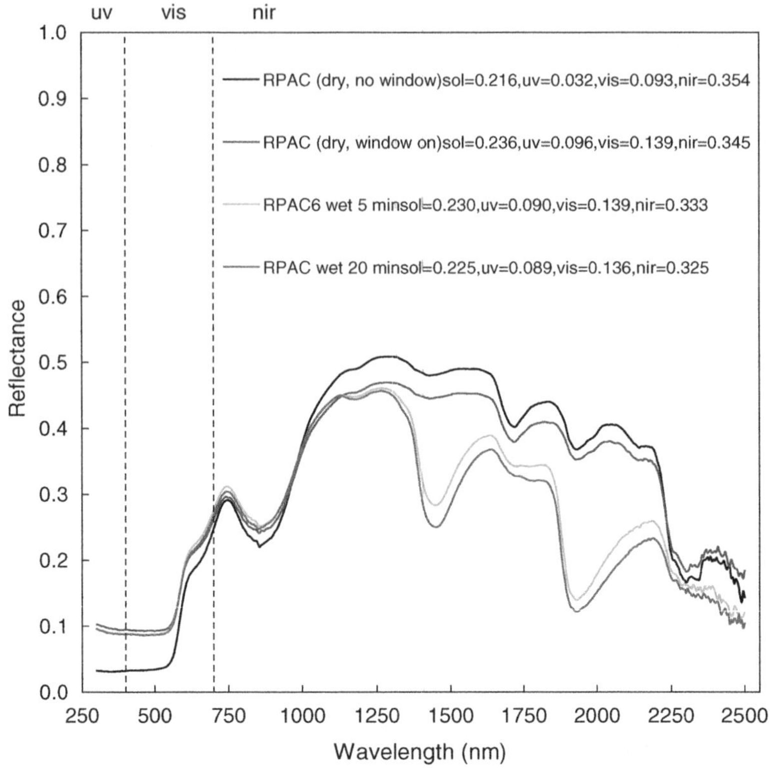

Fig. 18 Spectral reflectance of the top coat when dry (measured with or without the quartz glass window), and after 5 and 20 min of partial immersion in water

reflectance for higher values of dry reflectance, for porous materials (Cresswell et al. [40] measured the solar reflectance of soils decreasing from 0.40, when dry, to 0.20, when wet).

5 Thermal Performances Variability with Ageing and Moisture

As previously introduced, we performed laboratory tests to measure the hygrothermal performances through ageing times: thermal resistance [12], and time shift and decrement factor [41].

5.1 Steady State Thermal Performance

The thermal resistance of the main section profile, for the entire specimen, the substrate, and the ETICS was measured according to ISO 9869 [34]. Data in Table 10 and Table 11 refer to the sample non-aged (T0), and after exposure to two macro-cycles (T1, and T2). Condition T2+ concerns ageing time T2 after a rain cycle (spraying 1 lt m^{-2} h^{-1} for 2 h). With a correction to the mean heat flow density (using the storage effect method detailed in ISO 9869), we derived thermal resistance (Fig. 19) and water content of each layer from an iterative numerical solution, starting from an estimate of moisture and temperature profiles in the sample, considering, for each layer, the thermal capacity, including moisture contribution.

The curve of thermal conductivity versus water content for the EPS has been computed by interpolation (the interpolation has been validated by comparison with the curve at 20 °C included in the WUFI database), including the contribution of air and water at different temperatures (see Fig. 20), since the thermal conductivity of water presents different values for solid and liquid phase (about 2.30 W m^{-1} K^{-1} for ice at −10 °C, 2.20 W m^{-1} K^{-1} for melting ice, and about 0.61 W m^{-1} K^{-1} for liquid water at 20 °C).

Results highlight the difference between thermal resistance values calculated with standard reference and measured ones and clearly show the performance decay of the thermal insulation. At the same ageing conditions, after a rain cycle a great loss in thermal resistance can be observed.

We note a limited reduction in thermal resistance of the substrate that is supposed to be due to increase in moisture content because of water permeated through ETICS. Construction water is estimated to be digested at tests beginning (6 months later than substrate been built). On the other hand, thermal resistance of the ETICS suffers a pronounced loss, that could be explained with an increase in moisture content in outer layers of EPS (even setting as zero thermal resistances of base coat and finishing coat, this insulation decay could not be reached), where water is in solid phase (most exterior EPS section) and in physical transition phase (EPS middle section).

Performing a test with ice water or water in transition phase does not introduce a useless complication in data analysis, but allows reading clearly the evolution of hygrothermal performances, and, especially, it considers the conditions for which the thermal insulation is designed. Without ice presence, could be hard to separate the thermal resistance loss from measurement errors and uncertainties.

5.2 Infrared Thermography

A development of a relevant thermal bridge in coincidence with joints between insulation panels, especially horizontal one, can be noticed analysing data

Table 10 Estimated water content (w_c) and thermal conductivity (λ_{wet}) of the tested sample not aged (T0), and after ageing (T1, T2, and T2+). In the calculation, the EPS board is discretized in six layers of 1 cm

Layer	t (m)	w_c T0 (kg m^{-3})	w_c T1 (kg m^{-3})	w_c T2 (kg m^{-3})	w_c T2+ (kg m^{-3})	$\lambda_{wet\ T0}$ (W m^{-1} K^{-1})	$\lambda_{wet\ T1}$ (W m^{-1} K^{-1})	$\lambda_{wet\ T2}$ (W m^{-1} K^{-1})	$\lambda_{wet\ T2+}$ (W m^{-1} K^{-1})
Gypsum	0.005	4.35	3.59	5.46	4.09	0.208	0.207	0.210	0.208
Plaster	0.015	33.53	29.92	39.28	29.96	0.913	0.901	0.932	0.901
Masonry	0.12	25.00	29.11	33.72	52.36	0.375	0.413	0.415	0.431
Plaster	0.015	49.38	61.56	54.72	66.3	0.966	1.007	0.984	1.023
Binder	0.003	62.52	61.56	54.72	66.35	1.010	1.007	0.984	1.023
EPS 1	0.01	5.00	5.00	5.00	5.00	0.0339	0.0335	0.0337	0.0337
EPS 2	0.01	7.50	10.00	15.00	20.00	0.0335	0.0334	0.0338	0.0339
EPS 3	0.01	10.00	15.00	20.00	37.50	0.0332	0.0365	0.0358	0.0382
EPS 4	0.01	12.50	17.50	25.00	40.00	0.0358	0.0371	0.0392	0.0435
EPS 5	0.01	15.00	20.00	27.50	42.50	0.0360	0.0374	0.0396	0.0435
EPS 6	0.01	23.59	22.36	29.83	43.04	0.0380	0.0378	0.0399	0.0436
Base coat	0.005	43.71	39.05	47.45	30.00	0.947	0.931	0.960	0.901
Top coat	0.0015	13.17	8.37	16.36	3.98	0.870	0.870	0.870	0.870

Table 11 Thermal performances of the tested sample not aged (T0), after ageing (T1, and T2), and at ageing time T2 after a rain cycle (T2+)

Thermal Performance		T0	T1	T2	T2+	Unit
Substrate resistance	$R_{cd,12}$	0.376	0.346	0.344	0.334	[m^2 K W^{-1}]
ETICS resistance	$R_{cd,23}$	1.725	1.684	1.640	1.553	[m^2 K W^{-1}]
Conductive resistance	$R_{cd,T}$	2.100	2.030	1.984	1.887	[m^2 K W^{-1}]
Conductance	Λ	0.476	0.493	0.504	0.530	[W m^{-2} K^{-1}]
Tot resistance	R_T	2.270	2.200	2.154	2.057	[m^2 K W^{-1}]
Transmittance[a]	U	0.440	0.455	0.464	0.486	[W m^{-2} K^{-1}]

[a] The surface resistances considered in this case are the standard ones (according to ISO 6946), not those estimated in the laboratory, just to provide with a simple reference number for design conditions: 0.04 m^2 K W^{-1} the exterior, and 0.13 m^2 K W^{-1} the interior

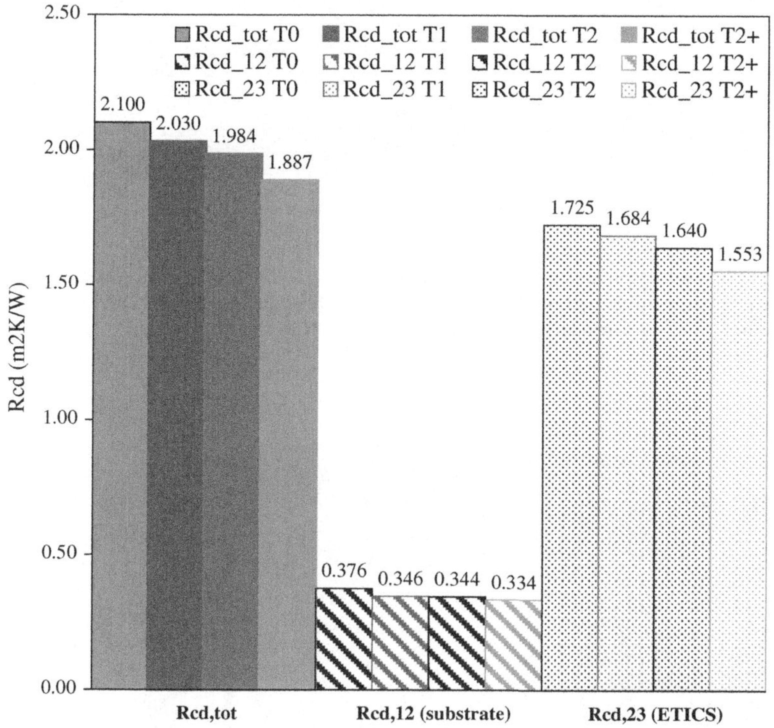

Fig. 19 Total conductive thermal resistance (Rcd,tot), thermal resistance of the masonry wall (Rcd,12) and thermal resistance of the ETICS (Rcd,23) at ageing time T0 (initial condition), T1, T2, and T2+ (the same as T2, but after a rain cycle)

provided by the infrared thermography performed at ageing time T2+ (see Fig. 21). Over the vertical joint the trend of water is to spread and not to rest only close to the joint line.

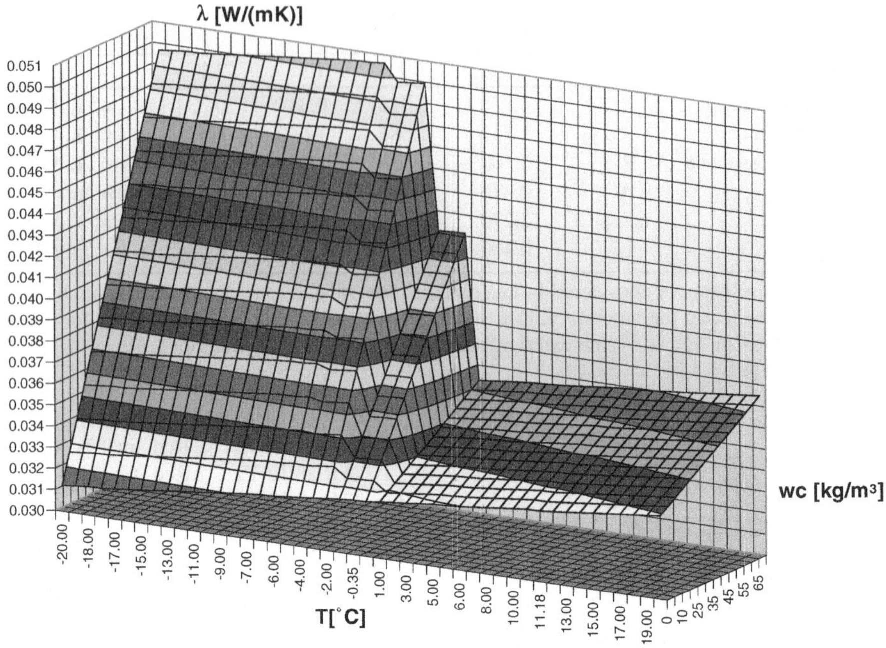

Fig. 20 Thermal conductivity of EPS at different temperatures for different moisture contents (in the figure only the initial part of the curve is presented, for moisture contents up to 70 kg m^{-3}). In the interpolation, we consider water within the pores all in the same phase

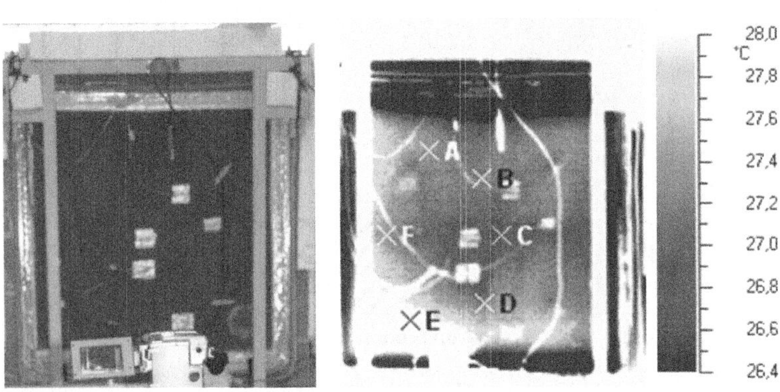

Point	T[°C]
A	27.2
B	27.3
C	26.8
D	27.1
E	28.4
F	27.2

Fig. 21 Infrared thermography at ageing time T2+. Laboratory air temperature $T_{A,LAB} = 29.0$ °C. Surface emissivity $\varepsilon = 0.90$. Thermocamera distance from specimen: 3 m

We also point out that, as demonstrated in the section discussing the optical and radiative properties, porous materials, when moist, offer lower solar reflectance than when they are dry. Thus, during thermal shock events, given the different capillary absorption of the main section and of the joints between insulation panels, there might also be a different radiative response over the façade. The different capillary absorption of the main section and the joints is due to the fact that the joint behaves like a capillary. For ETICS made out of the lab this effect is even stronger, since in the lab we achieved a perfect planar surface for the substrate, and there were no gaps between the insulation boards. Normally, in the construction yard, the panels are not matched perfectly, and the mortar of the base coat fills the gaps (especially were the insulation panels have been sawed, namely in coincidence with building details); this produces even stronger thermal bridges and "moisture bridges" than those we recorded in the laboratory.

5.3 Unsteady State Thermal Performance

To assess the evolution of time shift and decrement factor (as they are defined in ISO 13786 [42]), we exposed the sample to harmonic variations in air temperature (inside the weathering chamber), with the hygrothermal cycles previously identified as SINa (reproducing the air temperature in Milano in summer conditions), and SINb (producing the surface temperature, which the sample would have facing South in a clear sky summer day) (Table 12).

Results for T0 (non-aged sample) are not available, since these tests were introduced only later. For highly insulated samples,[2] we acknowledge some issues in the measurement procedure, namely:

- With a low sinusoidal temperature wave (min 20 °C, max 30 °C) it is difficult to read φ, and f_a, since the very small thermal gradient between laboratory (20 ± 2 °C) and climatic chamber;
- A non perfectly sinusoidal temperature wave representing the sol–air temperature has a higher maximum, but the same minimum value in solicitation, and the same problem in measuring f_a;
- Adopting higher values for the temperature waves, to enhance readability of the test, would alter the thermal properties of water, and the overall hygro-thermal response; and
- For well-insulated samples, it is difficult to measure f_a, since the error in estimation of results is of the same magnitude of the heat-flowmeter resolution. In our case, we could not read f_a.

Since these constraints, we adopted the selected testing forcing, despite issues in readability of results. However, we stress upon the need of standardisation in terms of sensors resolution, laboratory conditions, curing time between tests, and complementary side tests to check the main measurement.

---

[2] For highly insulated samples recent studies [43] showed that φ, and f_a may not be the best parameters to be used in order to asses the comfort in buildings.

Table 12 Thermal performances of the tested sample not aged (T0), after ageing (T1, and T2), and at ageing time T2 after a rain cycle (T2+)

Thermal Performance		T0	T1	T2	T2+	Unit
Substrate heat capacity	C_{12}	162.9	165.4	168.0	177.5	[kJ m^{-2} K^{-1}]
ETICS heat capacity	C_{23}	18.7	19.0	19.9	21.3	[kJ m^{-2} K^{-1}]
Computed[a] time shift	ΔT	8 h 54′	8 h 45′	8 h 49′	8 h 58′	[hh:mm]
Computed[a] decrement factor	f_a	0.271	0.278	0.274	0.263	[–]
Time shift SINa	ΔT	n.a.	10 h 48′	06 h 34′	n.a.	[hh:mm]
Time shift SINb	ΔT	n.a.	11 h 07′	05 h 09′	n.a.	[hh:mm]

[a] These values have been computed considering the water contents, the thermal capacities, the volumic masses, and the thermal resistances derived with the steady state tests, considering two layers: substrate (including masonry wall and the plasters) and ETICS

5.3.1 Error Risk Analyses

Given the risk of uncertainty in measurement results, and the risk of error in the calculation of dynamic performance due to uncertainty in input, we performed a risk analysis. We assumed a probability distribution (Gauss) for thermal conductivity, density, and specific heat capacity of each layer. Then, for each parameter, we considered that a variability comprised between −20 % and +20 % (centred on a value selected as average or most probable) should represent the 99.7 % of all the possible case ($\pm 3\sigma$). Only for surface heat resistances (see Fig. 22), we assumed a flat probability distribution—any value has the same probability to be correct—since they are not the result of an industrial process, and may assume diverse values for the same building component. As centre value for surface heat resistances we assumed 0.17 m^2 K W^{-1} (average value estimated from measurements). For all other layers, instead, we assumed a normal probability distribution (see Fig. 23).

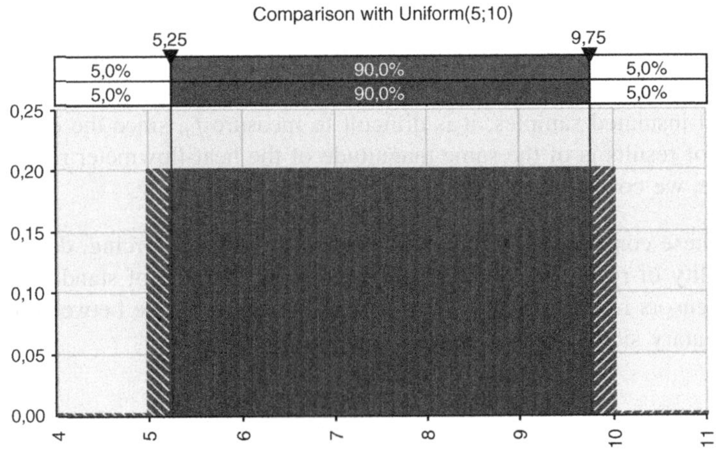

Fig. 22 Probability distribution assigned to surface heat resistance

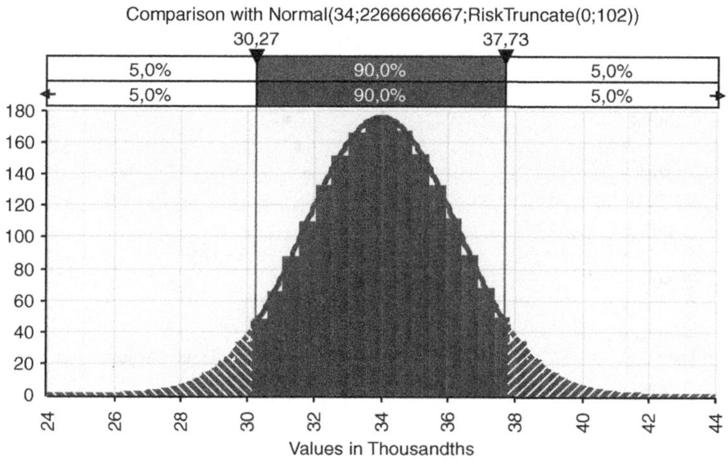

Fig. 23 Normal probability distribution assumed for material properties (here for the thermal conductivity of EPS)

Risk Analysis: Three moisture Content Conditions

We computed Y_{12} (see Fig. 24), f_a (see Fig. 25), and φ (see Fig. 26) for three conditions of moisture content:

- Case A—all layers are dry;
- Case B—all layers present with their typical moisture content; and
- Case C—all layers present the moisture content in condition of free saturation.

Results are displayed versus cumulative distribution functions (by percentile), to assess the stochastic reliability of computed thermal performances. With reference to periodic solicitation, the dynamic thermal response (Y_{12}, φ, f_a) of building components—according to ISO 13786—can be related to the periodic penetration depth δ (where $\delta = (\lambda T\ \pi^{-1}\ \rho^{-1}\ c_p^{-1})^{0.5}$, with T the period of solicitation in seconds), and hence to thermal conductivity, and the product of heat capacity and density. These quantities depend on water content (and temperature); thus, moisture has a dual effect: it decreases the thermal conductivity of a layer, and it increases the heat capacity and density of the layer. The risk analysis highlights that with typical moisture content (case B), in comparison with the dry condition (case A), φ decreases (at 50 %-ile from 8 h 25′ in case A, to 7 h 57′ in case B), while increase both Y_{12} (at 50 %-ile from 0.106 W m^{-2} K^{-1} in case A, to 0.129 W m^{-2} K^{-1} in case B) and f_a (at 50 %-ile from 0.29 in case A, to 0.32 in case B). Then, for high water contents (free saturation, case C) φ reaches the maximum (at 50 %-ile 10 h 28′), while Y_{12} and f_a show the minimum values (at the 50 %-ile respectively 0.19 W m^{-2} K^{-1}, and 0.22).

For the materials used for the aged sample, we did not measure the moisture storage function, the moisture diffusivity, and thermal conductivity depending on

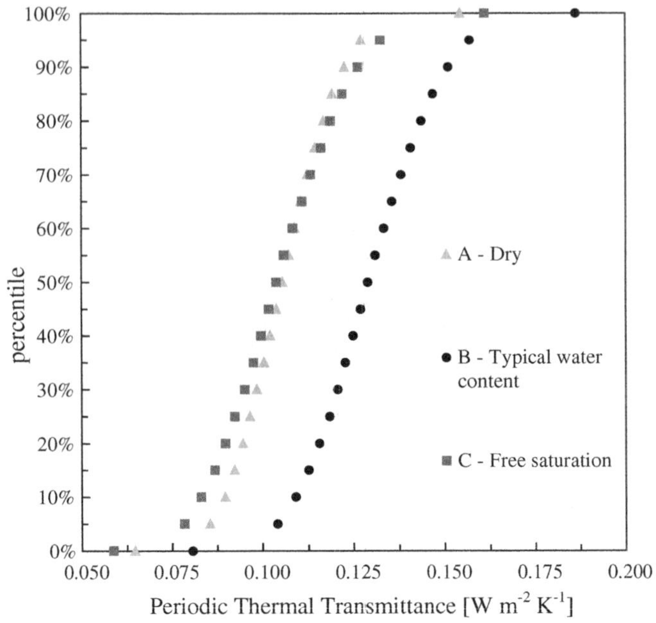

Fig. 24 Periodic Thermal Transmittance when *A* all the layers are dry, or they present *B* typical moisture content (specific for each material), or *C* free saturation water content

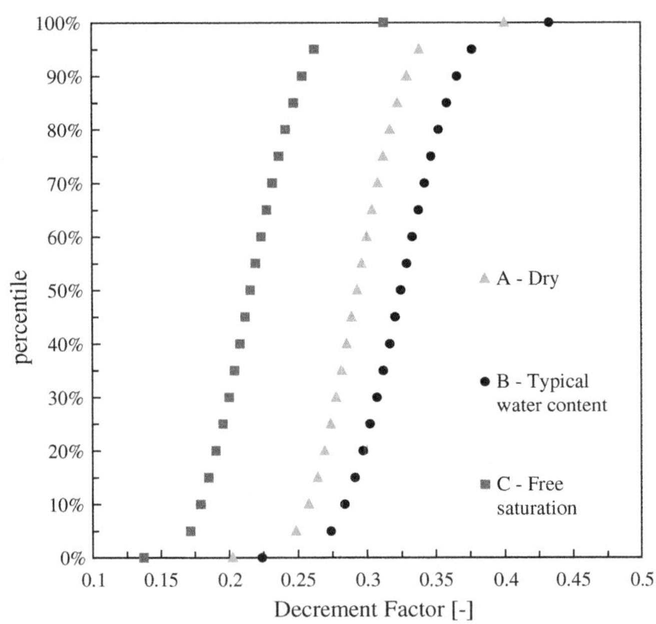

Fig. 25 Decrement Factor when *A* all the layers are dry, or they present *B* typical moisture content (specific for each material), or *C* free saturation water content

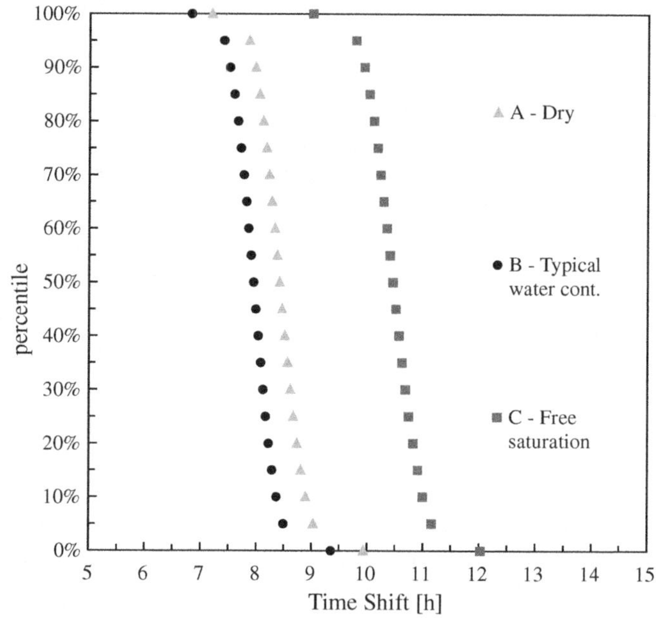

Fig. 26 Time shift for conditions *A* (dry materials), *B* (materials with their typical moisture content), and *C* (materials with their maximum moisture content in free saturation). Typical and maximum moisture content according to Künzel [44]

moisture content, but we assumed—as trial values for the stochastic risk analysis—the curves given in the literature [44, 45]. We adopted this approach since we note that, for many materials, properties may be highly influenced by laying conditions. For instance, a plaster may offer different properties when applied to a masonry substrate or by itself, manufacturing a small laboratory sample. We also note that, as yet, there is no test method to measure these properties depending also on temperature. In this work, we focus on dynamic parameters; other studies analyse the influence of variability of hygrothermal parameters values on moisture content, and they confirm the large influence that these parameters have on the result [46].

Risk Analysis: Laboratory Testing and Stochastic Monte Carlo Simulation

We performed a second error risk analysis, with the same procedure, assuming as central values the thermal conductivities, specific heat capacities, and densities that we measured (and derived depending on the moisture content computed as discussed above) for the ageing conditions T0 (non aged), T1, T2, and T2+ (i.e. T2 immediately after a rain cycle).

In our experiment, we observed a loss in thermal resistance (see Table 11 and Fig. 19), with increasing moisture content due to rain penetration occurred during ageing cycles, and decrease in time shift (from roughly 11 h at T1, to about 6 h at T2). We explain the decrease in time shift, considering that the increase in moisture content in the insulation board caused a reduction of the thermal resistance, more influent than the increase of the heat capacity for the same layer. Moreover, we did not observe a relevant increase in moisture content in the masonry substrate (except at T2+), meaning that the most of water penetrated and concentrated in the thermal insulator (see Table 10). Comparing measurement results with computer simulations, we observed that with laboratory testing we obtain a larger variation in time shift for the same moisture contents: with tests, φ is 10 h 48′ at T1, and 6 h 34′ at T2, while with calculations φ varies (at the 50 %-ile) from 7 h 26′ (T1) to 7 h 35′ (T2). We explain this noting that the moisture content was derived for the main section, while the most relevant rain penetration occurs through the joint between insulation boards. In addition, we note that the time shift measured in free saturation condition (10 h 28', curve C - 50 %-ile, Fig. 26) corresponds roughly to the time shift measured at T1 (10 h 48', Fig. 27). Furthermore, water is subject also to draining forced by gravity, and we expect the largest moisture content to be at the base of the sample. Moreover, we note a large influence on results of the surfaces heat resistances, which are difficult to characterise in the laboratory and may vary test by test (depending on the sample and the thermal gradient).

The surface temperatures on the main section and on the joint between insulation boards, during SINb cycles (see Fig. 28), for instance, show differences of

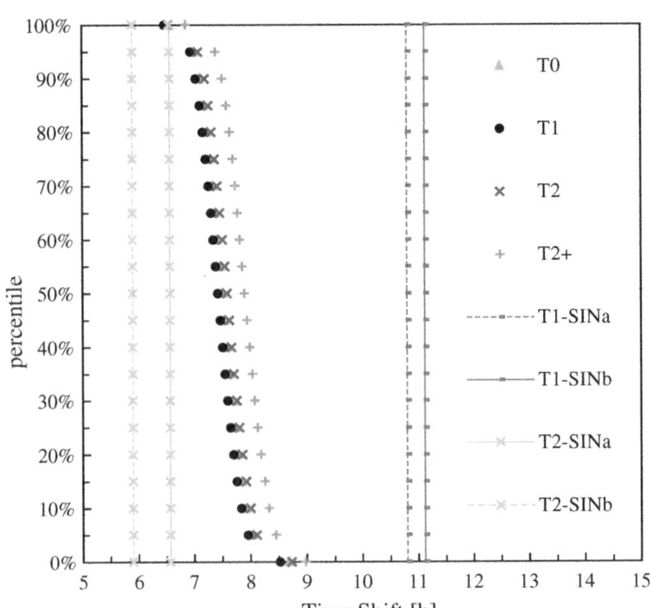

Fig. 27 Time shift for ageing times: measured (*lines*) and modelled (*plots*)

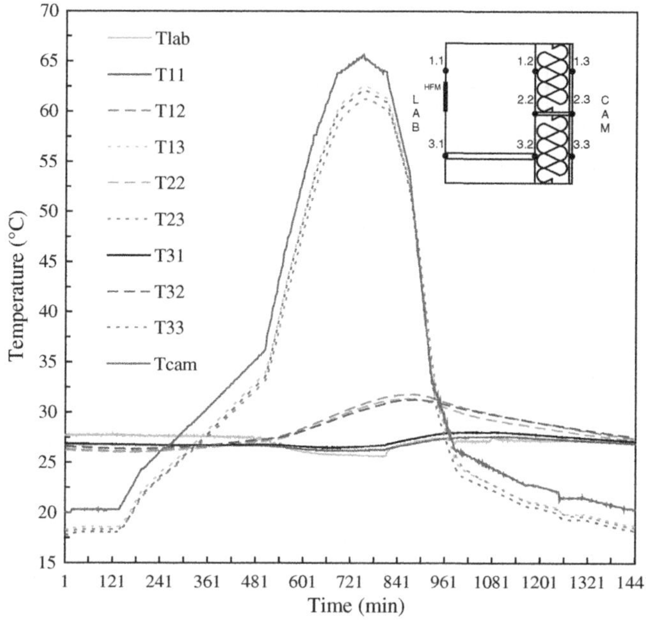

Fig. 28 Measurement of the temperatures within the section during test SINb, at ageing time T2

the order of 1 °C at peaks, while at the interface between the insulation and the masonry wall the differences are of the order of 0.5 °C (with a time shift between the peaks in the main section and on the joint between insulation boards of roughly 20 min). Qualitatively the same behaviour can be observed during the hygro-thermal cycles to assess the thermal inertia of the system (see Fig. 29). In this latter case, we observed that after roughly 8 h the temperatures at the interior surface reached the 90 % of the value achieved at stability. We note that the temperature variation at the interface between the ETICS and the masonry substrate is initially rapid, then perfect stability is reached only 24 h after the initial impulse, namely when the thermal capacitor is completely full. (We point out that the pink solid line in Figs. 28 and 29 does not properly represent the laboratory conditions: it actually indicates the air temperature at the boundary layer on the interior surface, which reflects the impulses of temperatures transported across the sample.)

6 Degradation Evolution

At a first level of qualitative observation of the degradation evolution, while performing the accelerated ageing exposure, we noted that at the end of UV and winter cycles the surfaces appeared substantially unchanged. Only a just

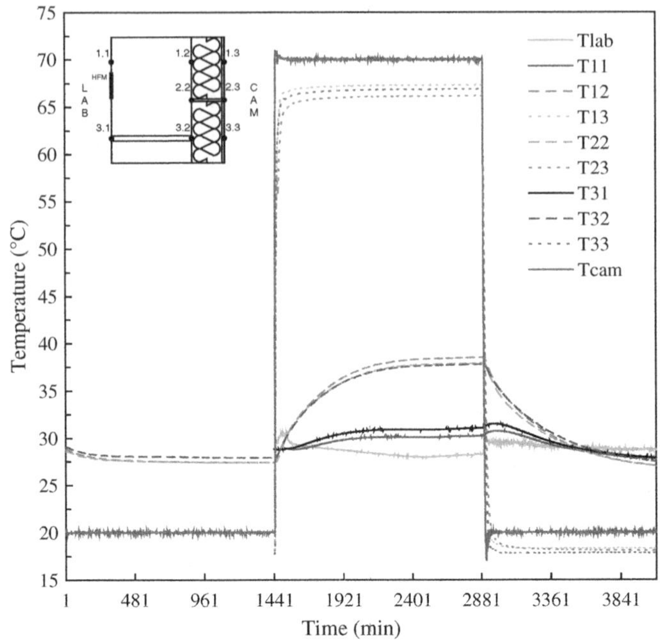

Fig. 29 Measurement of the temperatures within the section during test TI at ageing time T2

perceivable change in colour was noted . Then, we observed relevant changes right after the summer thermal shocks cycles (comparing degradation at T0 in Fig. 30 and at T1 in Fig. 31): localized blisters showed all over the surface of the sample, while the colour still did not change significantly.

Blisters had diameters ranging from a few millimitres up to roughly 9 cm, more diffuse in the upper left corner of the sample, and over the horizontal joint between the insulation boards. Moreover, Fig. 31, taken immediately after the end of ageing macro-cycle T1, shows white halos in several areas of the sample, which we regard as limestone deposits, and not as subflorescences phenomena, considered the nature of the finishing coat.

Proceeding with the accelerated ageing exposure, we did not note any remarkable variation in the intensity and distribution of blisters (see Fig. 32). The colour of the finishing, also, did not suffer large variations. Then, with magnifications of the images, we noted that the finishing coat presented an opening of the pores, increasingly with ageing. For instance for the observation point P3 (i.e. the central point of the sample), at ageing time T1.1 (see Fig. 33)—namely during ageing time T1 after UV and winter cycles and before thermal shock cycles—the surfaces appeared to be almost unchanged, with the pores of the finishing coat still closed, and without any perceivable degradation indicator. Then, at ageing time T2.1 (i.e. after UV and winter cycles and before thermal shock cycles during

Fig. 30 The big sample (1 m^2) applied as door to the weathering chamber before ageing (T0)

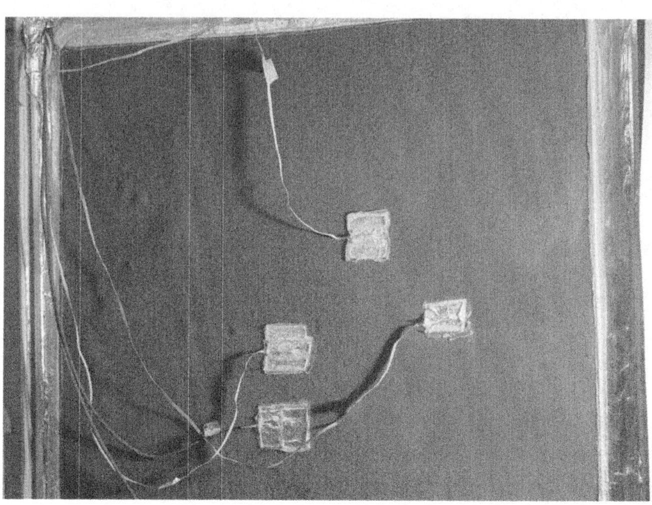

Fig. 31 The big sample (1 m^2) after the first ageing macro-cycle (T1)

Fig. 32 The big sample (1 m^2) at the end of ageing macro-cycle T2

Fig. 33 Observation point P3—Opening of the pores of the finishing coat at ageing time T1.1 (i.e. after UV and winter cycles, but before summer thermal shock cycles)

ageing macro-cycle T2) we noted more clearly the effect of rain-wash on the colour finishing, and the pores of the top coat appeared more open (see Fig. 34).

At a second stage of the analysis, as previously described, to assess the degradation evolution we developed a degradation scale, based on the method

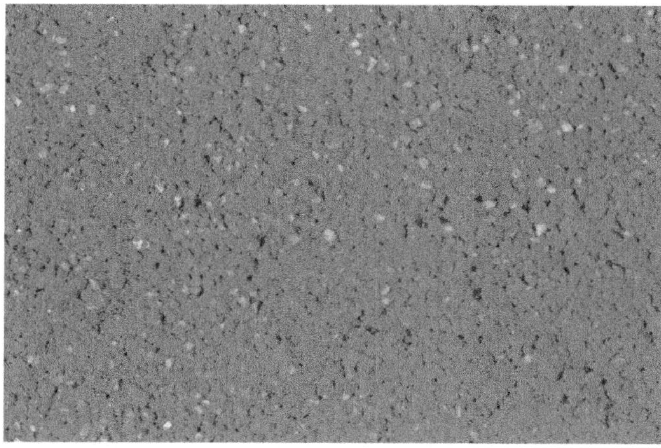

Fig. 34 Observation point P3—Opening of the pores of the finishing coat at ageing time T2.1 (i.e. after UV and winter cycles, but before summer thermal shock cycles during ageing time T2)

Table 13 Levels of blistering with regard to quantity, sizem and intensity of changes

Level	Quantity (n° of blisters/m^2)	Size (blisters surfaces in cm^2)	Intensity of changes
0	No detectable defects	Non visible under × 10 magnification	Unchanged
1	Very few (1–20)	Only visible under magnification up to × 10	Blistering
2	Small but significant number (20–40)	Just visible with normal corrected vision	Bulging
3	Moderate quantity (40–60)	Clearly visible up to 10 cm^2	Bulging, and detachments of portions of finishing
4	Remarkable quantity (60–80)	10–50 cm^2	
5	Dense pattern (>80): complete failure	Larger than 50 cm^2	

identified in ISO 4628 [32, 47], taking into account quantity, size, and intensity of the anomalies over a surveyed surface. In detail, we specialized the ranking method for blistering (see Table 13), and for quantity and size we considered six levels, from 0 (no defects/anomalies) to 5 (i.e. most serious defects/anomalies), while with regard to intensity we distinguished only four levels, from 0 to 3, identifying the physical stages of the degradation observed in case studies (analysed in the preliminary phases of the experimental programme).

Once defined the method, we identified the position of the blisters over the surface of the sample (see Fig. 35) with a grid of 10 × 10 cm (a stripe 5 cm wide at the edges was excluded from the analysis). Moreover, the dimension of each blister was surveyed, and the extension was measured.

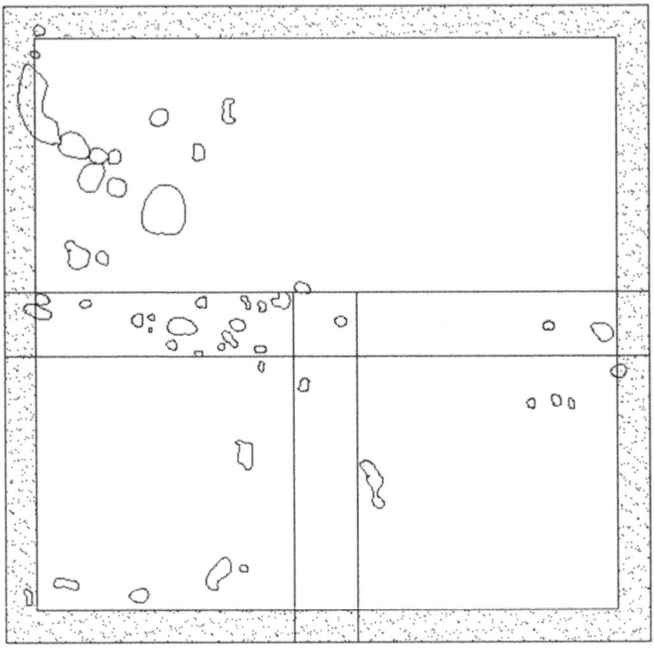

Fig. 35 Distribution of blisters over the surface of the sample at ageing time T2

We recorded 49 blisters over the whole surface, of which 20 over the horizontal joint between insulation boards, 2 over the vertical joint, 7 on the lower left panel and 5 on the lower right panel, and 15 on the upper panel.

Subsequently, the surface of each blister was computed: the most of blisters have diameter comprised between 1 and 1.5 cm and 1.5 and 2 cm, and the maximum surface area detected was of 58.5 cm^2 (diameter of 8.63 cm), while the smallest blister was of roughly 0.71 cm^2 (diameter of 0.95 cm). The total surface coverage of the blister was of roughly 3 % of the entire surface. Hence, according to the proposed classification we describe the degradation at ageing time T2 at level 3, with regard to quantity, and with regard to the intensity of changes, while, concerning the size of the blisters, we detected that level 5, corresponding to failure, was reached.

We explain the fact that no detachment was observed with the elasticity of the top coat, which had an acrylic binder with PVC equal to 80 %, thus the 20 % of resin content. Moreover, after the evaporation of water, the acrylic resin tends to film, providing additional protection against ageing.

7 Concluding Remarks

According to the implemented analysis and first degradation causes hypothesis, thermal performance decay should be mainly due to water content increase in EPS (so a relevant rise in apparent thermal conductivity of the layer is registered). A first hypothesis of ageing evolution could be the following (developed by different contributes given by different phases of ageing cycle):

- UV cycles: UV degradation of polymeric binder of finishing coat (progressive chain scission), increase in capillary cracking and in penetration depth of UV rays, thus changes in the colour of the finishing;
- Winter cycles: base coat and finishing are subject to tension force, capillary cracks width increases and water penetration depth in rain phase is allowed, that causes rain-wash. Wet rendering has a reduced traction resistance and during freeze–thaw cycles contribute to cement matrix degradation.
- Summer thermal shock cycles: base coat and finishing are subject to compression and an adhesion crisis between base coat and finishing is encountered. Due to this phenomenon finishing is deformed and moisture storing is allowed in this room, so water vapour pressure during heating phase cause blistering of finishing system.

We monitored the hygrothermal performances of a sample of an ETICS over masonry wall subject to accelerated ageing. We observed decreasing thermal resistance—almost only for the ETICS—due to increase of water content caused by rain penetration. Then, performing dynamic tests with sinusoidal temperature solicitation, a decrease in time shift was observed. We also computed the time shift (together with decrement factor and periodic thermal transmittance) for the sample with moisture content derived at each ageing time. We observed large differences in values of computed and measured time shift. We impute these differences to the fact that we measured the moisture content for the main profile section, while we estimate (thanks to IR thermography) that most relevant water penetration occurs through the joint between insulation boards.

We conclude that ageing, and moisture, whose content within layers is increased by ageing, affect dynamic thermal performances. Moreover, we argue that considering the main section profile is not sufficient to capture hygro-thermal dynamic performances of ETICS. We also point out the need of standardisation of un-steady state thermal performance testing methods.

The survey of degradation with photos has shown the most significant degradation evolution at ageing time T1, with the development of blisters on finishing coat surface. During all ageing time, an increase in the dimension of pores of finishing coat has been observed.

The aim of this experimental work is to predict the service life of an ETICS over a masonry wall, designing an ageing procedure based on ISO 15686-2. More precisely, our aim is to achieve performance over time functions, the observation of degradation over time, and correlations between degradation and performance loss. Knowledge regarding the performances over time may allow to compute the service life of a component in use conditions, since different buildings may have different requirements regarding the several performances of a building component, and any of these might determine the end of the service life, depending on the users' needs.

To assess in use conditions different from those simulated in the laboratory one might estimate the service life with simplified tools such as the factor method proposed in ISO 15686 (an application to ETICS is presented in [48, 49]), or with more refined instruments, such as heat and moisture transport simulations coupled with Monte Carlo method, to consider the variability of input parameters.

Acknowledgments We acknowledge the Italian Ministry of Research (MIUR) for the initial funding of the experimental programme, with the research contract "PRIN2003–2005: Methodologies for design and assessment of durability of building components in sustainable building processes: standard experimental evaluation and correction suitable for the utilisation in specified use conditions aimed to maintenance planning for buildings". We also want to thank for their precious help the BSc and MSc students in Building Engineering: Filippo Caimi, Andrea Dalcò, Alberto Fumagalli, Ilde Ferrara, Andrea Ragozzino, Francesco Rotella, Mattia Sala, and Matteo Tansi. We wish to thank Dr. Ronnen Levinson and all the Heat Island Group of the Lawrence Berkeley National Laboratory (Berkeley, California, USA) for having granted access to Riccardo Paolini to their laboratories and instrumentation with which the measurements of the optical and radiative properties were performed. We thankfully acknowledge Dr. H.M. Künzel of IBP-Fraunhofer for providing us with the software WUFI, which was helpful to better read the results of the experiments and design further tests.

References

1. EOTA: European Organization of Technical Approvals, ETAG 004—Edition March 2000. Guideline for European Technical Approval of external thermal insulation composite systems with rendering. Bruxelles, Belgium (2000)
2. Daniotti B., Paolini R.: Durability Design of External Thermal Insulation Composite Systems with Rendering. 10th DBMC, Lyon 17–20 April 2005
3. Hens, H., Carmeliet, J.: Performance Prediction for Masonry Walls with EIFS Using Calculation Procedures and Laboratory Testing. J. Therm. Envelope Build. Sci. **25**, 167–186 (2002)
4. Künzel, H., Künzel, H.M., Sedlbauer, K.: Long-term Performance of External Thermal Insulation Systems (ETICS). Acta Archit. **5**(1), 11–24 (2006)
5. Sahal, N., Lacasse, M.A.: Water entry function of a hardboard siding-clad wood stud wall. Build. Environ. **40**, 1479–1491 (2005)
6. Beaulieu, P., Bomberg, M., Cornick, S., Dalgliesh, A., Desmarais, G., Djebbar, R., Kumaran, K., Lacasse, M., Lackey, J., Maref, W., Mukhopadhyaya, P., Nofal, M., Normandin, N., Nicholls, M., O'Connor, T., Quirt, D., Rousseau, M., Said, N., Swinton, M., Tariku, F.: Final Report from Task 8 of MEWS Project (T8-03)—Hygrothermal Response of Exterior Wall Systems to Climate Loading: Methodology and Interpretation of Results for Stucco, EIFS, Masonry and Siding-Clad Wood-Frame Walls. Institute for Research in Construction, National Research Council Canada, Ottawa (2002)

7. Zirkelbach D., Holm A., Künzel H.M.: Influence of Temperature and Relative Humidity on the Durability of Mineral Wool in ETICS. 10th International Conference On Durability of Building Materials and Components, Lyon, France, 17-20 April 2005
8. Abuku, M., Janssen, H., Poesen, J., Roels, S.: Impact, absorption and evaporation of raindrops on building facades. Build. Environ. **44**, 113–124 (2009). doi:10.1016/j.buildenv.2008.02.001
9. de Freitas, V.P., Abrantes, V., Crausse, P.: Moisture Migration in Building Walls—Analysis of the Interface Phenomena. Build. Environ. **31**(2), 99–108 (1996)
10. Bronski M.B.: Wall Cladding System Durability Lessons Learned from the Premature Deterioration of Wood-Framed Construction Clad with Exterior Insulation and Finish Systems (EIFS) in the U.S., Proceedings 10th International Conference on Durability of Building Materials and Components, Lyon, France, 17–20 April 2005
11. Barreira E., de Freitas V.P.: Importance of Thermography in the Study of ETICS Finishing Coatings Degradation due to Algae and Mildew Growth. Proceedings 10th International Conference on Durability of Building Materials and Components, Lyon, France, 17–20 April 2005
12. Daniotti B., Paolini R.: Evolution of Degradation and Decay in Performance of ETICS. Proceedings 11th International Conference on Durability of Building Materials and Components, Istanbul, Turkey, 11–14 May 2008, Vol. 4, pp. 1523–1530, (2008)
13. ISO, ISO 15686-2: Building and constructed assets. Service life planning. Part 2. Service life prediction principles. International Organization for Standardization, Geneva, CH (2000)
14. Daniotti B., Re Cecconi F.: CIB W080: Test Methods for Service life Prediction—State of the Art Report on Accelerated Laboratory Test Procedures and Correlation between Laboratory Tests and Service Life Data, CIB Publication 331, CIB—International Council for Research and Innovation in Building and Construction, Rotterdam, The Netherlands, (2010) ISBN: 978-90-6363-062-1—Available online at: http://cibworld.xs4all.nl/dl/publications/w080_wg3_report.pdf
15. EOTA: European Organisation for Technical Approvals 1999, Assessment of working life of Products, EOTA, Brussels, EOTA Guidance document 003/edition December 1999
16. Re Cecconi F., De Angelis E.: Guasti in edilizia: ammaloramenti dell'edificio, suggerimenti di ripristino e di prevenzione. Banca Dati dei casi di guasto. Rimini, Maggioli editore (2008)
17. Sleiman M., Destaillats H., Kirchstetter T., Gilbert H., Berdahl P., Akbari H., Marlot L., Quelen S., Preble C., Spears M., Ban-Weiss G., Levinson R.: Accelerated Aging Protocols for Roofing Materials: Version 1.0. In: Proceedings of International Workshop on Advances in Cool Roof Research, Berkeley, CA—July 28 & 29, 2011
18. Künzel H.M., Krus M., Fitz C., Hofbauer W., Scherer C., Breuer K.: Accelerated Test Procedure to Assess the Microbial Growth Resistance of Exterior Finishes. 12th International Conference on Durability of Building Materials and Components, Porto, Portugal, 12-15th April, 2011
19. UEAtc: Directives for the Assessment of External Insulation Systems for Walls (Expanded Polystyrene Insulation Faced with a Thin Rendering), June 1988. Union Européenne pour l'Agrément technique dans la construction (European Union of Agrement, 1988)
20. UEAtc: Technical Guide for the Assessment of External Wall Insulation Systems Faced with Mineral Render. Union Européenne pour l'Agrément technique dans la construction (European Union of Agrement, 1992)
21. ISO, ISO 15927-1: Hygrothermal performance of buildings—Calculation and presentation of climatic data—Part 1: Monthly means of single meteorological elements. International Organization for Standardization, Geneva, CH (2003)
22. UNI, UNI 10349: Riscaldamento e raffrescamento degli edifici—Dati climatici (Heating and cooling of buildings. Climatic data). Ente Nazionale Italino di unificazione, Milano, Italy (1993)
23. ISO, ISO 15927-4: Hygrothermal performance of buildings—Calculation and presentation of climatic data—Part 4: Hourly data for assessing the annual energy use for heating and cooling. International Organization for Standardization, Geneva, CH (2005)

24. Maggi P.N., Rejna M.G., Daniotti B., Re Cecconi F., Poli T., Rigamonti B., Jornet A., Teruzzi T.: Experimental program to evaluate building elements service life: first results on brickwork in Proceedings of 8th "Durability of Building Materials & Components" Conference, Vancouver (1999)
25. Daniotti B., Iacono P.: Evaluating the service life of external walls. A comparison between long-term and short-term exposure. 10th DBMC, Lyon 17–20 April 2005
26. Daniotti B., Lupica Spagnolo S., Paolini R.: Climatic Data Analysis to Define Accelerated Ageing for Reference Service Life Evaluation. Proceedings 11th International Conference on Durability of Building Materials and Components, Istanbul, Turkey, 11-14 May 2008, vol. 3, pp. 1343–1350 (2008)
27. CEN, EN 196-1: Methods of testing cement. Determination of strength. European Committee for Standardization, Bruxelles, Belgium (1996)
28. CEN, EN 13163: Thermal insulation products for buildings—Factory made products of expanded polystyrene (EPS)—Specification. European Committee for Standardization, Bruxelles, Belgium (2003)
29. Berdahl, P., Akbari, H., Jacobs, J., Klink, F.: Surface roughness effects on the solar reflectance of cool asphalt shingles. Sol. Energy Mater. Sol. Cells **92**, 482–489 (2008)
30. ISO, ISO 10456: Building materials and products—Hygrothermal properties—Tabulated design values and procedures for determining declared and design thermal values. International Organization for Standardization, Geneva, CH (2007)
31. ASTM, ASTM American Society for Testing and Materials, E 312-06: Standard Practice for Description and Selection of Conditions for Photographing Specimens Using Analog Cameras and Digital Still Cameras. American Society for Testing and Materials, West Conshohocken, PA, USA (2006)
32. ISO, ISO 4628-1: Paints and varnishes—Evaluation of degradation of coatings—Designation of quantity and size of defects, and of intensity of uniform changes in appearance. Part 1: Part 1: General introduction and designation system. International Organization for Standardization, Geneva, CH (2003)
33. NORMaL, NORMal 44–93: Assorbimento d'acqua a bassa pressione. Milano, Italy (1993)
34. ISO, ISO 9869: Thermal insulation—Building elements—In-situ measurements of thermal resistance and thermal transmittance. International Organization for Standardization, Geneva, CH (1994)
35. ISO, ISO 15148: Hygrothermal performance of building materials and products—Determination of water absorption coefficient by partial immersion. International Organization for Standardization, Geneva, CH (2003)
36. ASTM, ASTM American Society for Testing and Materials, E 903-96: Standard test method for solar absorptance, reflectance, and transmittance of materials using integrating spheres. American Society for Testing and Materials, West Conshohocken, PA, USA (1996)
37. ASTM, ASTM American Society for Testing and Materials, C 1549-02: Standard test method for determination of solar reflectance near ambient temperature using a portable solar reflectometer. American Society for Testing and Materials, West Conshohocken, PA, USA (2002)
38. ASTM, ASTM American Society for Testing and Materials, C 1371-04: Standard test method for determination of Emittance of Materials Near Room Temperature Using Portable Emissometers. American Society for Testing and Materials, West Conshohocken, PA, USA (2004)
39. ASTM, ASTM American Society for Testing and Materials, G 173-03: Reference Solar Spectral Irradiances: Direct Normal and Hemispherical on 37° Tilted Surface. American Society for Testing and Materials, West Conshohocken, PA, USA (2003)
40. Cresswell, H.P., Painter, D.J., Cameron, K.C.: Tillage and water content effects on surface soil hydraulic properties and shortwave albedo. Soil Sci. Soc. Am. J. **57**, 816–824 (1993)
41. Daniotti B., Re Cecconi F., Paolini R.: Effects of ageing and moisture on dynamic thermal performance of ETICS cladding, Proceedings 12th International Conference on Durability of

Building Materials and Components, Porto, Portugal, 12–15 April 2011, vol. 1, pp. 217–224 (2011)
42. ISO, ISO 13786: Thermal performance of building components—Dynamic thermal characteristics—Calculation methods. International Organization for Standardization, Geneva, CH (2007)
43. Di Perna, C., Stazi, F., Ursini Casalena, A., D'Orazio, M.: Influence of the internal inertia of the building envelope on summertime comfort in buildings with high internal heat loads. Energy Buildings **43**(1), 200–206 (2011)
44. Künzel, H.M.: Simultaneous Heat and Moisture Transport in Building Components (One- and two-dimensional calculation using simple parameters). Fraunhofer Institute of Building Physics. Fraunhofer Verlag, Stuttgart, Germany (1995)
45. IEA: International Energy Agency 1991, Condensation and Energy, Catalogue of Material Properties, Report Annex XIV, Volume 3 International Energy Agency, Condensation and Energy, Catalogue of Material Properties, Report Annex XIV, Volume 3, 1991
46. Salonvaara M., Karagiozis A., Holm A.: Stochastic Building Envelope Modeling—The Influence of Material Properties. Proceedings VIII Conference of Performance of Exterior Envelopes of Whole Buildings, 2–7 December 2001, Clearwater Beach, Florida, USA (2001)
47. ISO, ISO 4628-2: Paints and varnishes—Evaluation of degradation of coatings—Designation of quantity and size of defects, and of intensity of uniform changes in appearance. Part 2: Assessment of degree of blistering. International Organization for Standardization, Geneva, CH (2003)
48. Daniotti B., Lupica Spagnolo S., Paolini R.: Factor Method Application Using Factors' Grids'. Proceedings of the 11th International Conference on Durability of Building Materials and Components. 11–14th May 2008 vol. 3, pp. 1449-1457 (2008)
49. Daniotti B., Lupica Spagnolo S., Paolini R.: Service life estimation of building components: methods for durability assessment in use conditions. Proceedings of International conference on construction & Building research, 24–26th June 2009, Madrid (Spain). pp. 130–140 (2009)
50. ISO, ISO 12572: Hygrothermal performance of building materials and products—Determination of water vapour transmission properties. International Organization for Standardization, Geneva, CH (2001)
51. CEN, EN 13494: Thermal insulation products for building applications—Determination of the tensile bond strength of the adhesive and of the base coat to the thermal insulation material. European Committee for Standardization, Bruxelles, Belgium (2003)

Durability Assessment of Adhesive Systems for Bonding Ceramic Tiles on Façades: The Research and the Practice

Vasco Peixoto de Freitas, Helena Corvacho, Marisa Quintela and J. M. P. Q. Delgado

Abstract The aim of this work is to give a contribution for the assessment of the durability of adhesive systems for bonding ceramic tiles on façades, based on experimental works carried out at the Building Physics Laboratory, Faculty of Engineering, University of Porto, Portugal (LFC—FEUP), in the perspective of both the research and the possible practical applications. The durability assessment of any construction material is not an easy task. The researchers must choose between several different approaches, each of which has advantages and limitations. While, in a building in use, the materials or components are subjected to varying actions in an environment where the great majority of the degradation agents are not controllable by man, in a laboratory, one can choose to expose the materials to artificial and controlled conditions, using climatic chambers for accelerated ageing. The main difficulty of this type of test lies in the interpretation of the results, in what concerns their correspondence to real time. Researchers on durability have been discussing this problem for a long time: how to get a valuable correlation between the results of accelerated ageing tests and the one issued from natural exposure. This work is presented in two different stages. In a first stage, a discussion on durability assessment is presented based on an experimental research study, concerning the performance over time of different types of cementitious adhesives exposed to accelerated ageing tests and also to natural ageing. The main advances and the main difficulties in implementing a service life prediction model

V. P. de Freitas (✉) · H. Corvacho · M. Quintela · J. M. P. Q. Delgado
LFC—Building Physics Laboratory, Civil Engineering Department, University of Porto, 4200-465 Porto, Portugal
e-mail: vpfreita@fe.up.pt

H. Corvacho
e-mail: corvacho@fe.up.pt

M. Quintela
e-mail: betaniaquintela@gmail.com

J. M. P. Q. Delgado
e-mail: jdelgado@fe.up.pt

V. P. de Freitas and J. M. P. Q. Delgado (eds.), *Durability of Building Materials and Components*, Building Pathology and Rehabilitation 3,
DOI: 10.1007/978-3-642-37475-3_7,

are identified. Suitable strategies are discussed for the future development of this approach. In a second stage, an extensive case study is presented. The main goal of this case study is to evaluate the durability of alternative adhesive systems for bonding ceramic tiles on the façades of a building, located near the sea. For this purpose, accelerated ageing tests are performed following two different ageing procedures, allowing the comparison of the performance over time of the systems under analysis.

Keywords Durability · Performance · Sustainable construction · Accelerated ageing tests · Natural ageing · Adhesive systems · Prediction model

1 Introduction

The objective of any service life prediction of a material, system or component integrated into a building is to evaluate its ability to satisfactorily perform its operations throughout the service life of the building or throughout the period considered reasonable for its replacement or repair.

The preliminary experimental studies that are presented here lead to the proposal of a model for service life prediction applied to cementitious adhesives for ceramic tiles which show a clear degradation of the adhesive strength over time.

A reflection is presented on the factors that could influence the service life prediction, on the structure of the model and on the possibility of its generalized application.

In the complete case study presented ahead, a comparative durability assessment is made. The main goal of this case study was to compare the durability of several alternative adhesive systems, based on the variation over time of their adhesion to the support. The evaluation of the adhesive strength of the different solutions was carried out by performing pull-off tests, on specimens, at the initial conditions (after 28 days of curing) and after submitting the specimens to different cycles of accelerated ageing.

An efficient adhesive bonding of exterior ceramic tiles applied on façades is an obvious important factor to ensure the safety and the durability of the façade. The failure of adhesive bonding has immediate consequences and therefore is a common concern of the building industry and of building owners.

This case study focuses on the behaviour of ceramic tiles applied on an exterior surface where the substrate is concrete C35/40. This type of envelope is frequently subjected to pathologies that can have different causes. The pull-off test is a very popular way of evaluating the adhesive strength of renderings or adhesives for ceramic tiles and it is frequently used as a tool for a correct diagnosis of causes for degradation of the façades exterior layers.

Several standards are available to frame the evaluation of adhesive strength by means of the pull-off test, namely EN 1348 [1] *Adhesives for tiles—Determination of tensile adhesion strength for cementitious adhesives* and EN 12004 [2] *Adhesives For Tiles—Requirements, Evaluation Of Conformity, Classification And Designation.*

2 Service Life Prediction: The Research

2.1 Service Life Prediction Model Adopted by LFC-FEUP

Experimental studies carried out in the LFC-FEUP, were based on the general methodology proposed by ISO 15686 [3] standard and lead to the proposal of the model that is presented here [4]. Test campaigns were carried out for accelerated ageing. Natural ageing stations where the materials were exposed to the external environment were simultaneously set up.

In this way, according to the diagram in Fig. 1, when in possession of the test results, referring to accelerated ageing and to natural ageing, given that the type of degradation resulting from either one or the other test is seen to be similar, it is possible to interpret these results and develop a model of service life prediction of the products and systems being studied through the establishment of a correlation between the number of cycles of artificial ageing and the real time of natural ageing (see Fig. 2).

Knowing the degradation curve of a given characteristic (as an example, the Fig. 2 considers the adhesive strength between the cementitious adhesive and its substrate) it will be possible to establish a correspondence between the number of accelerated artificial ageing cycles and the number of years in real time [5].

2.2 Critical Analysis of the Difficulties Associated to the Chosen Model

The model presupposes the identification of a degradation curve as a function of time, obtained in accelerated ageing tests. This procurement can be met with several difficulties. Firstly, the most relevant agents of degradation for the material being studied have to be rigorously identified and their variation known. For this identification, the performances of the preliminary tests become important. Secondly, performance characteristics to be evaluated must be selected, opting for those that are shown to be critical for durability. This choice is very important and could be based, on the one hand, on the results of the preliminary tests and, on the other hand, on the historical pathologies normally associated with the material to be evaluated. If the identified degradation agents or the selected performance characteristics were not the most adequate, the tests could be shown to be inconclusive and the model would not be able to be applied.

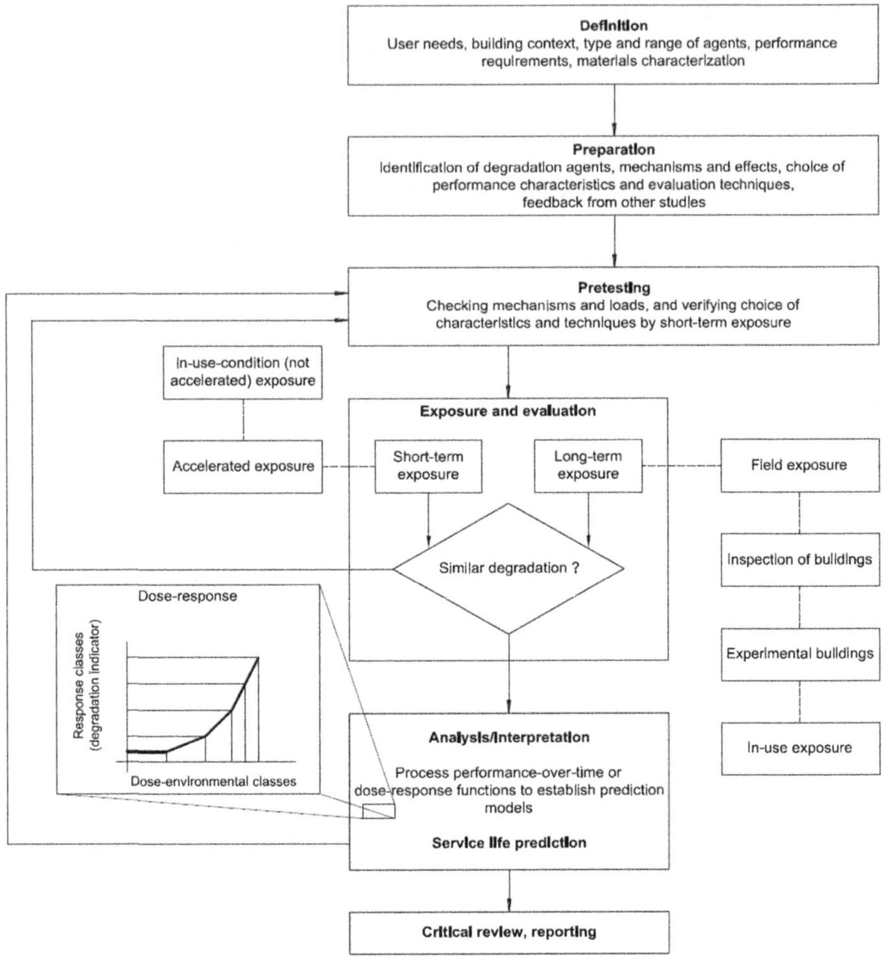

Fig. 1 Systematic methodology for service life prediction of building components (ISO 15686)

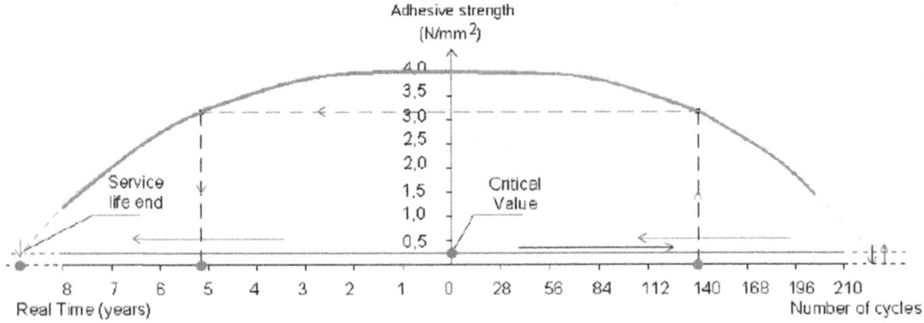

Fig. 2 Correlation between the numbers of accelerated artificial ageing cycles and real time natural exposure—prediction model

With the adequate performance characteristics chosen and the necessary artificial accelerated ageing tests performed, another problem could still arise: the results obtained for different stages of ageing do not reveal any definable tendency, making it impossible to draw a degradation curve. In this case, the hypothesis that there are other aspects that are more relevant to ageing of the material or system in study becomes plausible: the agents of degradation being considered may not be very relevant for the variation of the characteristic being tested as it is overall dependent on, for example, the execution, on the interaction between different elements of the construction (interfaces), on the type of in-use conditions, etc. In this case, the application of the model as presented becomes unviable.

It can also be observed that the model assumes the occurrence of a degradation curve that is continuous in time and does not consider the interventions hypothesis along this timeline that restore or improve a given performance characteristic, which could obviously take place in reality. However, the objective of the model is to obtain a prediction of a reference value for the service life. With this value determined, it will then be possible to take into account the effects of modifying factors related to the various aspects present in the life cycle of each system or material, with these factors being related to the design, the execution, the use, the specific environments to which they will be exposed, maintenance measures, etc.

Regarding the service life prediction starting from a specific number of tests in a natural ageing scheme, the conclusions made will be dependent on the ease or the difficulty of gauging the deterioration observed and its comparison with the results of the accelerated ageing tests. In this way, for products or systems with a history of applications, it will also be possible to complement the information with a study of its record with respect to pathologies and the repair measures, being able to more easily evaluate the significance of the observed deterioration in the early stages of the experimental installation. In addition to this, the tests under in-use conditions, performed in buildings of different ages, could substitute or complement the tests made on the experimental installations of natural ageing, as long as it is possible to retrieve results in a sufficient number to be statistically significant and as long as there is information about the type of products applied, the date of application and the conditions to which they have been subjected to over the years.

2.3 Preliminary Experimental Studies

2.3.1 Brief Description

In a preliminary study [6] the durability of cementitious adhesive in bonded ceramic coverings of external walls was evaluated. This kind of covering has a long tradition in Portuguese construction. It continues, in modern times, to be widely used, both for its aesthetic aspect as well as for its increased durability and satisfactory functional performance. Added to this is the relative ease of cleaning and maintenance.

The system under study is basically composed of three essential elements: the substrate, the bonding material (in this case, the cementitious adhesive) and the ceramic tile. The evaluation of durability of one of the elements must take into account, by necessity, the durability of the system and the compatibilities between the different elements.

The substrate for the ceramic coverings is defined as a function of the nature of its constituent materials. In the present study, only concrete walls covered with extremely hardened mortar were considered. The designation of "polymer cementitious adhesive" applies to a hydraulic adhesive mortar with a cement base, with aggregates (fine granulometry) and additives that improve water retention, plasticity and adherence.

The ceramic tiles are thin plates made from clay and/or other organic raw materials and are generally used as coverings for walls and floors. They are, usually, formed by extrusion or pressing at room temperature but can also be moulded by other processes. Next they are dried and finally cooked at a temperature appropriate for obtaining the required characteristics.

2.3.2 Performance Indicators Evaluated and Criteria for Durability

In the evaluation of the durability of the adherent ceramic covering system, the criterion selected for its characterization was the value of the adhesive strength to its substrate. It was considered that this characteristic would be the most important one for the service life prediction of the system. The set critical value was that of 0.3 N/mm^2. Over a period of time, adhesion tests were performed in order to evaluate the adhesive strength, at different moments of the accelerated ageing process, with the aim of characterizing its influence on the performance of the cementitious adhesive.

Figure 3 and Table 1 show us the characteristics of the specimens used in this preliminary study.

Of the various mechanisms and agents of degradation responsible for the ageing of the study systems, those of hygrothermic nature are considered, namely: variation in temperature; variation in relative humidity; incidence of solar radiation and precipitation. These actions have a determining role in the ageing of the exterior coverings. Therefore, the specimens under analysis were submitted to accelerated ageing cycles that reproduce these actions.

Accelerated Ageing Procedure

The cycles performed in the study of cementitious adhesive consist of the exposure of the specimens to extreme conditions of usage, with harsh variations in temperature, relative humidity and radiation in such a way as to provoke the rapid degradation of the system of covering, Fig. 4 portrays the adopted cycle.

Fig. 3 Scheme of the specimens used

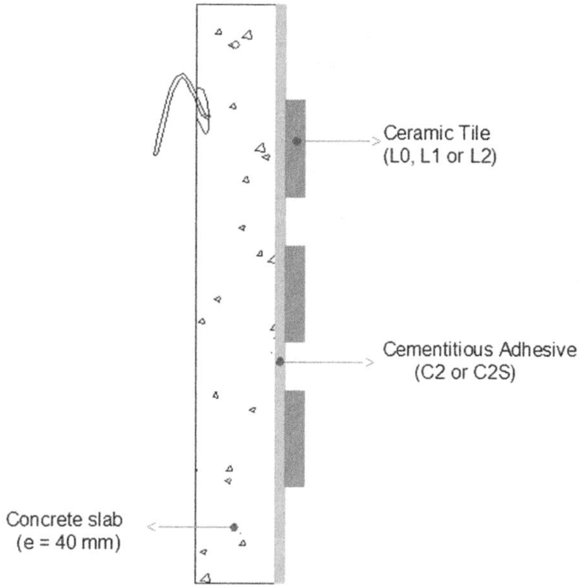

Table 1 Ceramic tiles, cementitious adhesive and specimen's identification

Specimens	Ceramic tile	Cementitious adhesive
PE0	L0	C2
PE1	L1	C2
PE2	L1	C2S
PE3	L2	C2
PE4	L2	C2S
Designation		Group[a]
Ceramic tiles	L0	BIa
	L1	AI
	L2	BIIa
Designation		Class[b]
Cementitious adhesive	C2	C2
	C2S	C2S

[a] EN 14411 [7]
[b] CSTB [8]

Simultaneously, a station was set up for natural ageing consisting of a wall covered with different ceramic tiles (see Fig. 5) and, exposed to the external environment. Similarly to that performed in the laboratory, adhesion tests were performed on these walls.

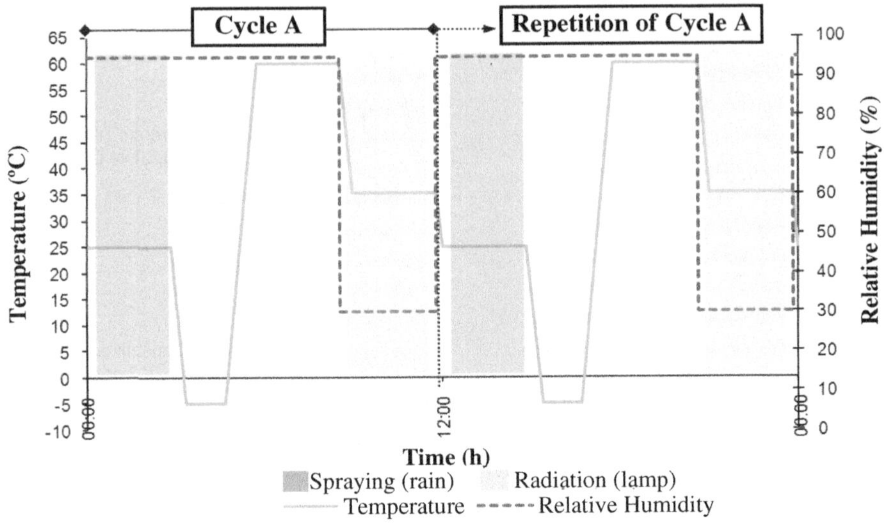

Fig. 4 Complete cycles performed in 24 h

Fig. 5 Natural ageing station—ceramic covering

Pull-Off Test

The adhesive strength is determined as the maximum tensile strength applied by a direct load perpendicular to the surface being tested [9–11]. The pull-off test is classified as a near-to-surface, partially destructive method which is able to measure tensile adhesive strength of various materials. The tensile load is applied by means of a defined pull-head plate glued to the test area. The adhesive strength is the ratio between the failure load and the test area.

Pre-cut tiles with 50 × 50 mm were used. The test machine for direct pull tensile force test is in accordance with standard requirements, with the additional

Fig. 6 Pull-off test machine

feature of automatic control of the applied force provided by an attached electrical engine, as presented in Fig. 6.

As mentioned in the introduction, several standards are available to frame the evaluation of adhesive strength by means of the pull-off test, namely EN 1348 [1] and EN 12004 [2] which were used in this study. In the latter, the definition of types of failures can be found, as sketched in Fig. 7.

2.3.3 Experimental Results

As regards the durability of cementitious adhesive, the graphic in Fig. 8 presents the variation in the adhesive strength as a function of the number of accelerated ageing cycles to which the specimens designated as PE0, PE1 and PE3 had been subjected. In these three specimens cementitious adhesive of type C2 EN 12004 [2] was used along with three different types of ceramic tiles. There is a very clear decreasing tendency and it can be predicted that, independent of the type of ceramic tile used, in the set of tests performed on the specimens with class C2 cementitious adhesive, 140 cycles would be sufficient to drop below the critical adhesive strength value of 0.30 N/mm^2 and thereby reach the service life term of the product.

Regarding the tests performed on the wall of the natural ageing station, given the little time of exposure that passed by until the performance of the first tests, it was only intended to demonstrate the methodology to follow to establish the correlation between the results obtained from accelerated ageing and those obtained with natural ageing.

It must be said that the type of failure is strongly dependent on the type of ceramic tile. The failure that occurs in the PE0 tests, concerning tiles with a low absorption coefficient (0.02 %) was of the adhesive type. Then, the measured values of the adhesive strength correspond, in fact, to the adhesive tension between the cementitious adhesive and its substrate. In the other tests (PE1 and PE3), where

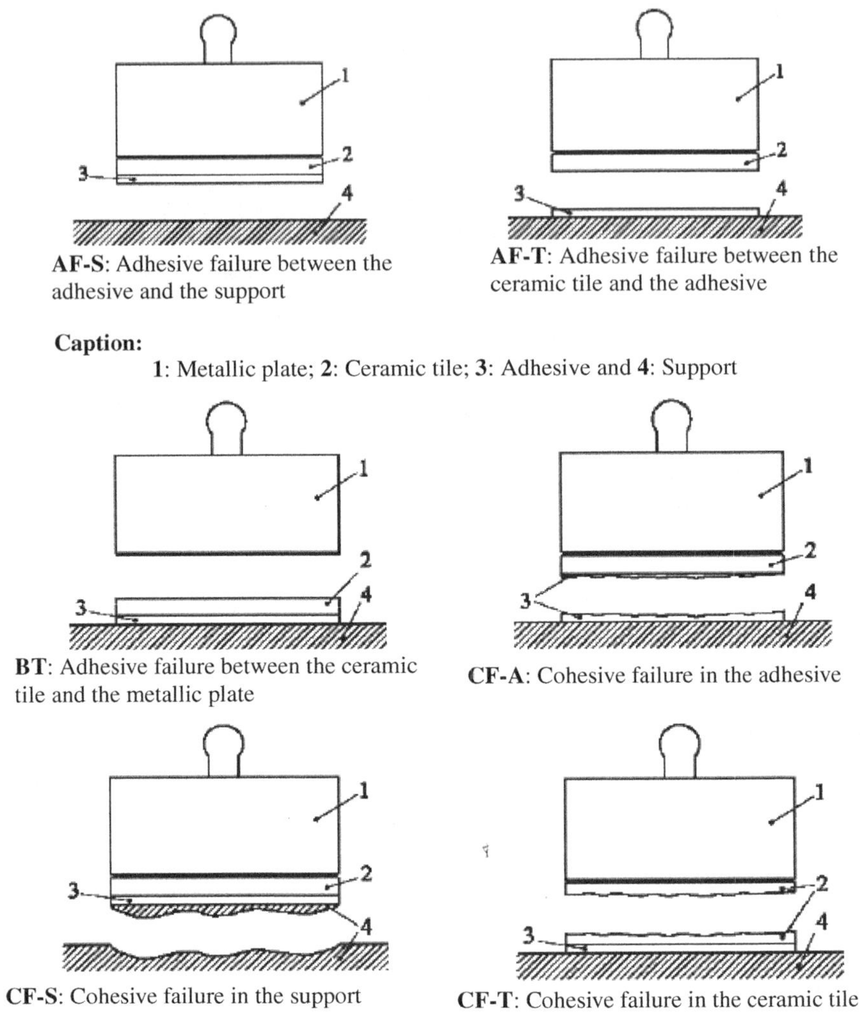

Fig. 7 Types of failures (EN 12004 [2])

the ceramic tiles used have a higher absorption coefficient (>0.5 %), the failure was of the cohesive type, which means that it occurred within the thickness of the layer of the cementitious product. In these circumstances, the measured values do not correspond to the adhesive tension between the cementitious adhesive and its substrate but to one of an inferior limit of the searched parameter.

In this study, cementitious adhesive of type C2S (EN 12004 [2]) was also tested and, in this case, the failure was always of the cohesive type. The degradation curves for C2S product have the same configuration as the ones for C2 product but,

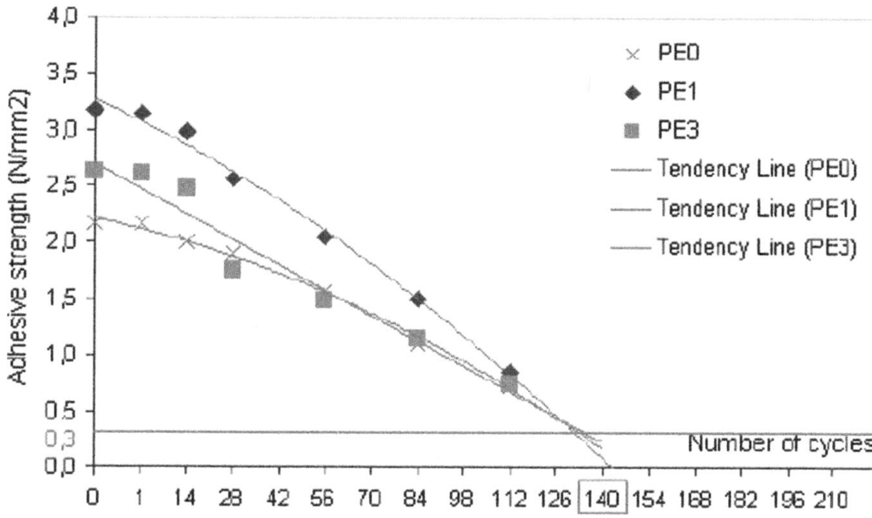

Fig. 8 Service life term prediction for C2 cementitious adhesive

as expected, a higher number of ageing cycles was necessary for reaching service life term.

Regarding the natural ageing station, the substrate was a rendered brick wall, unlike in the specimens used in accelerated ageing of the cementitious adhesive where the substrate was concrete, as can be seen in Fig. 3. In the adhesion tests performed at the natural ageing station, the rupture was always of the cohesive type and occurred within the render layer and not within the cementitious adhesive layer. Therefore, the only possible evaluation concerning the adhesive strength of the studied product over time, at the natural ageing station, was that its values are higher than the values measured in the tests.

2.4 Considerations Regarding the Applicability of the Model in Durability Assessment

2.4.1 Main Problems

After carrying out the experimental research, it was possible to identify clearly the main difficulties for the development of the model. They are the following:

- Adequate identification of the degradation agents;
- Definition of the parameters to measure;
- Definition of criteria to be adopted;
- Configuration of accelerated ageing tests;
- To execute samples that reproduce the in-site applied products;

- To obtain interesting and identifiable degradation curves;
- To obtain results in a sufficient number to be statistically significant;
- How to take into account the variability of the execution;
- To manage adequately the interrelationship of the different parts of the systems (beginning on the substrate wall).

As regards the cementitious adhesive for bonding ceramic tiles, once it was possible to obtain a degradation curve in the accelerated ageing tests, to be able to reliably apply the model, it became necessary to perform the tests in the natural ageing station, for defined time periods, in order to be able to establish a relationship between the short term tests and the long term tests.

It must be noticed also that for products with a reduced application experience or for new products, it is difficult to avoid the necessity of waiting for a relatively long time for the long term tests. If a service life prediction in absolute terms is intended, performance data for these products for in-use conditions are not available which would serve to evaluate the significance of the tests in which they were obtained. But this is not the case for cementitious adhesive, for which there is a long experience of application. It will be possible to collect the performance data for in-use conditions, although total characterization would not always be easy in terms of historical application and use. In Fig. 9 a comparison between the results for accelerated artificial ageing and for natural ageing is presented.

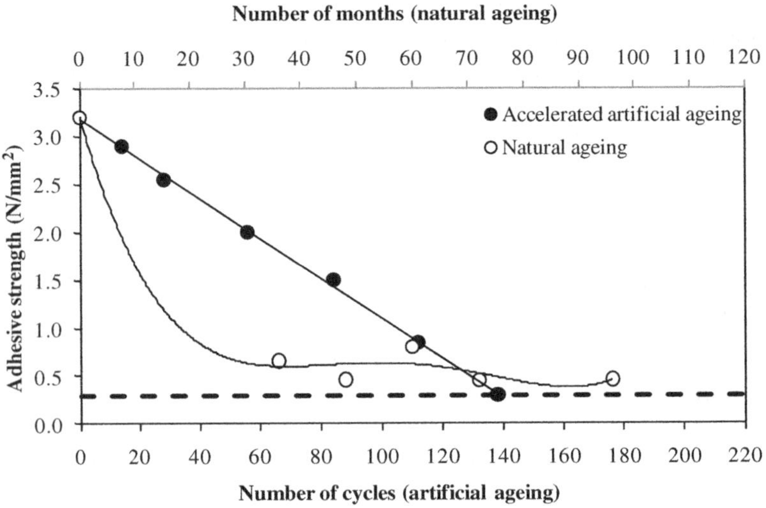

Fig. 9 Application of the model to cementitious adhesive

2.4.2 Future Strategies

In Fig. 9, it can be observed that the degradation curves are not similar although both of them show a clear decreasing trend. This leads to the need of a deeper reflection on the mechanisms of degradation and on the dose-response assessment. Simultaneously, the reproducibility of the type of degradation curve obtained for natural ageing must be confirmed.

For the development of the prediction model and its future application to different systems and materials, campaigns of accelerated ageing tests must be performed, accompanied by natural ageing tests and the systematic collection of data from existing buildings. For the adopted strategies to be actually efficient, a rigorous identification of the relevant characteristics to be tested (those that are seen to be critical to durability) is very important and also the performance of preliminary tests for the gauging of the expected variations (dose-response assessment) and for the identification of the most influential agents.

After a service life prediction has been obtained for specified conditions, where the predicted service life will work as a reference service life, it becomes necessary to evaluate the influence of the specific conditions of each real situation. For this, for example, the factorial method proposed in the ISO standard 15686 or other equivalent method could be adopted. The method takes into account, for each situation in the project, the conditioning factors of an estimated service life. They are: the intrinsic quality of the material or system; the quality of the design; the quality of the execution; the environmental, interior and exterior characteristics; the type of use and the level of maintenance.

3 Case Study: The Practice

This extensive case study presents the results of a work carried out by request of the Port Authority of "Porto de Leixões" to analyse the hygrothermal performance and the durability of several solutions for bonded ceramic tiles to waterproof the façades of the planned new Cruise Terminal (see Fig. 10).

Fig. 10 New cruise terminal (images provided by the designer, architect Luis Pedro Silva)

In what concerns durability, the main goal of this work was to evaluate the performance over time in terms of adhesion of some alternative adhesive systems for ceramic tiles to be applied on the façades of the new Cruise Terminal. The evaluation of the adhesive systems performance was carried out, on specimens, by measuring the adhesive strength to the support in the initial conditions (after 28 days of curing) and after submitting the specimens to different cycles of accelerated ageing.

3.1 Experimental Set-Up

The specimens used in this case study were made with concrete C35/40 (support), with the application of a waterproofing and adhesive system, as sketched in Fig. 11. For each of the 6 bonding systems tested (see Table 2) 10 specimens were used [12].

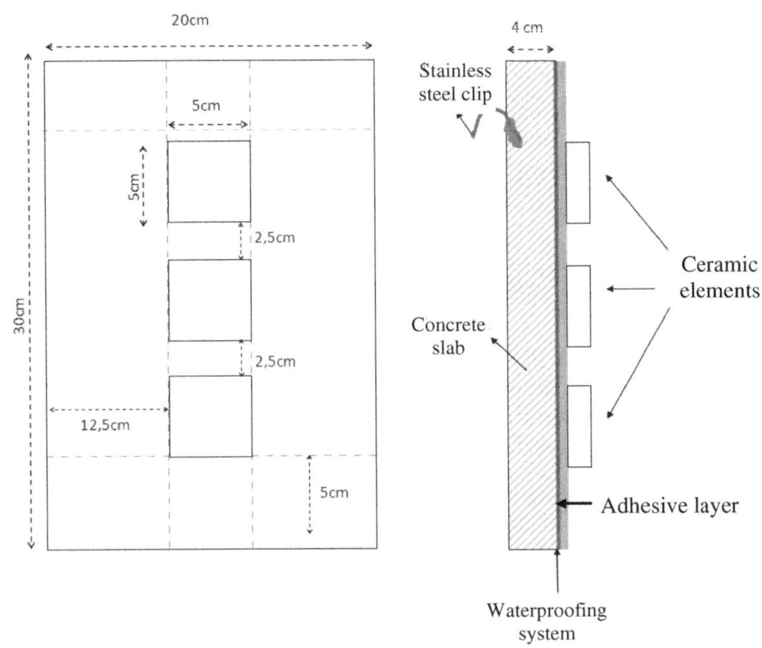

Fig. 11 Sketch of the specimens subjected to tests

Table 2 Brief description of the bonding systems tested

Bonding system	Description	Layers
BA1	A waterproofing product (W1) and a reaction resin adhesive (RA1)[a]	2
BA2	A waterproofing product (W2) and the same reaction resin adhesive as BA1 (RA1)[a]	2
K	A waterproofing product (W3) and a reaction resin adhesive (RA2)[a]	2
M	Waterproofing and adhesion provided by a reaction resin adhesive (RA3)[a]	1
S	Waterproofing and adhesion provided by a cementitious adhesive (CA)	1
T	Waterproofing and adhesion provided by a reaction resin adhesive (RA4)[a]	1

[a] All the reaction resin adhesives tested are classified as R2T according to NP EN 12004

3.1.1 Accelerated Ageing Procedures

First Ageing Procedure: Cycle A

The first accelerated ageing procedure, designed as cycle A, consists of the variation of conditions of temperature and relative humidity, in a climatic chamber, associated with the effects of rain (spraying with water) and of solar radiation (Xenon arc lamp), having a duration base of 12 h. Figure 4 shows the climatic conditions of this cycle.

For this ageing procedure, the performance of the six bonding systems was evaluated at four different moments in time: after 14, 28, 42 and 75 days of exposure in the climatic chamber subjected to repeated cycles of 12 h.

Second Ageing Procedure: Cycle B

The second ageing procedure, designed as cycle B and based on EN 1015-21 [13], is described in Table 3. It consists of the repetition of a heating/freezing cycle (Cycle B1) and a humidification/freezing cycle (Cycle B2). Each of the cycles, B1 and B2, has duration of 24 h. In the transition from cycle B1 to cycle B2, the

Table 3 Cycles combination B

Cycle B1 Heating/ Freezing				Standard environment (laboratory environment)				Cycle B2 Humidification/ Freezing		
T (°C)	RH (%)	D (h)		T (°C)	RH (%)	D (h)		T (°C)	RH (%)	D (h)
60	n.c.	8						Immersed in water		8
20	65	0.5	+	20	65	48	+	20	65	0.5
−20	n.c.	15						−15	n.c.	15
20	65	0.5						20	65	0.5

T temperature, *RH* Relative Humidity, *D* duration, *n.c.* Not controlled

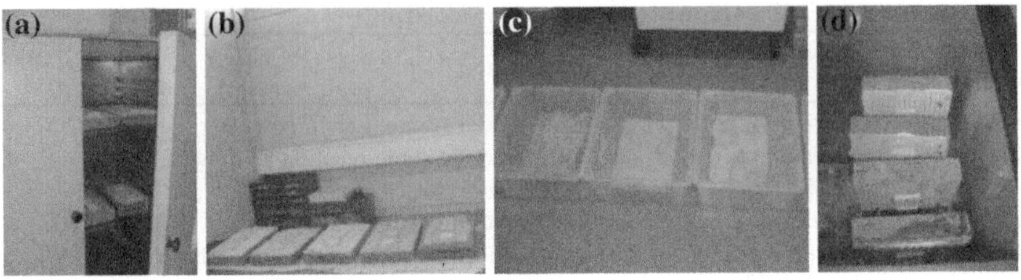

Fig. 12 The four different stages of the second ageing procedure: **a** heating, **b** transition period in laboratory environment, **c** immersion and **d** freezing

specimens remain for 48 h in a controlled environment (at the laboratory). Figure 12 shows the different stages of the second ageing procedure.

For this specific building, since it will be subjected to particularly harsh environmental conditions, in close contact with the sea, the immersion of the specimens in Cycle B2 was performed with sea water.

For this second ageing procedure, the performance of the six bonding systems was evaluated at three different moments in time: T1, after the exposure to (2 × cycle B1 + standard environment + 2 × cycle B2); T2, after the exposure to (4 × cycle B1 + standard environment + 4 × cycle B2) and T3, after the exposure to (8 × cycle B1 + standard environment + 8 × cycle B2).

3.2 Results

The experimental results are presented for each bonding system tested. The first series of results is the reference series for the initial conditions which means that no accelerated ageing procedure was applied. The second and third series present the results of the accelerated ageing procedures, cycle A and cycle B.

3.2.1 Bonding System BA1

The following tables and figures present the results related to bonding system BA1 for the first series without accelerated ageing (Table 4) and the results of the accelerated ageing procedures, cycle A (Table 5 and Figs. 13 and 15) and cycle B (Table 6 and Figs. 14 and 15).

As given in Table 2, the bonding system BA1 is constituted by a waterproofing product (W1) and a reaction resin adhesive (RA1) of class R2T according to NP EN 12004.

Table 4 Pull-off test results without accelerated ageing, reference value (t0)

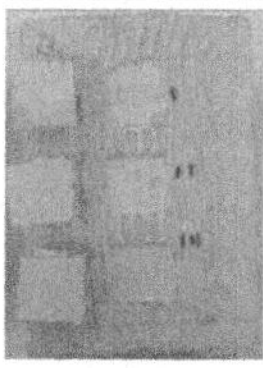

Bonding system	Sample	Adhesive strength (N/mm^2)	Type of failure	ΔAS (%)
BA1_t0	I	1.20	CF-sA (I)	+9
	II	1.10	CF-sA (I)	−0
	III	1.10	CF-sA (I)	

ΔAS Positive and negative maximum differences from the average value of adhesive strength
CF-sA (I) Cohesive failure by the waterproofing layer

Table 5 Pull-off test results with accelerated ageing (Cycle A)—t14, t28, t42 and t75

Bonding system	Sample	Adhesive strength (N/mm^2)	Type of failure	ΔAS (%)
BA1_A_t14	I	0.95	CF-sA (I)	+8
	II	0.85	CF-sA (I)	−8
	III	1.00	CF-sA (I)	
BA1_A_t28	I	0.80	AF-sS	+0
	II	0.90	AF-sS	−11
	III	0.90	AF-sS	
BA1_A_t42	I	1.10	CF-sA (I)	+5
	II	0.95	CF-sA (I)	−10
	III	1.05	CF-sA (I)	
BA1_A_t75	I	0.80	CF-sA (I)	+22
	II	1.10	CF-sA (I)	−11
	III	0.80	CF-sA (I)	

ΔAS Positive and negative maximum differences from the average value of adhesive strength
CF-sA (I) Cohesive failure by the waterproofing layer
AF-sS Adhesive failure between the waterproofing layer and the concrete support

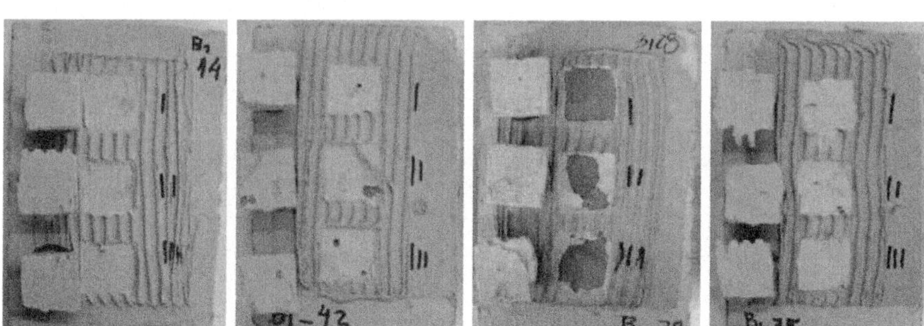

BA1_A_t14 BA1_A_t42 BA1_A_t28 BA1_A_t75

Fig. 13 Type of failure—Cycle A (bonding system BA1)

Table 6 Pull-off test results with accelerated ageing (Cycle B)—t1, t2 and t3

Bonding system	Sample	Adhesive strength (N/mm^2)	Type of failure	ΔAS (%)
BA1_B_t1	I	1.10	CF-sA (I)	+10
	II	1.10	CF-sA (I)	−15
	III	0.85	CF-sA (I)	
BA1_B_t2	I	0.95	CF-sA (I)	+5
	II	1.15	CF-sA (I)	−14
	III	1.10	CF-sA (I)	
BA1_B_t3	I	0.95	AF-sS	+6
	II	0.20	CF-T[a]	−11
	III	0.80	AF-sS	

ΔAS Positive and negative maximum differences from the average value of adhesive strength
CF-sA (I) Cohesive failure by the waterproofing layer
AF-sS Adhesive failure between the waterproofing layer and the concrete support
CF-T Cohesive failure by the ceramic tile
[a] This value was not considered given its very significant deviation from the average value. A defect in the tile was found

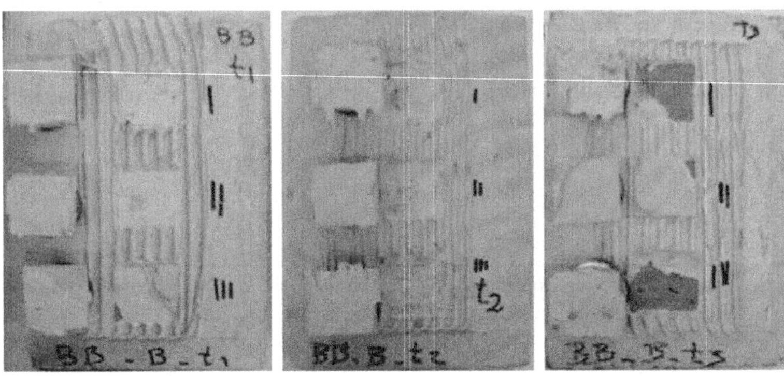

BA1_B_t1 BA1_B_t2 BA1_B_t3

Fig. 14 Type of failure—Cycle B (bonding system BA1)

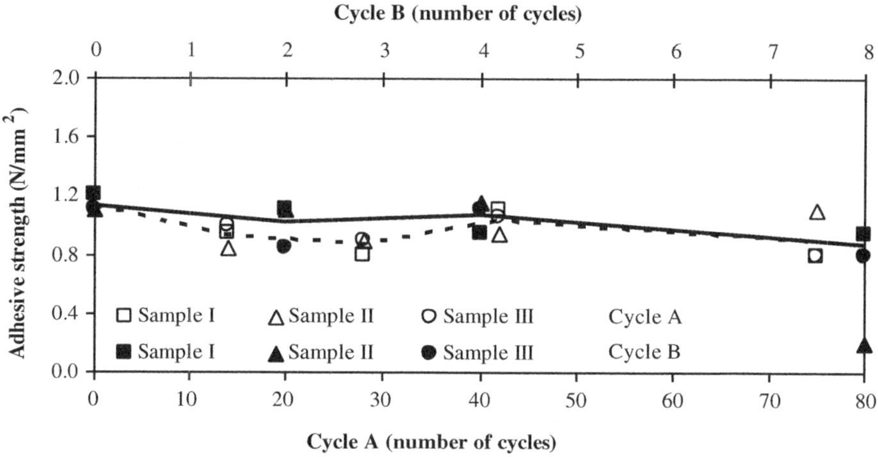

Fig. 15 Pull-off test results for the two ageing procedures (bonding system BA1)

3.2.2 Bonding System BA2

The following tables and figures present the results related to bonding system BA2 for the first series without accelerated ageing (Table 7) and the results of the accelerated ageing procedures, cycle A (Table 8 and Figs. 16 and 18) and cycle B (Table 9 and Figs. 17 and 18).

As given in Table 2, the bonding system BA2 is constituted by a waterproofing product (W2) different from the one of system BA1 and the same reaction resin adhesive (RA1).

Table 7 Pull-off test results without accelerated ageing, reference value (t0)

	Bonding system	Sample	Adhesive strength (N/mm^2)	Type of failure	ΔAS (%)
	BA2_t0	I	2.00	AF-sA(I)	+5
		II	2.20	AF-sA(I)	-24
		III	1.45	AF-T	

ΔAS Positive and negative maximum differences from the average value of adhesive strength
AF-sA (I) Adhesive failure between the waterproofing layer and the adhesive
AF-T Adhesive failure between the ceramic tile and the adhesive

Table 8 Pull-off test results with accelerated ageing (Cycle A)—t14, t28, t42 and t75

Bonding system	Sample	Adhesive strength (N/mm^2)	Type of failure	ΔAS (%)
BA2_A_t14	I	2.00	AF-sS	+11
	II	1.65	AF-sS	−13
	III	2.10	AF-sA(I)	
BA2_A_t28	I	1.60	AF-sS	+12
	II	1.45	AF-sS	−15
	III	1.90	AF-sS	
BA2_A_t42	I	1.45	AF-sA(I)	+24
	II	1.55	AF-T	−40
	III	0.75	AF-sS	
BA2_A_t75	I	0.20	AF-sS	+40
	II	0.35	AF-sS	−20
	III	0.25	AF-sS	

ΔAS Positive and negative maximum differences from the average value of adhesive strength
AF-sS Adhesive failure between the waterproofing layer and the concrete support
AF-sA (I) Adhesive failure between the waterproofing layer and the adhesive
AF-T Adhesive failure between the ceramic tile and the adhesive

Fig. 16 Type of failure—Cycle A (bonding system BA2)

Table 9 Pull-off test results with accelerated ageing (Cycle B)—t1, t2 and t3

Bonding system	Sample	Adhesive strength (N/mm^2)	Type of failure	ΔAS (%)
BA2_B_t1	I	2.10	AF-T	+5
	II	2.00	AF-sA (I)	−10
	III	1.80	AF-sA (I)	
BA2_B_t2	I	1.90	AF-sA (I)	+10
	II	2.20	AF-sA (I)	−5
	III	1.95	AF-sA (I)	
BA2_B_t3	I	2.30	AF-sA (I)	+5
	II	2.10	AF-sA (I)	−5
	III	1.15	[a]	

ΔAS Positive and negative maximum differences from the average value of adhesive strength
AF-T Adhesive failure between the ceramic tile and the adhesive
AF-sA (I) Adhesive failure between the waterproofing layer and the adhesive
[a] This value was not considered given its very significant deviation from the average value. A defect in the tile was found

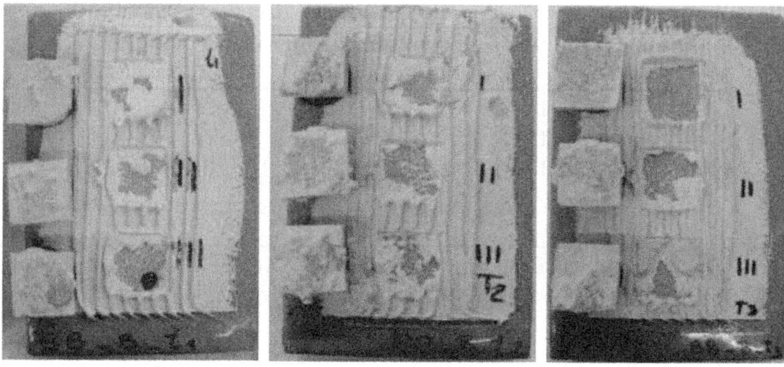

BA2_B_t1 BA2_B_t2 BA2_B_t3

Fig. 17 Type of failure—Cycle B (bonding system BA2)

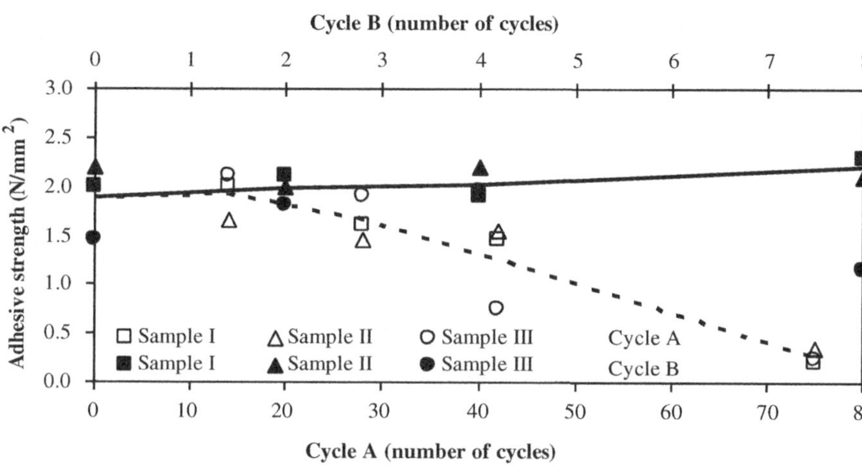

Fig. 18 Pull-off test results for the two ageing procedures (bonding system BA2)

3.2.3 Bonding System K

The following tables and figures present the results related to bonding system K for the first series without accelerated ageing (Table 10) and the results of the accelerated ageing procedures, cycle A (Table 11 and Figs. 19 and 21) and cycle B (Table 12 and Figs. 20 and 21).

As given in Table 2, the bonding system K is constituted by a waterproofing product (W3) and a reaction resin adhesive (RA2) of class R2T according to NP EN 12004, both different from the previous tested systems (BA1 and BA2).

Table 10 Pull-off test results without accelerated ageing, reference value (t0)

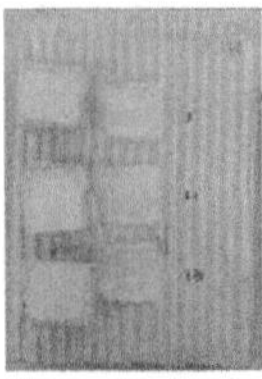

Bonding system	Sample	Adhesive strength (N/mm^2)	Type of failure	ΔAS (%)
K_t0	I	1.35	AF-sA(I)	+4
	II	1.15	AF-sA(I)	−12
	III	1.35	AF-sA(I)	

ΔAS Positive and negative maximum differences from the average value of adhesive strength
AF-sA (I) Adhesive failure between the first and second waterproofing layers

Table 11 Pull-off test results with accelerated ageing (Cycle A)—t14, t28, t42 and t75

Bonding system	Sample	Adhesive strength (N/mm^2)	Type of failure	ΔAS (%)
K_A_t14	I	0.80	CF-sA(I)	+0
	II	0.80	CF-sA(I)	−13
	III	0.70	AF-sS	
K_A_t28	I	0.75	CF-sA(I)	+7
	II	0.75	CF-sA(I)	−7
	III	0.65	CF-sA(I)	
K_A_t42	I	0.55	AF-sS	+9
	II	0.60	AF-sS	−0
	III	0.55	AF-sS	
K_A_t75	I	0.40	AF-sS	+0
	II	0.45	CF-sA(I)	−11
	III	0.45	CF-sA(I)	

ΔAS Positive and negative maximum differences from the average value of adhesive strength
CF-sA(I) Cohesive failure by the waterproofing layer
AF-sS Adhesive failure between the waterproofing layer and the concrete support

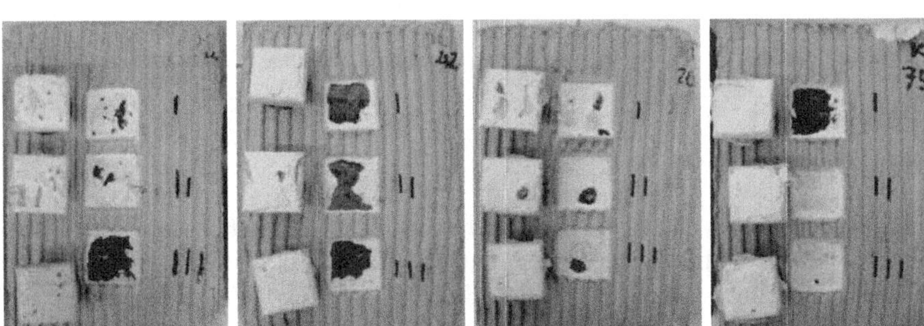

K_A_t14 K_A_t42 K_A_t28 K_A_t75

Fig. 19 Type of failure—Cycle A (bonding system K)

Table 12 Pull-off test results with accelerated ageing (Cycle B)—t1, t2 and t3

Bonding system	Sample	Adhesive strength (N/mm^2)	Type of failure	ΔAS (%)
K_B_t1	I	1.55	AF-sA(I)	+11
	II	1.40	AF-sA(I)	−11
	III	1.25	AF-sA(I)	
K_B_t2	I	1.30	AF-sA(I)	+8
	II	1.20	AF-sA(I)	−4
	III	1.15	AF-sA(I)	
K_B_t3	I	1.25	AF-sS	+17
	II	1.40	AF-sS	−13
	III	1.05	AF-sS	

ΔAS Positive and negative maximum differences from the average value of adhesive strength
AF-sA(I) Adhesive failure between the first and the second waterproofing layers
AF-sS Adhesive failure between the waterproofing layer and the concrete support

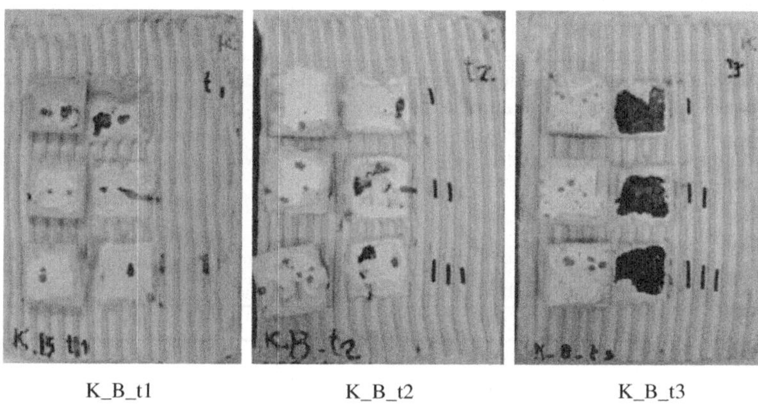

Fig. 20 Type of failure—Cycle B (bonding system K)

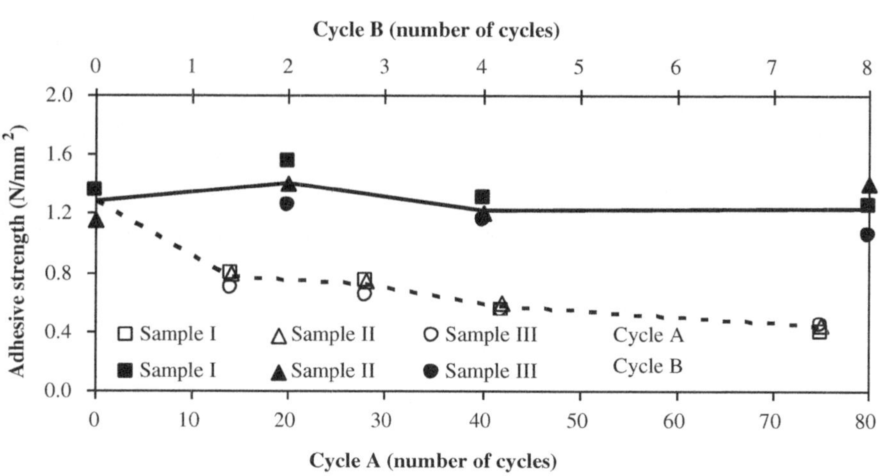

Fig. 21 Pull-off test results for the two ageing procedures (bonding system K)

3.2.4 Bonding System M

The following tables and figures present the results related to bonding system M for the first series without accelerated ageing (Table 13) and the results of the accelerated ageing procedures, cycle A (Table 14 and Figs. 22 and 24) and cycle B (Table 15 and Figs. 23 and 24).

As given in Table 2, the bonding system M is constituted by only one product, a reaction resin adhesive (RA3) which provides both waterproofing and adhesion.

Table 13 Pull-off test results without accelerated ageing, reference value (t0)

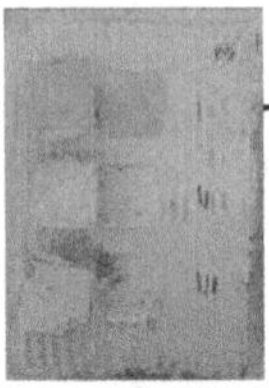

Bonding system	Sample	Adhesive strength (N/mm^2)	Type of failure	ΔAS (%)
M_t0	I	0.85	CF-T[a]	+10
	II	2.20	CF-A	−5
	III	1.90	CF-A	

ΔAS Positive and negative maximum differences from the average value of adhesive strength
CF-T Cohesive failure by the ceramic tile
CF-A Cohesive failure by the adhesive
[a] This value was not considered given its very significant deviation from the average value. A defect in the tile was found

Table 14 Pull-off test results with accelerated ageing (Cycle A)—t14, t28, t42 and t75

Bonding system	Sample	Adhesive strength (N/mm^2)	Type of failure	ΔAS (%)
M_A_t14	I	1.00	CF-A	+5
	II	1.15	CF-A	−9
	III	1.10	CF-A	
M_A_t28	I	1.00	AF-S	+0
	II	1.00	AF-S	−10
	III	0.90	AF-S	
M_A_t42	I	1.10	AF-S	+5
	II	1.05	AF-S	−0
	III	1.05	AF-S	
M_A_t75	I	1.15	AF-S	+16
	II	1.45	AF-S	−8
	III	1.20	CF-T	

ΔAS Positive and negative maximum differences from the average value of adhesive strength
CF-A Cohesive failure by the adhesive
AF-S Adhesive failure between the concrete support and the adhesive
CF-T Cohesive failure by the ceramic tile

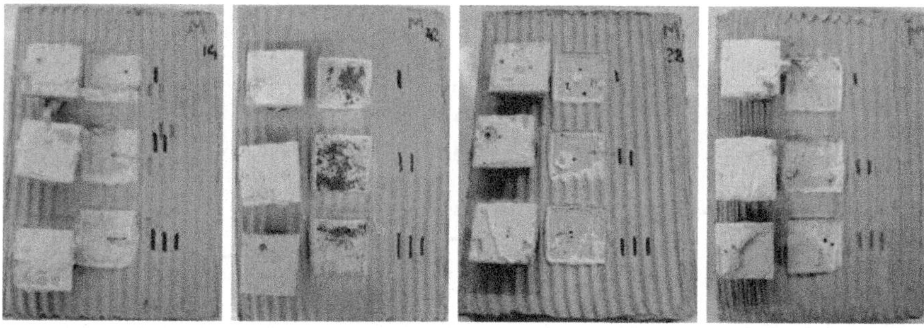

M_A_t14 M_A_t42 M_A_t28 M_A_t75

Fig. 22 Type of failure—Cycle A (bonding system M)

Table 15 Pull-off test results with accelerated ageing (Cycle B)—t1, t2 and t3

Bonding system	Sample	Adhesive strength (N/mm^2)	Type of failure	ΔAS (%)
M_B_t1	I	1.55	CF-A	+3
	II	1.60	CF-A	−3
	III	1.65	AF-S	
M_B_t2	I	1.35	CF-A	+7
	II	1.60	CF-A	−10
	III	1.55	CF-A	
M_B_t3	I	1.15	CF-A	±0
	II	1.15	AF-S	
	III	1.15	AF-S	

ΔAS Positive and negative maximum differences from the average value of adhesive strength
CF-A Cohesive failure by the adhesive
AF-S Adhesive failure between the concrete support and the adhesive

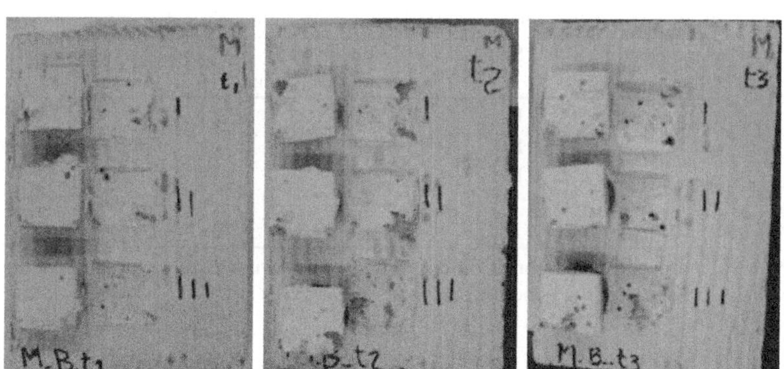

M_B_t1 M_B_t2 M_B_t3

Fig. 23 Type of failure—Cycle B (bonding system M)

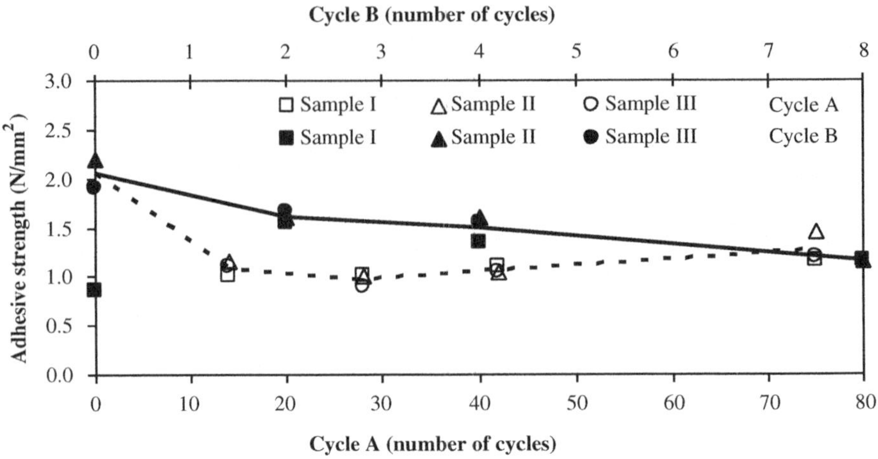

Fig. 24 Pull-off test results for the two ageing procedures (bonding system M)

3.2.5 Bonding System S

The following tables and figures present the results related to bonding system S for the first series without accelerated ageing (Table 16) and the results of the accelerated ageing procedures, cycle A (Table 17 and Figs. 25 and 27) and cycle B (Table 18 and Figs. 26 and 27).

As given in Table 2, the bonding system S is constituted by only one product, a cementitious adhesive (CA), which provides both waterproofing and adhesion.

Table 16 Pull-off test results without accelerated ageing, reference value (t0)

	Bonding system	Sample	Adhesive strength (N/mm^2)	Type of failure	ΔAS (%)
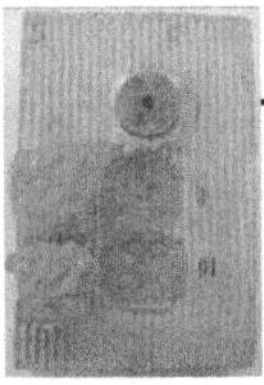	S_t0	I	4.00	[a]	+3
		II	4.00	CF-S	−6
		III	3.65	CF-S	

ΔAS Positive and negative maximum differences from the average value of adhesive strength
CF-S Cohesive failure by the concrete support
[a] without failure

Table 17 Pull-off test results with accelerated ageing (Cycle A)—t14, t28, t42 and t75

Bonding system	Sample	Adhesive strength (N/mm^2)	Type of failure	ΔAS (%)
S_A_t14	I	2.40	CF-T	+5
	II	3.20	CF-T	−17
	III	3.05	CF-T	
S_A_t28	I	1.40	CF-T	+111
	II	4.00	[a]	−87
	III	0.25	AF-S	
S_A_t42	I	0.30	AF-S	+50
	II	0.15	AF-S	−25
	III	0.15	AF-S	
S_A_t75	I	4.00	[a]	±0
	II	4.00	[a]	
	III	4.00	[a]	

ΔAS Positive and negative maximum differences from the average value of adhesive strength
CF-T Cohesive failure by the ceramic tile
AF-S Adhesive failure between the concrete support and the adhesive
[a] without failure

S_A_t14 S_A_t42 S_A_t28 S_A_t75

Fig. 25 Type of failure—Cycle A (bonding system S)

Table 18 Pull-off test results with accelerated ageing (Cycle B)—t1, t2 and t3

Bonding system	Sample	Adhesive strength (N/mm^2)	Type of failure	ΔAS (%)
S_B_t1	I	4.00	[a]	±0
	II	4.00	[a]	
	III	4.00	[a]	
S_B_t2	I	4.00	CF-S	+0
	II	3.95	CF-S	−1
	III	4.00	[a]	
S_B_t3	I	2.80	CF-S	+33
	II	1.70	CF-S	−19
	III	1.85	CF-S	

ΔAS Positive and negative maximum differences from the average value of adhesive strength
CF-S Cohesive failure by the concrete support
[a] without failure

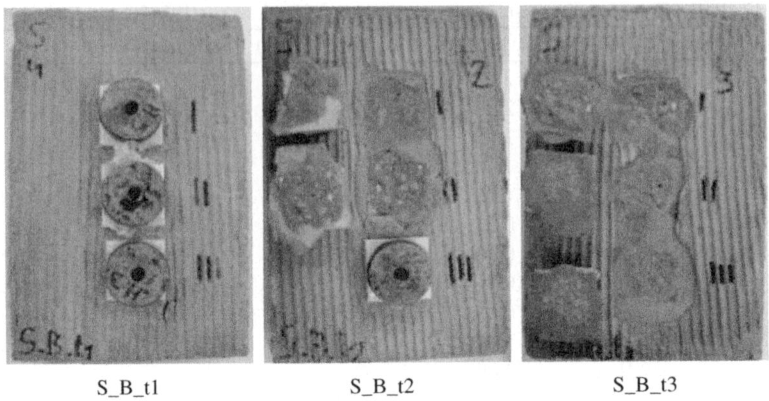

Fig. 26 Type of failure—Cycle B (bonding system S)

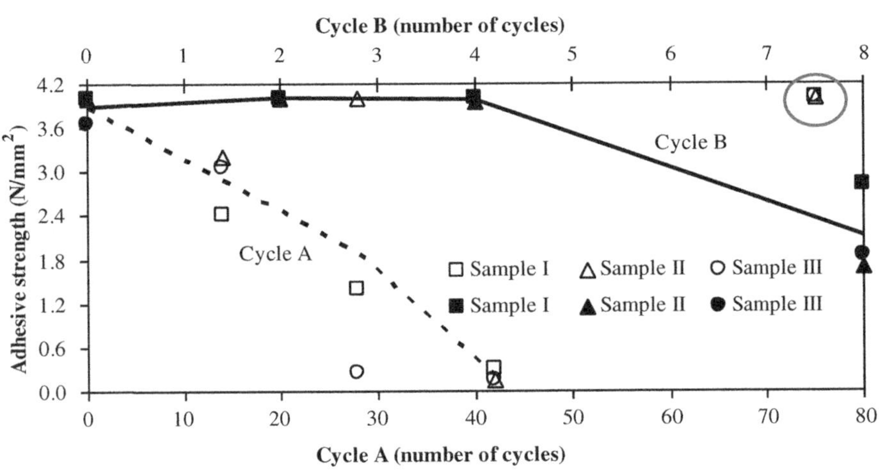

Fig. 27 Pull-off test results for the two ageing procedures (bonding system S)

Table 19 Pull-off test results without accelerated ageing, reference value (t0)

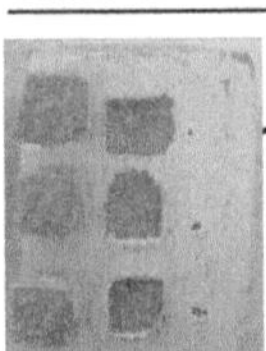

Bonding system	Sample	Adhesive strength (N/mm^2)	Type of failure	ΔAS (%)
T_t0	I	2.90	CF-S	+7
	II	3.00	CF-S	−7
	III	2.60	CF-S	

ΔAS Positive/negative maximum differences from the average value of adhesive strength
CF-S Cohesive failure by the concrete support

Table 20 Pull-off test results with accelerated ageing (Cycle A)—t14, t28, t42 and t75

Bonding system	Sample	Adhesive strength (N/mm^2)	Type of failure	ΔAS (%)
T_A_t14	I	1.55	AF-S	+29
	II	1.35	AF-S	−21
	III	2.20	CF-A	
T_A_t28	I	1.20	AF-S	+21
	II	2.35	CF-A	−38
	III	2.25	CF-A	
T_A_t42	I	1.90	CF-A	+21
	II	1.55	AF-S	−21
	III	2.35	AF-S	
T_A_t75	I	1.90	AF-S	+3
	II	1.95	AF-S	−3
	III	1.85	AF-S	

ΔAS Positive and negative maximum differences from the average value of adhesive strength
AF-S Adhesive failure between the concrete support and the adhesive
CF-A Cohesive failure by the adhesive

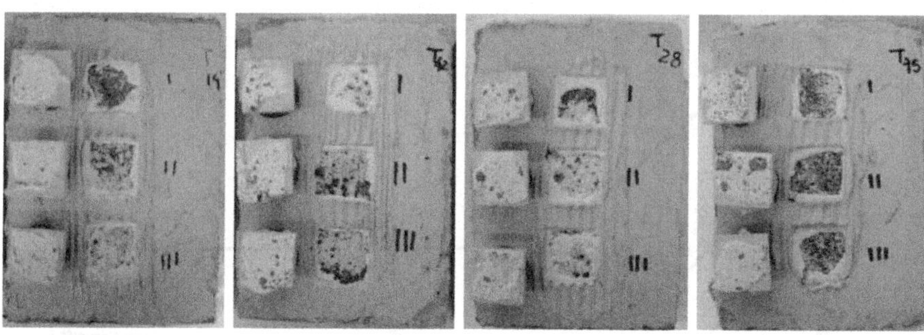

T_A_t14 T_A_t42 T_A_t28 T_A_t75

Fig. 28 Type of failure—Cycle A (bonding system T)

Table 21 Pull-off test results with accelerated ageing (Cycle B)—t1, t2 and t3

Bonding system	Sample	Adhesive strength (N/mm^2)	Type of failure	ΔAS (%)
T_B_t1	I	3.10	CF-S	+15
	II	3.10	CF-S	−28
	III	1.95	CF-S	
T_B_t2	I	1.60	[a]	±0
	II	1.80	[a]	
	III	2.5	CF-S	
T_B_t3	I	0.75	[a]	+11
	II	2.05	CF-A	−11
	III	2.55	CF-A	

ΔAS Positive and negative maximum differences from the average value of adhesive strength
CF-S Cohesive failure by the concrete support
CF-A Cohesive failure by the adhesive
[a] This value was not considered given its very significant deviation from the average value. A defect in the tile was found

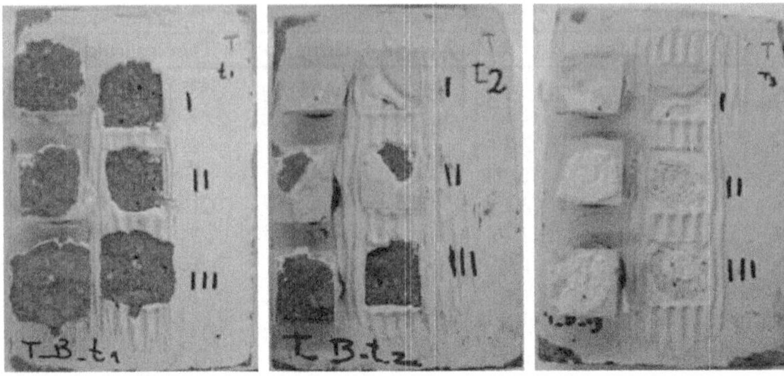

T_B_t1 T_B_t2 T_B_t3

Fig. 29 Type of failure—Cycle B (bonding system T)

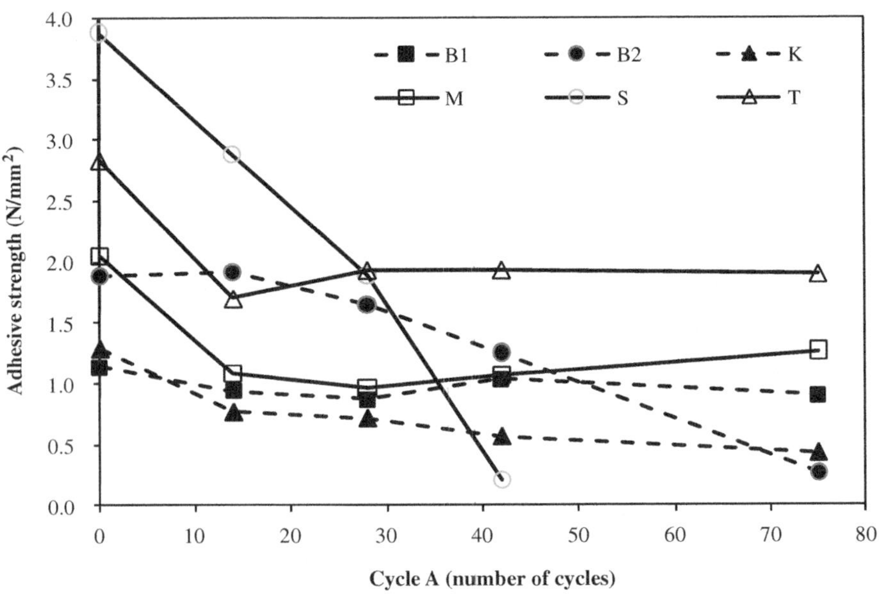

Fig. 30 Pull-off test results for the two ageing procedures (bonding system T)

3.2.6 Bonding System T

The following tables and figures present the results related to bonding system T for the first series without accelerated ageing (Table 19) and the results of the accelerated ageing procedures, cycle A (Table 20 and Figs. 28 and 30) and cycle B (Table 21 and Figs. 29 and 30).

As given in Table 2, the bonding system T is composed by only one product, a reaction resin adhesive (RA4), which provides both waterproofing an adhesion.

Finally, Fig. 31 and Table 22 show a summary of the experimental results described in detail above. It is possible to observe, that the cementitious adhesive presented a marked degradation curve, due to the first accelerated ageing

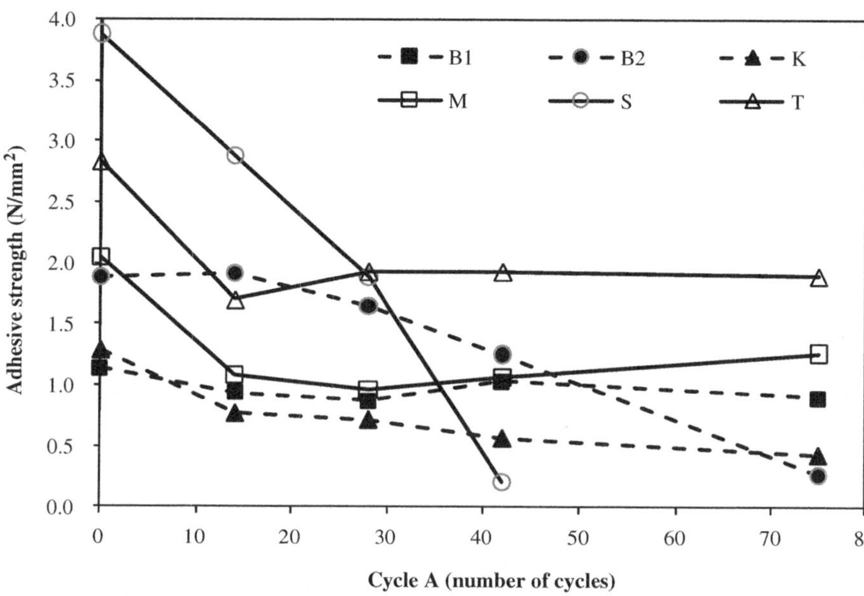

Fig. 31 Summary of Pull-off test results for the first ageing procedure (Cycle A)

Table 22 Summary of the pull-off test results

Adhesive strength average values (N/mm²)				
Bonding system	Initial reference value (t0)	Minimum (test time)	Accelerated ageing procedure	
			Cycle A	Cycle B
BA1	1.1	0.90 (several)	No significant variation	No significant variation
BA2	1.9	0.30 (A_t75)	Decreasing trend	No significant variation
K	1.3	0.40 (A_t75)	Decreasing trend	No significant variation
M	2.0	0.90 (A_t28)	Negligible decreasing trend	Decreasing trend
S	3.9	0.15 (A_t42)	Minimum value too low	Decreasing trend
T	2.8	1.20 (A_t28)	Decreases in the 1st accelerated ageing time and maintains thereafter	Negligible decreasing trend

procedure (Cycle A), not observed with reaction resin adhesives. This conclusion is in accordance with the preliminary results presented in Fig. 9. In the case of the use of reaction resin adhesives the results obtained are less expressive, with no significant variation of the adhesive strength in some of the tests (see Table 22).

4 Conclusions

In this work, a model for the service life prediction of a material or system of construction based on accelerated artificial ageing tests and on natural ageing tests was presented. Based on an experimental research work on the durability of cementitious adhesives for bonding ceramic tiles on façades, the development of the model was discussed and the main difficulties were identified in order to establish the best future strategies for its improvement and future applications. The results of that experimental research work are still not enough to fully confirm the proposed model but some encouraging results could be added from the practice, namely from the presented case study.

A test campaign with the aim of characterizing the performance over time of different bonding systems for ceramic tiles to be applied on the façades of a large building, the new Cruise Terminal of “Porto de Leixões”, was presented as a case study. This comparative analysis took, from the research, the accelerated ageing procedures already tested and confirmed their relevance. The kind of experiments carried out, although inconclusive in what concerns a durability assessment in absolute terms, allows a relevant comparison between several solutions, in this case between the adhesive strength over time of different bonding systems.

In what concerns cementitious adhesive the results obtained in the case study are consistent with the one of the preliminary research work. The cementitious adhesive presents a marked degradation curve, due to the accelerated ageing procedure named Cycle A. This clear trend was not observed in the case of reaction resin adhesives.

The degradation effects obtained with the accelerated ageing procedure named Cycle B were less expressive and, for some of the bonding systems tested, there was no significant variation of their adhesive strength.

Acknowledgments The authors would like to thank Ana Sá and Luís Silva (Weber Portugal) for their invaluable help during the experimental work development. J.M.P.Q. Delgado would like to thank Fundação para a Ciência e a Tecnologia (FCT) for financial support through the grant SFRH/BPD/84377/2012.

References

1. EN 1348: Adhesives for Tiles—Determination of Tensile Adhesion Strength for Cementitious Adhesives. European Committee for Standardization, Brussels (2007)
2. EN 12004: Adhesives for Tiles—Definitions and Specifications. European Committee for Standardization, Brussels (2008)
3. ISO 15686-1: Buildings and Constructed Assets—Service Life Planning—Part 1: General Principles. International Organization for Standardization, Brussels (2001)
4. Quintela, M.A.: Durability of one-coat rendering mortar for external use. MSc. Thesis, Faculty of Engineering of the University of Porto, Portugal (in Portuguese) (2006)
5. Freitas, V.P., Corvacho, H., Vaz Sá, A., Quintela, M.: Discussing the durability assessment of cement mortars—a contribution for a prediction model. In: Proceedings of the 11th International Conference on Durability of Building Materials and Components (DBMC-2008), Istanbul, Turkey 11–14 May 2008
6. Sá, A.V.: Durability of cementitious adhesive in ceramic adherent coverings of walls. MSc. Thesis, Faculty of Engineering of the University of Porto, Portugal (in Portuguese) (2005)
7. EN 14411: Ceramic Tiles—Definitions, Classification, Characteristics and Marking. European Committee for Standardization, Brussels (2003)
8. Centre Scientifique et technique du Bâtiment (CSTB), Classification des colles à car-relage—Définitions et spécifications, Cahier 3264 du CSTB (2000)
9. Flores-Colen, I., Brito, J., Freitas, V.P.: Expected render performance assessment based on impact resistance in situ determination. Constr. Build. Mater. **23**(9), 2997–3004 (2009)
10. Maranhão, F.L., Loh, K., John, V.M.: The influence of moisture on the deformability of cement–polymer adhesive mortar. Constr. Build. Mater. **25**(6), 2948–2954 (2011)
11. Ramos, N.M.M., Simões, M.L., Delgado, J.M.P.Q., Freitas, V.P.: Reliability of the pull-off test for in situ evaluation of adhesion strength. Constr. Build. Mater. **31**(1), 86–93 (2012)
12. RILEM: Recommendation MDT D 3. Determination in situ of the adhesive strength of rendering and plastering mortars to their substrate. Mater. Struct. **37**, 488–490 (2004)
13. EN 1015-21: Methods of Test for Mortar for Masonry—Part 21: Determination of the Compatibility of One-Coat Rendering Mortars with Substrates. European Committee for Standardization, Brussels (2002)

Behavior to Salt Crystallization of Repointing by Ready-Mix Mortars: Experimental Data and Application of a Probabilistic Model

Luigia Binda, Elsa Garavaglia and Cristina Tedeschi

Abstract During their life Cultural Heritage buildings suffer different attacks due to the changes induced to the surrounding environment along the years. Each attack causes damages of different magnitude which are, often, time dependent. Salt crystallization is one of the highest causes of damage in masonry connected with an aggressive environment. Mortar joints as brick and stone can be a vehicle of the water and salts coming from capillary rise or sea spray. The maintenance against damages consists in replacing (re-pointing) the joint using a new mortar. The commercial mortars for repair, present nowadays on the market, give rise to a question: how durable is the mortar once it has been used to repair masonry? In the chapter, to investigate durability of these materials, and the eventual damage that they induce at the interface mortar/brick, a procedure based on salt crystallization tests is described and the development of a simple probabilistic damage model introduced. The salt crystallization tests were carried out on brick masonry specimens prepared with the same type of bricks and six types of commercial mortars based on natural hydraulic binders (NHL). During the test the loss of material from the surface was measured and assumed as parameter of damage. The loss is referred to the vertical section of the specimen and quantified through a computer code starting from successive readings of the surface decay along defined profiles using a laser profilometer device. The randomness affecting the damage due to the salt crystallization and the consequent continuing loss of material suggests studying the deterioration under a probabilistic point of view where the continuous deterioration of specimens can be assumed to be a stochastic process.

L. Binda (✉) · E. Garavaglia · C. Tedeschi
Department of Structural Engineering, Politecnico di Milano, P.Za L. da Vinci 32, 20133 Milan, Italy
e-mail: binda@stru.polimi.it

E. Garavaglia
e-mail: garava@stru.polimi.it

C. Tedeschi
e-mail: tedeschi@stru.polimi.it

V. P. de Freitas and J. M. P. Q. Delgado (eds.), *Durability of Building Materials and Components*, Building Pathology and Rehabilitation 3,
DOI: 10.1007/978-3-642-37475-3_8,

Keywords Probabilistic model · Fragility curves approach · Accelerated crystallization tests · Masonry durability

1 Introduction

Structures subjected to an aggressive environment may suffer degradation of their component materials during their service life. The deterioration of material due to environmental attacks also affects load bearing structures, where the main aggressors are natural frost-defrost cycles, temperature and humidity variation [1–3]. In the presence of moisture and/or capillary rise these cycles may produce the crystallization of salts inside the component materials (bricks, stones, and mortar), with the resulting loss of surface material. In historic buildings and monuments the loss of surface material has, of course, to be considered an important damage. The decay due to salts is a typical damage affecting the Mediterranean historic heritage and it can seriously compromise the conservation of the exposed surfaces of the historic masonry buildings.

The related decay mechanisms have been discussed in several studies, as well as the influence of many factors affecting salt movements, their crystallization location and the degradation rate in masonry building [4, 5]. The research has pointed out the great importance of the compositional and microstructural features of porous materials on their salt-decay attitude and the need for experimental work taking into account the masonry substrate.

The maintenance of historic masonry requires actions aimed at the rehabilitation and repairing of damages happened over time, with products compatible with the original material, able to consolidate the structure and increase its reliability. Traditionally the repairing occurs with the use of hydrated lime-based mortars. Nowadays, new products for masonry repair are available on the market and they represent a convenient alternative to the traditional hydrated lime-based mortars.

Often the products present on the market are well classified in terms of stress and strain capability, physical and chemical characteristics, simplicity of use and application, but it is difficult to know, a priori, the possible damage induced by the new material to the ancient one and the durability of the new material after it is set in use. In order to investigate these important aspects, that influence the choice of a repairing material or technique rather than another, in last ten years appropriate laboratory accelerated tests were developed by Binda's research group. The laboratory tests were here applied to natural hydraulic lime (NHL) binders and to commercial ready to mix mortars which are proposed by the producers as materials for the conservation of the cultural heritage (CH). A complete characterization of the compositional, mineralogical, physical and mechanical properties has been performed on the studied mortars by means of a wide range of techniques, including optical microscopy, XRD, FTIR, SEM and reported in [6–8].

Since soluble salt, especially sulphates, can cause the highest damages, salt crystallization tests have been carried out on masonry specimens built with one type of soft mud bricks and six types of NHL mortars following the RILEM pre-standard procedure [9] for the evaluation of the material resistance to sulfates. The damage evolution of the wallettes has been monitored by visual observation and quantified as the material loss from the masonry surface after several crystallization cycles. Remarkable differences in the damage extent and in the decay patterns have been observed, depending on the characteristic features of the different mortars.

The damage induced by accelerated tests must be quantified with parameters which are significant also for the real structures. Therefore, on the basis of the recorded experimental data, a suitable damage parameter describing the material deterioration process has been chosen. The damage measurements have been made through a laser device along chosen profiles on the upper surface of the mortar/brick system. The loss is quantified as the variation of the profile depth over the time, defined as the area between two subsequent profiles measured every 4 weeks during the test. The parameter is then reported in damage diagrams as a percentage of the initial area of the specimen section.

The progressive decay of a given material and the consequent decreasing of its durability can be performed only by studying the behavior of the system over time. The randomness present in mortar, bricks and interfaces suggests to deal with the deterioration process as a stochastic process of the random variable ℓ (where ℓ is the measured loss of area). Therefore, the deterioration process is interpreted with a process, $L(c, \ell)$, function of the time (or cycles) c and of the damage ℓ. Modeling has been implemented following a Log Normal distribution. Assuming some significant damage thresholds $\overline{\ell}$ and following the procedure developed by Garavaglia et al. [10–12] the determination of the fragility curve describing the probability to reach or exceed $\overline{\ell}$ over time (number of cycles), has been, here, approached. The model of the theoretical fragility curves follows the typical approach of a reliability problem and the distribution chosen in the modeling was a Weibull distribution; its parameters have been estimated through a computer code which was implemented by an optimization procedure and by some significant statistical tests.

The aim of the research was the development of a simple model able, for each type of repairing material present on the market, to forecast the magnitude of the expected damage and its occurrence time. The studied methodology can be in the future a support for the decision makers when the maintenance planning of a historic building is developed.

2 Laboratory Tests

Four ready-mix commercial mortars and two commercial binders were selected. All the products are proposed by the producers as materials for restoration purposes, and classified as cement-free, with a minimum content of soluble salts and are based on NHL binders. A series of laboratory tests were carried out on theses mortars and binders in order to obtain the characteristics which are influencing their service life.

2.1 Mortar Characterization

The products here called MA, MB, MC, MD refer to the ready-mix mortars, while the ones called LA and LB refer to commercial binders later on mixed with sand and water to obtain mortar. The materials characterization of each product was performed both on anhydrous raw materials and on hardened mortars samples by means of: stereomicroscopy; X-ray diffraction; FT-IR spectroscopy; ionic chromatography; optical microscopy of thin sections observed under visible light and UV light; SEM + EDX analysis. The related results have been thoroughly reported in previous works [6–8] and are here only summarized.

Concerning the powdered raw materials, the main binder phase in most of the analyzed samples is Larnite. It is a di-calcium silicate (Ca_2SiO_4) analogous to C_2S, which is typical of NHL compositions and responsible for the hydraulic behavior of the binders. Larnite is associated to portlandite ($Ca(OH)_2$) in the composition of all specimens except for MA. Sample MA, differently from all the others, only shows the presence of portlandite while no Larnite is traced. Aggregates of the ready-mixed products are quite heterogeneous: quartz and calcite are the most diffused mineral phases, which are present together with plagioclase, mica and dolomite. The analysis of samples from the hardened mortars provided further compositional and micro-structural information (Figs. 1a, b).

The mortar MA has residual non hydrated C_2S grains uniformly dispersed within the matrix and a high amount of fine grained calcite is present as carbonatic filler. The hardened matrix of MB is very compact and very rich in partially reacted slag fragments which contribute to the hydraulic behavior. C_2S is the main hydraulic phase traceable in the binding matrix of mortar MC, but few C_3S grains are present as well. The presence of an air-entraining additive in this mix leads to a peculiar microstructure with small and regular rounded shape pores. MD has a compact binder with only few residual non hydrated C_2S grains dispersed in the matrix. Samples LA and LB, which are prepared mixing together commercial binders and standard quartz-siliceous aggregate, show predominant C_2S and few C_3S grains in the hardened mortars. Moreover, in LB binder composition a carbonatic filler is present. On the hardened mortar specimens also porosimetry analyses were carried out by mercury porosimeter. In Table 1 the total porosity and the average pore radius in volume are reported.

Salt crystallization tests were carried out on the hardened mortars. Two weekly cycles (4 daily cycle/week) were sufficient to induce severe damage in most of the mortars. The evaluation of the mass uptake (percentage variation between initial and final dry mass of specimens) confirms that most of the sodium sulfate accumulation occurs during the first imbibition cycles and that in several cases

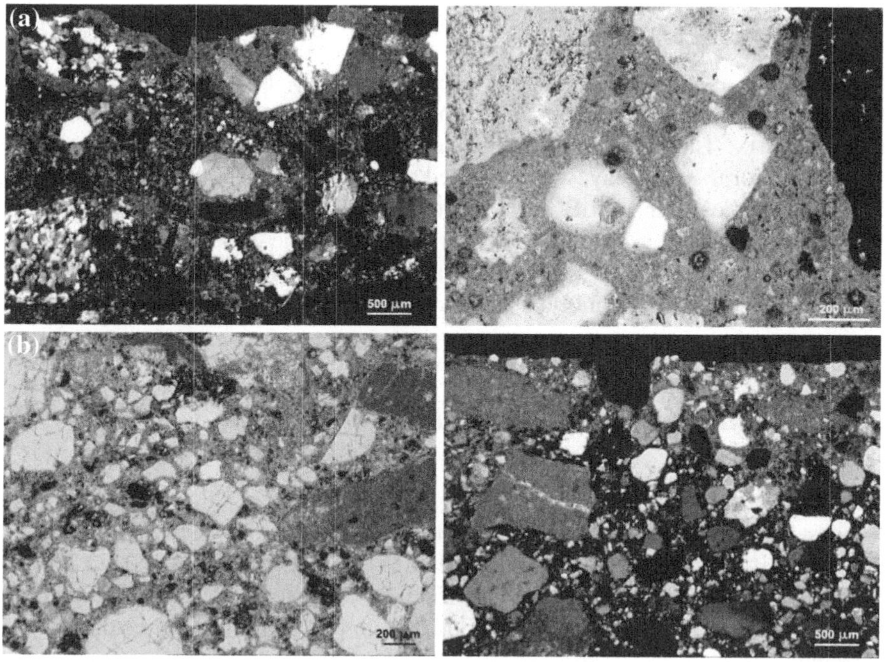

Fig. 1 **a** Mortar MA: carbonation front and irregular shaped porosity (*left*, CPL); not reacted C_2S particles dispersed in the binder's matrix (*right*, PPL). **b** Mortar MD: quartz and limestone aggregate and non homogeneous distribution of C_2S particles (*left*, PPL); carbonation front along the border (*right*, CPL)

Table 1 Porosity and median pore radius (Volume)

Type of mortar	Porosity (%)	Median pore radius (Volume [µm])
MA	30.37	0.05
MB	26.48	0.16
MC	27.08	0.43
MD	34.26	0.66
LA	22.59	0.27
LB	21.42	0.20

(MB, LAS, LB) any further salt penetration is prevented after the first weekly cycle [7]. On the contrary accumulation in MA samples keeps going on since the very last cycle and determines the highest final mass increase for this mortars, while MB has the smallest one.

As a matter of fact, MA, which has the highest compressive strength value, is the most damaged by sodium sulfate crystallization. On the other hand, MB is totally undamaged at the end of the test despite its low mechanical characteristic. The outstanding resistance of MB is most probably due to its very compact microstructure, deriving from the slag lime-binder, that does not allow massive imbibition as in the case of MA. The rather good performance of MC can be influenced both by its good mechanical strength and by the presence of an air-entraining agent in its composition. The enhanced porosity in this case can provide further empty spaces where crystallization occurs without damaging effects to the mortar matrix. The two commercial binders, LA and LB, show a rather comparable behavior to salt decay, with a limited to medium damage level. The presence of a carbonatic filler in LB composition seems to not affect significantly its resistance, compared to the one of LA. In conclusion, the durability of the different mortars to salt decay is a function of a number of parameters acting simultaneously, especially when mortars with so different characteristics are considered.

2.2 *Procedure of the Salt Crystallisation Test for Masonry Specimens*

The test was carried out by the authors on the basis of previous research [4, 5, 13, 14] and according to the Recommendation RILEM MS A.1 of RILEM TC 127-MS [9]. Each wallette consisted of three courses of bricks with two horizontal bed joints and a vertical one (Fig. 2). A red soft-mud brick of high porosity, used for restoration was selected.

Anhydrous ready-mix products were just added with the correct amount of pure water according to the indication of technical data sheets to prepare the MA, MB, MC, MD mortars. The Commercial NHL binders LA, LB were mixed with clean standard aggregate with a 1:3 binder/aggregate ratio (w/w) and then pure water was added according to the standard requirements. All mixing operations were conducted under controlled conditions with the aid of a mortar mixer. From Table 1 it can be seen that there is no direct correlation between the total porosity and the average pore radius.

The masonry specimens were called MA, MB, MC, MD, LA and LB, as the mortars used for joints.

The wallettes (approx. 250 × 200 × 120 mm) were put in contact with their back side with a 10 % (w %) Na_2SO_4 solution (anhydrous Na_2SO_4 reagent grade,

Fig. 2 Surface of MA wallette after 7 weekly cycles: (**a**) after brushing (**b–c**) details of MA specimen after 7 weekly cycles: severe scaling of the mortar can be observed

Fluka). Then they were stored over a layer of dry gravel in a plastic container open at the top, sealed along the borders (Fig. 3a); the upper face was exposed to a laboratory controlled environment (20 °C and 50 % R.H.).

After 4 weeks the first crystallization cycle (or, better, step) was concluded, the wallettes were subjected to (1) visual inspection, (2) cleaning from efflorescence and detached materials with a soft brush and a vacuum cleaner, (3) photographic survey, (4) description of the observed damage, (5) reading of the chosen surface profiles by means of a laser profilometer which allowed quantifying the damage. In Fig. 2a–c pictures of steps (2–4), are shown, referred to the specimens made with mortar MA and also called MA.

De-mineralised water was then added to the containers and a new 4 weeks cycle began.

3 Damage Measurement by a Laser Profilometer Device

A laser profilometer was used to monitor the damage (Fig. 3b, c) [13, 14]. The use of the laser profilometer allows to measure, with a very good resolution, the loss of material from the exposed surface calculated at subsequent times.

Profiles recorded at the end of each 4 weeks step show how the surface is changing along the time due to progressing of the decay and the loss of material can be measured as the area of the vertical section of the specimen lost at each profile (Fig. 4) [12]. The presence of profile bulging or swelling (Fig. 4) means that the detachment of material has started due to the delamination caused by salt crystallization underneath the masonry surface as crypto florescence. The results can be used for probabilistic modeling of the progressive damage and of the life-cycle assessment of the material [10–12].

Visual inspection continuously made during the tests revealed some common behavior of the mortar joints. Since the first days, soft efflorescence, in the shape of needles, appeared on the surface of MA (Fig. 2a). In other cases efflorescence appeared at the interface brick–mortar joint. Later on, one month after the starting of the test, some bricks suffered delamination in the specimen MA. During the first and the second cycle of the test the highest loss of material, as powdering or

Fig. 3 **a** wallettes in the boxes, **b** laser profilometer, **c** scheme of the measurement

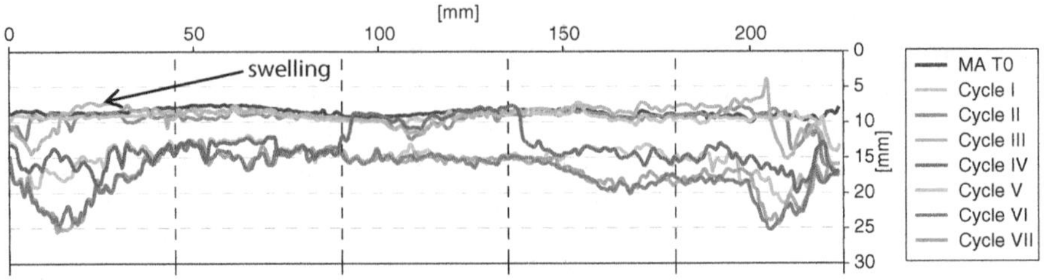

Fig. 4 Example of damage evolution over time: progressive loss of material recorded by laser profilometer

crumbling, occurred. At the end of the second cycle the damage seemed to stabilize.

The most critical situation is represented by the MA specimen, on which the swelling appears to be the most evident even at visual inspection (Fig. 2). This behavior could also be measured monthly by the laser profilometer as shown graphically in Fig. 4. Here only the results concerning the profiles taken in direction B–B are elaborated and discussed. They only report the damages measured along the horizontal mortar joint in Fig. 3c. A total number of 7 cycles has been carried out up to now, but the test will continue for at least other two months. In Fig. 4, the profiles in direction B–B from the first to the last cycle are reported for the specimen MA.

3.1 Surface Damage Definition and its Measurements

Since 1993, the laser profilometer is used to monitor the damage [13, 14] of wallettes submitted to different aggressive attacks. As said before, the device allows drawing plots of the wall profile in the chosen positions. Subsequent measurements show how the profile is changing in time; in this way it is possible to measure the material loss along the time. The resolution of the device is 0.01 mm. Figure 4 shows an example of profile for seven different measurements. Since bulging due to swelling phenomena is the previous step before detachment, it is possible to consider it as the starting point of damage (Fig. 4). Through a simple model, based on the Newton–Cotes first degree close formula, known as trapezoid role, the experimental measurements have been converted in new deterioration diagrams where the bulging has been eliminated (Fig. 5) [15].

The loss of material seems to be a good parameter to measure the masonry damage due to salt crystallization, which can be quantified using the new profiles of Fig. 5. Therefore, for each profile *i*, represented in Fig. 5 the loss ℓ_i of cross section of the wall (as a percentage of the lost area *A*), calculated at every time c^* of measurement ($c^* = 1, 7$ cycles), has been assumed as parameter of damage for the decay due to salt crystallization.

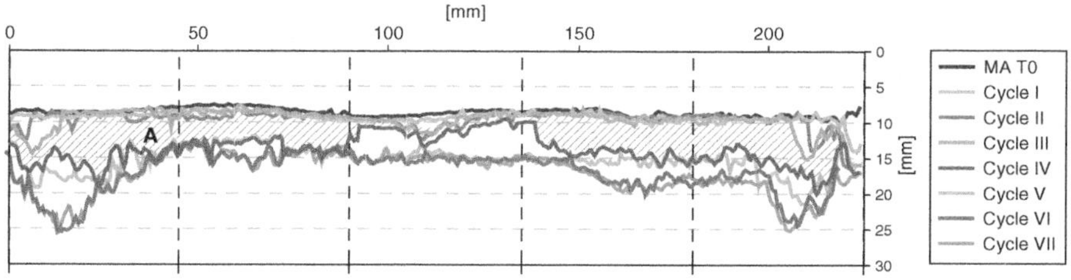

Fig. 5 Profiles of the progressive loss of material, cycle by cycle, after the swelling has been removed

As said above, in order to quantify ℓ_i, at every time c^*, the area included between two subsequent diagrams is assumed. This area is referred to the vertical section of specimen as, schematically, shown in Fig. 6a. The area is automatically calculated by the computer code studied to eliminate bulging [15]. To compare all the results obtained, the damage has been plotted in percentage:

$$\ell = \frac{\text{area, } \mathbf{A}\text{, lost}}{\text{area tranversal section}} * 100 \qquad (1)$$

To study the behavior of the mortar joint along the profile B–B (Fig. 6b) the profile was divided into 5 parts and for each part the lost area A (Fig. 5) and its percentage ℓ_i were calculated at each cycle in order to collect the variation of the damage along the profile.

To study the behavior of the mortar joint, bricks and interfaces, the profile A–A of Fig. 6b was divided into different portions each one corresponding to mortar joints, brick portions and interface portions; also in this case for each portion the lost area A (Fig. 5) and its percentage ℓ_i were calculated at each cycle in order to collect the variation of the damage along the profile. A simple linear interpolation of the experimental point permits to better read the behavior of the loss ℓ_i over time (linear splices). As an example, in Fig. 7 the loss ℓ_i (in %) referred to the brick portions of the specimen MB (profile A–A) against the number of cycles is plotted.

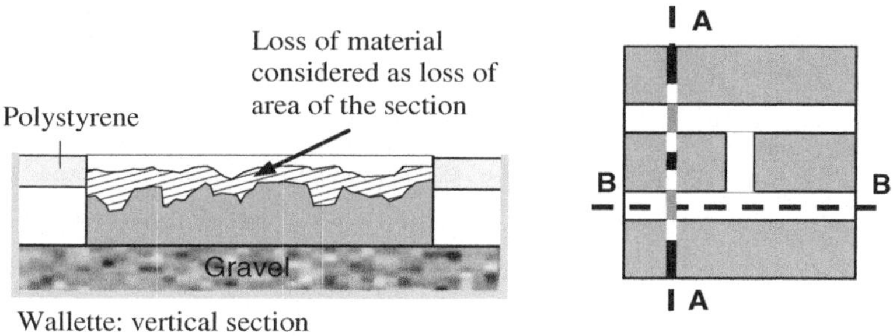

Fig. 6 **a** Schematic representation of the loss of material quantified cycle by cycle; **b** subdivision of profile A–A in mortar portions (*gray line*); brick portions (*black line*) and interface portions (*white lines*)

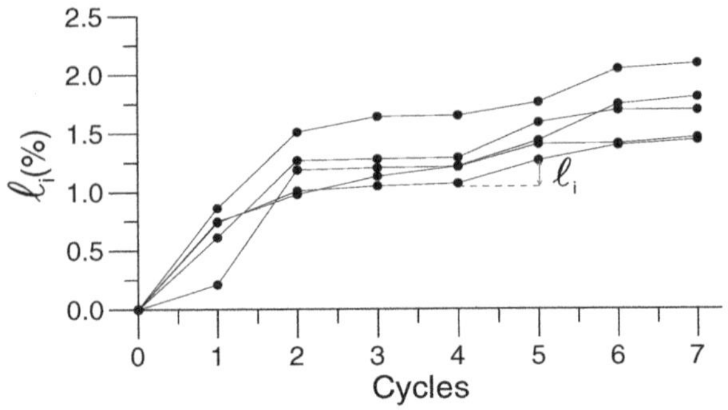

Fig. 7 Specimen MB-profile A–A, brick portions, deterioration versus cycles: example of behavior

4 Description of the Experimental Results

In Fig. 8, the average loss (in %) over the 5 points of measurement along the horizontal mortar joint (profile B–B) for all the wallettes is plotted, showing that mortar MA has the maximum damage followed by mortar MB, LB, LA, MC and MD respectively. In order to explain and better understand the reason for these different behaviors a tentative was made to correlate the loss of surface material with the median pore radius of the original mortars (see Table 1). It seems that the pore median parameter can be a good one to explain the behavior of the specimens to the crystallization test (Fig. 9). As it was said before in fact the salt damage highly depends from the porosity of the materials, or better from the pore size distribution: the lowest is the radius, the highest is the damage.

Figure 10 is a 3D representation of the damage as a function of median pore radius and of the number of steps, better confirming what is shown by Figs. 8, 9 and Table 1 (five damage levels).

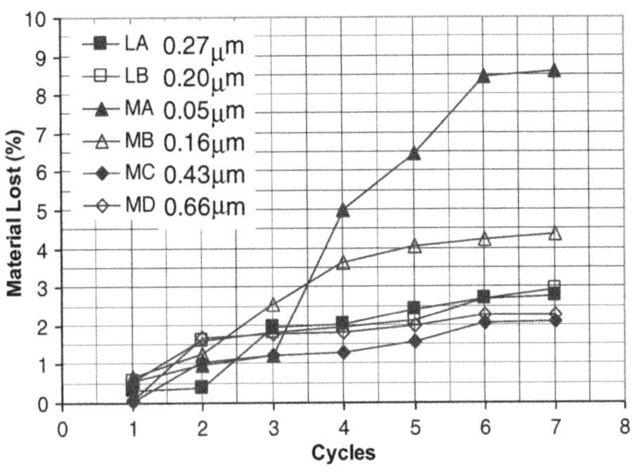

Fig. 8 Deterioration (mean values) versus steps for the *six* mortar types

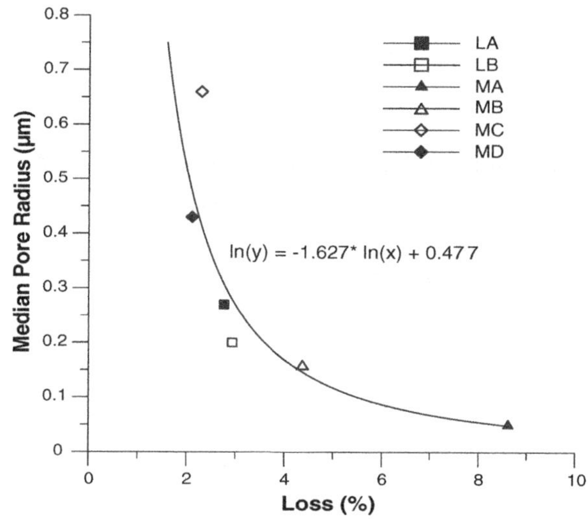

Fig. 9 Pores dimension versus maximum percentage of material area loss

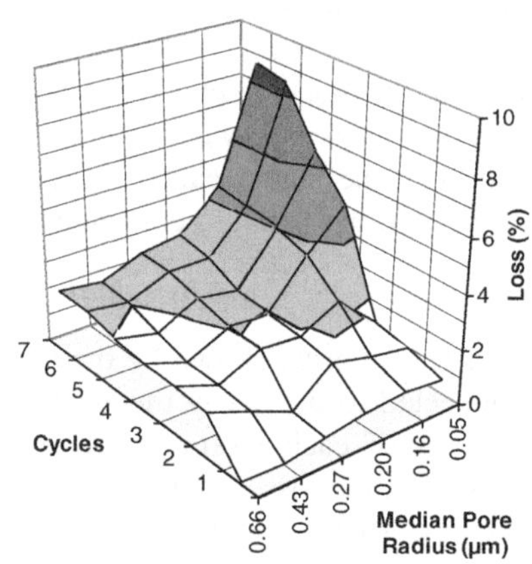

Fig. 10 Damage behavior as function of median pore radius and number of steps

5 The Assessment of Surface Damage Evolution Over Time

The deterioration of a historic masonry subjected to salts crystallization is a process that evolves in time and the stress induced into the system by the environmental crystallization attacks is largely uncertain, so as the damage produced. Therefore, it seems correct to interpret this kind of deterioration as a time dependent stochastic process where the random variable is the damage reached at each instant of the system life.

Following this hypothesis and starting from the laboratory test results, a probabilistic model has been developed to investigate the damage behavior, over time, of commercial mortars used in the maintenance of historic masonry and, consequently, to predict they durability.

5.1 Surface Deterioration Measurements: Modeling of Dispersion at Each Cycle

The laboratory test results presented in Sect. 3 have shown that the significant parameter describing the damage induced in the masonry by salt crystallization is the loss of surface material; this loss is function of step (cycle) c and damage ℓ. Following the hypothesis that the deterioration is a stochastic process the damage ℓ will be considered, here, to be a random variable (r.v.).

As illustrated in Sect. 3, during the laboratory test the damage measurements are made at each step c^*; by experience, the measurements, made at the same step, but on different samples, show scattering around an average value. Therefore, at each step c^* the deterioration process can be seen as function of the r.v. ℓ only, but the scattering of the measurements requires an appropriate modeling. The modeling of the measurements dispersion requires the choice of an appropriate density function $f_L(c*,\ell)$. The choice is not simple; it must be based on physical knowledge of the modeled phenomenon, but also on the characteristics of the distribution function in its tail, where often no experimental data can be collected. This last aspect of the matter can be investigated by analyzing the behavior of the hazard rate function $\phi_L(c*,\ell)$ connected with the chosen distribution function [11, 12]:

$$\phi_L(c*,\ell)dL = \Pr\{\ L<\ell\leq L+dL|\ell\geq\ L\} \tag{2}$$

The scattering shown by the laboratory data is probably due to the randomness connected to the decay mechanism in laboratory environment and to the characteristics of each mortar (porosity, different response to the salts attacks, etc.). However, the loss seems to be included in a certain range of values. Therefore, it seems correct to assume that, at a given step c^*, the probability of having a loss $(L<\ell\leq L+dL)$ decreases as the value L (magnitude of the loss) increases. The assumed hypothesis as a satisfied (but not unique) physical interpretation of the decay process leads to model the loss ℓ at the step c^* with a Lognormal distribution (Fig. 11) as follows:

$$L(c*,\ell) = \frac{1}{\ell\sqrt{2\pi\alpha}}\exp\left\{-\frac{[\log(\rho\ell)]^2}{2\alpha}\right\} \tag{3}$$

The estimation of the parameters α and ρ have been made through a computer code involving the maximum likelihood method. This modelling can be obtaining also through the function FMIN present in the MATLAB code.

This family of distributions presents the hazard rate function (2) decreasing as the value of L increases; this fact seems to respect the physical interpretation of the decay process previously proposed.

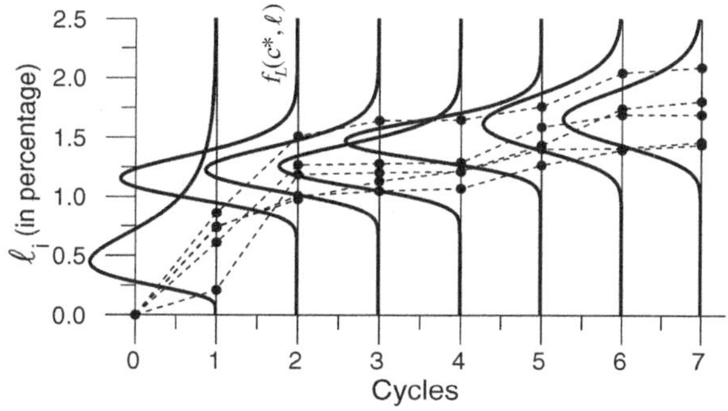

Fig. 11 MB specimen-profile A–A, brick portion, deterioration versus steps: model of dispersion with log-normal distributions

5.2 Probabilistic Prediction of Material Durability Over Time

If some significant damage thresholds $\overline{\ell}$ are considered and the time needed to exceed them must be predicted, then the deterioration process can be treated as a reliability problem.

Indeed, reliability *R*(*t*) is concerned with the performance of a system over time and it is defined as the probability that the system does not fail by time *t*. Here, this definition is extended by denoting with $\bar{R}(c)$ the probability that a system exceeds a given significant damage threshold $\overline{\ell}$ by the step *c*. The random variable that is used to quantify reliability is $\bar{C}$. $\bar{C}$ is the cycle in which the exceeding of the damage $\overline{\ell}$ can happen with a given level of probability. Thus, from this point of view, the reliability function is given by [10, 11]:

$$\bar{R}(c) = \Pr(\bar{C} > c) = 1 - F_{\overline{C}}(c) \tag{4}$$

where $F_{\overline{C}}(c)$ is the distribution function for $\bar{C}$.

Computing $F_{\overline{C}}(c)$ for different damage levels $\overline{\ell}$ allows for building the *fragility curve* for each $\overline{\ell}$.

A fragility curve describes the probability of reaching or exceeding a given damage $\overline{\ell}$ over time (or cycles). For a chosen damage level $\overline{\ell}$ at a given cycle c^*, the probability to reach $\overline{\ell}$ can be seen as the area *under* the threshold $\overline{\ell}$ and the probability of exceeding it can be seen as the area *over* the threshold $\overline{\ell}$ (Fig. 12: sketched area).

Indeed, the computed areas over different thresholds $\overline{\ell}$ provide the discrete exceeding probability used to fit the theoretical fragility curves (Fig. 13).

In order to model the experimental fragility curves and to evaluate $F_{\bar{C}}(c)$, a Weibull distribution has been chosen [11, 12, 16] as follows:

$$F_{\overline{C}}(c) = 1 - \exp[-(\rho c)^{\alpha}] \tag{5}$$

Fig. 12 MB specimen-profile A–A, brick portion, deterioration versus cycles: exceeding probability to cross the threshold $\bar{\ell}$

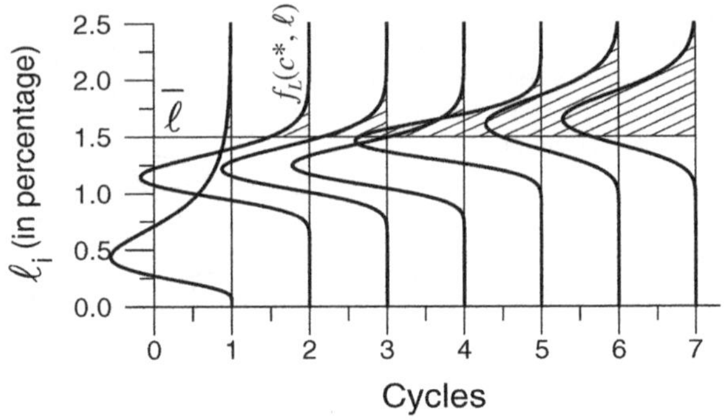

Fig. 13 MB specimen-profile A–A, brick portion, discrete exceeding probability defined for a given threshold $\bar{\ell}$

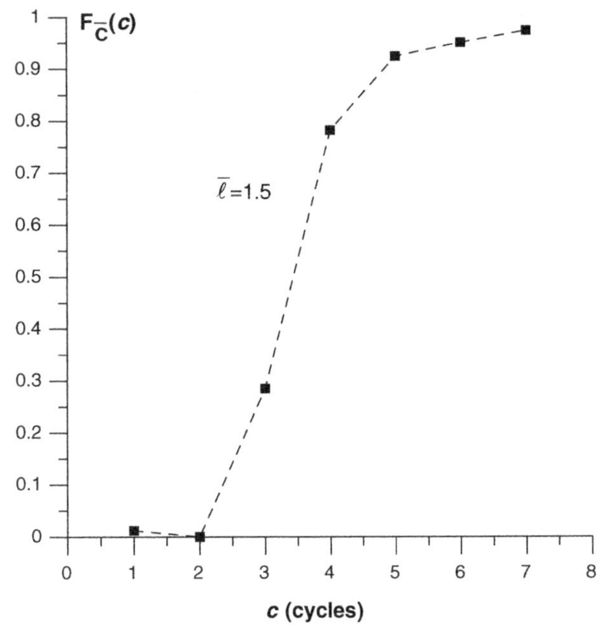

In fact this distribution seems to be a good interpretation of the physical phenomenon: the larger is the number of steps c needed to lose $\overline{L}$, the higher is the probability that the loss $\overline{L}$ will happen in the next interval $(c + dc)$. Therefore distributions with the hazard rate function $\phi_{\overline{C}}(c)$ increasing with c and tending to ∞ as $c \rightarrow \infty$ are needed. The Weibull distributions satisfy this requirement.

6 Results and Comments

The fitting of the experimental fragility curves with the distribution 5 has been made using an optimization procedure for the estimation of the parameters through a computer code involving the least squares method and implemented with Fortran IMSL Library. Since in this case the parameters estimation is made on cumulative distributions the least squares method is preferable than the maximum likelihood method.

In the following the fragility curves are presented for the profile B–B and for the profile A–A concerning mortar joints, interface mortar/brick and bricks.

6.1 Profile B–B

6.1.1 Mortar Joint

The evolution of the damage connected with the six mortars investigated is compared in Fig. 14 through the theoretical fragility curves$F_{\overline{C}}(c)$. It is evident that mortars characterized by higher median pore radius (MC and MD) present significant exceeding probability of high damage thresholds only for high numbers of cycles. Vice versa mortars characterized by small median pores radius (MA and MB) present high probability of damage already from the first cycles. The mortars LA and LB are characterized by an average median pores radius among all the other mortars and a small porosity; both these characteristics lead to average probability to exceed thresholds of significant damage.

6.2 Profile A–A

6.2.1 Mortar Joint

The damage level recorded on the mortar joint of profile A–A (Fig. 15) is smaller than the damage reported by the mortar joint along profile B–B (Fig. 14). This can be due to the dimension and location of the joint respect to the reading process but also to the fact that the interference with the bricks through the bond can increase in some way the damage. The mortar MA presents a high exceeding probability of high damage levels also for the first steps of the test, this can be due to the high porosity and the small median pore radius that characterize this type of mortar. The mortar MB is characterized by small median pore radius and average porosity so is LB, but they behave differently (MB shows an exceeding probability to a significant damage also in the first 7–10 cycles). It can be said that the mortar joint in profile A–A behave in a more dispersed way than for B–B.

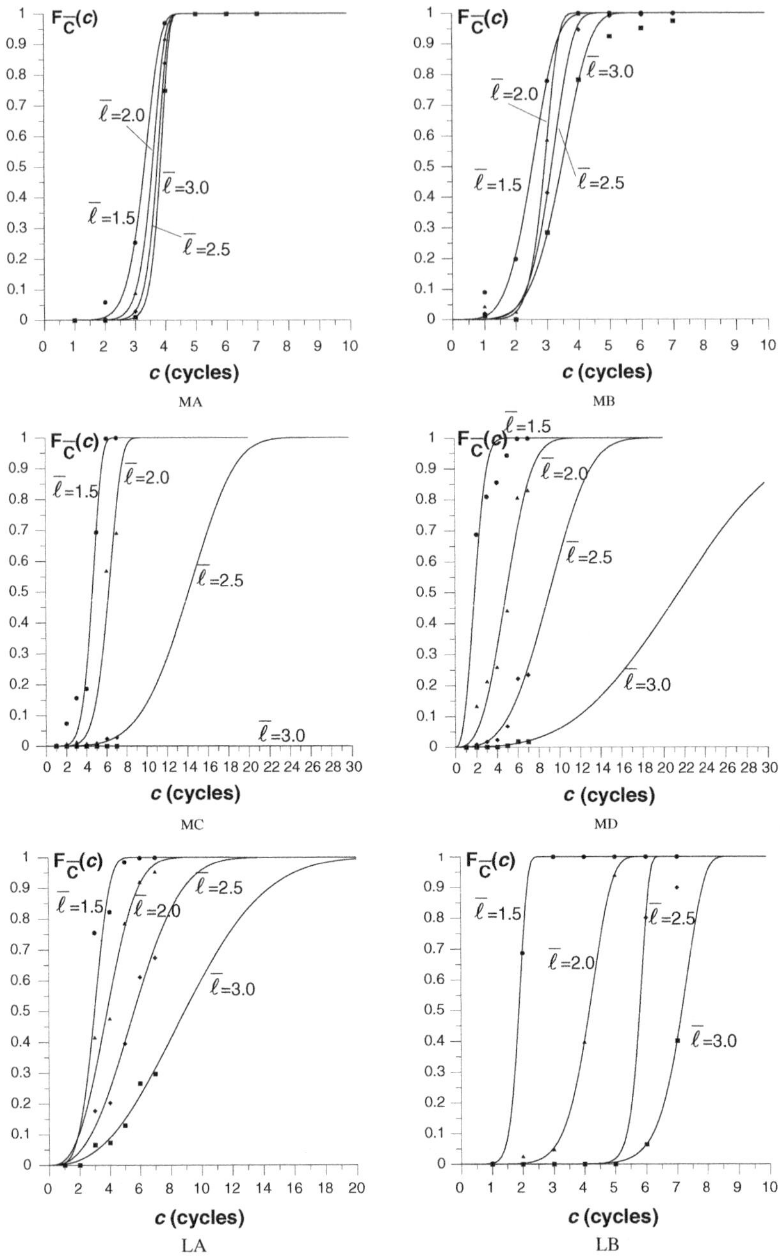

Fig. 14 Profile B–B, mortar joint: fragility curves for the different specimens and for different thresholds

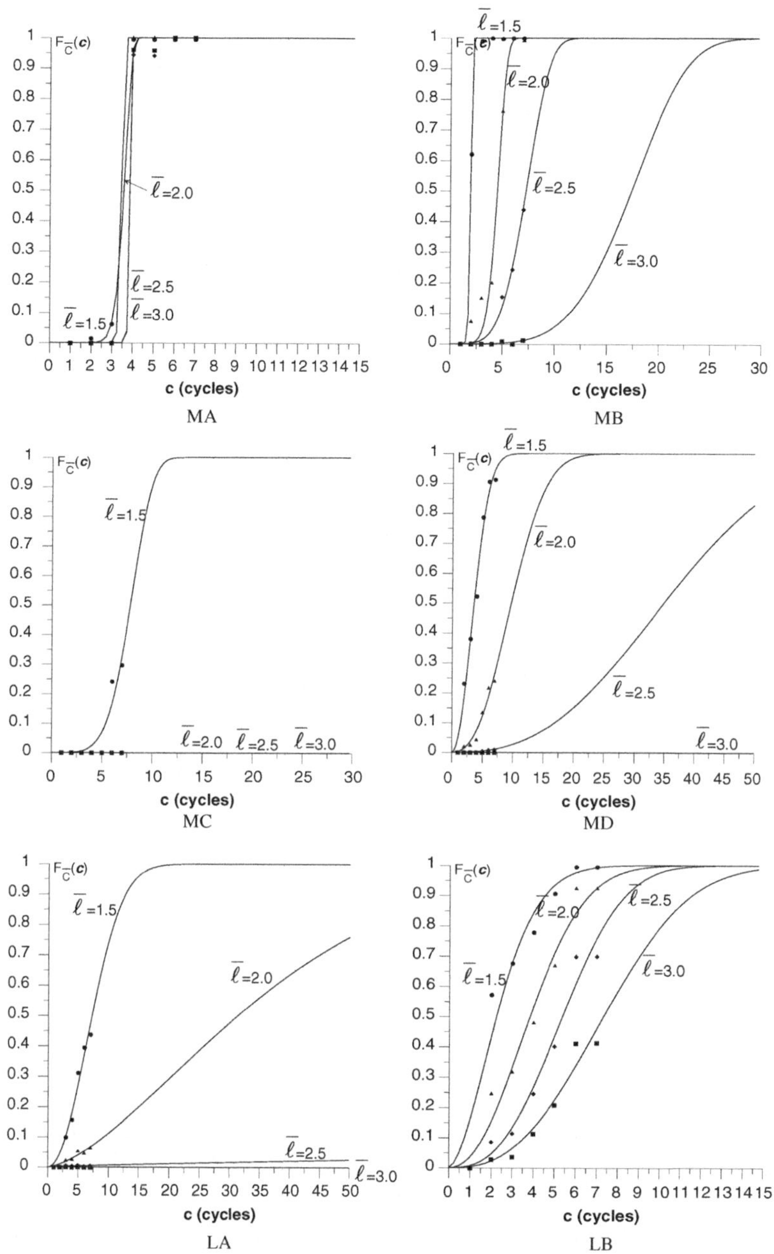

Fig. 15 Profile A–A, mortar joint: fragility curves for the different specimens and for different thresholds

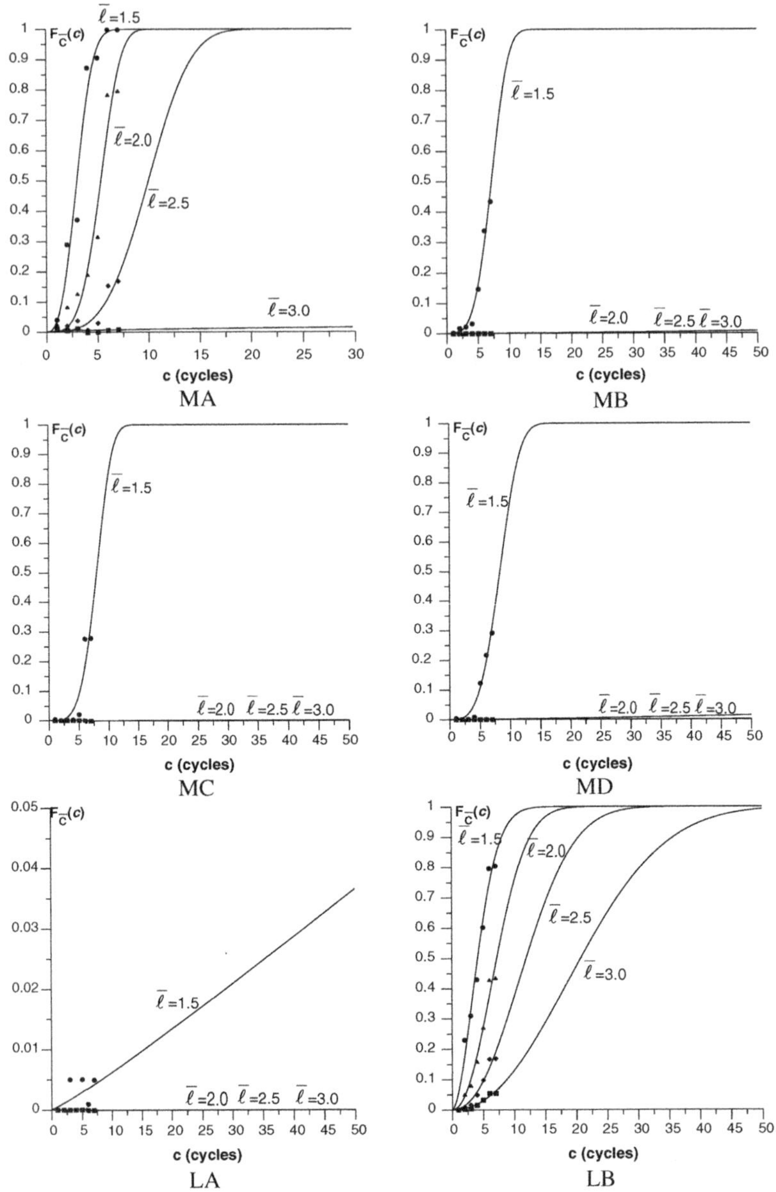

Fig. 16 Profile A–A, mortar/brick interface: fragility curves for the different specimens and for different thresholds

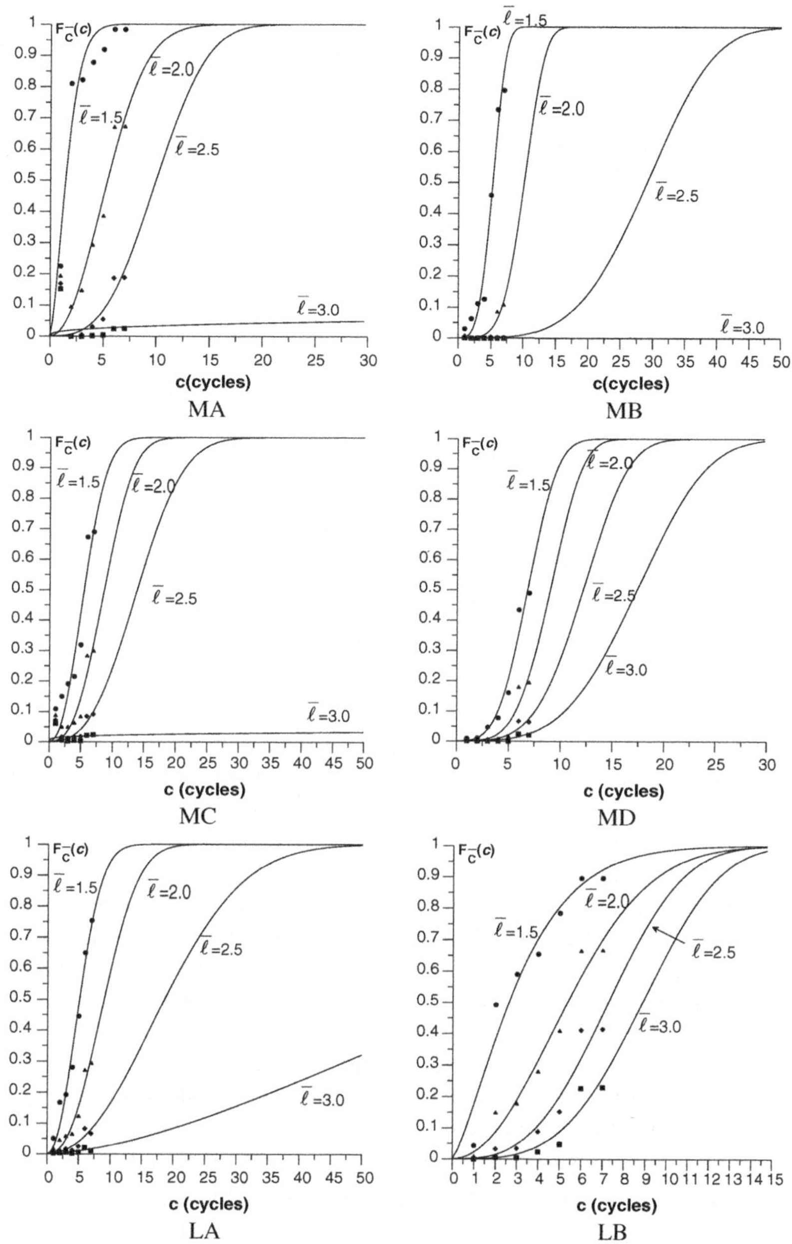

Fig. 17 Profile A–A, brick portions: fragility curves for the different specimens and for different thresholds

6.2.2 Mortar/Brick Interface

From the experimental results the damage suffered by the interfaces seems to be very small with respect to the damage suffered by the mortar joints. Therefore, as evident in Fig. 16, no predictions can be made concerning the exceeding probability of significant damage thresholds.

6.2.3 Bricks

The bricks used for the wallets are all of the same type; just the mortars present different mix. As evident in Fig. 17 the exceeding probability changes specimen by specimen; this result leads to think that the mortar influences also the behavior of the bricks areas far from the interfaces.

The mortars LA and MD significantly affect the damage of the type of brick here used, even if their porosity characteristics and median pore radius are different: LA has average median pore radius and small porosity; MD has high median pore radius and high porosity.

7 Conclusions

As well known, during their service life, masonry subjected to an aggressive environment may suffer degradation. The great randomness connected with the occurrence of critical attacks by salt crystallization suggests approaching the deterioration process of these materials by a probabilistic point of view. In the chapter has been shown that, by experimental evidence, the loss of surface material is a significant parameter in the damage measurement, cycle by cycle, and in the modeling of the damage evolution over time. The scattering around an average value, shown by measurements made at the same cycles but on different samples, suggest assuming the loss ℓ as a r.v.; the modeling of this behavior has been made with a Lognormal distribution.

In order to predict the evolution of the masonry damage over time the deterioration process can be seen as a reliability problem where the reliability function $R_{\overline{C}}$ has been defined as a function of the r.v. $\ell =$ loss of surface material; the failure process is seen as the probability of the system to exceed a given damage level $\overline{\ell}$. The fragility curves obtained allow defining the exceeding probability connected with each $\overline{\ell}$ chosen. This important information allows to appropriately planning of the maintenance and repair of the wall surface.

In this chapter the procedure has been applied on the prediction of the behavior, over time, of six different commercial mortars used in the repairing of historic masonry. The rather complex studied system, taking into account the damage of: bricks, mortar joint in longitudinal and transversal direction of the specimen, interfaces mortar/bricks, allowed to give a first detailed description of the effect of salt crystallization on masonry built with the chosen mortars.

The parameter chosen as a measure of the decay and the measurement technique seems to be once again appropriate to describe the damage along the time and to allow for a reliable comparison between the studied mortars. The durability to salt crystallization which affects many of the historic masonry surfaces can also be taken into account as a measure to choose the most appropriate mortar for repair, once the masonry texture and the existing bricks are known.

Concerning the studied mortars some conclusions can be made in detail for:

1. Mortars MA-MB-MC-MD, the damage of joint AA is always lower than BB. The interface shows a lower damage after 7 cycles. The brick damage is similar for all the specimens except for MA where it is higher;
2. Mortar LB; the damage is rather high everywhere then prevision is possible. The damage of joint BB starts before then AA. The interface shows apparently an earlier damage than the brick;
3. Mortar LA; the damage of the interface is rather low and constant for 7 cycles; more could be said in the next cycles. The brick is rather highly damaged laterally.

In conclusion the salt crystallisation test was effective in detecting the different durability of four commercial mortars and two binders when used as joints of a brick-masonry wall.

The worst behaviour was shown by mortar MA (lowest average pore radius) and the best by mortar MC (highest average pore radius); apparently the damage was higher for lower average pore radius. The mortar joint also can cause different damages at the mortar-brick interface and in some cases it can even cause damage to the bricks. This also means once again that the new commercial mortars should be carefully characterised and chosen before using them especially for conservation purposes.

Acknowledgments This paper received an award at the 12 DBMC Conference, Porto, 2011, the authors are grateful to Scientific Committee for this important recognition. Authors wish to thank D. Gulotta, M. Cucchi for their help in profilometer measurements and M. Iscandri for his help in laboratory tests; the research was carried out within the frame of EU NIKER Contract n.244123. A thank also to C. Arcadi and R. Cozza for their support in the contract management.

References

1. Charola, A.E.: Salts in the deterioration of porous materials: An overview. J. Am. Inst. Conserv. **39**, 327–343 (2000)
2. AA.VV.: Special issue on salt decay. Steiger, M., Siegesmund, S. (eds). Environmental Geology (52) (2007)
3. AA.VV. Salt weathering on building and stone sculptures, Proceedings from the international conference, 22–24 Oct. 2008, Technical University of Denmark, Copenhagen (2008)

4. Binda, L., Charola, A. E., Baronio, G.: Deterioration of porous materials due to salt crystallization under different thermohygrometric conditions, 5th International Conference on Deterioration and Conservation of Stone, Lausanne, Switzerland, 279–288 (1985)
5. Cardani, G., Tedeschi, C., Binda, L., Baronio, G.: Laboratory crystallization tests for damage evaluation maintenance of pointing in historic building: decay and replacement, Contract EV-CT98-0706, Final report (2001)
6. Gulotta, D., Toniolo, L., Binda, L. Tedeschi, C.: Investigation of commercial ready-mixed mortars for architectural heritage. 11th Conference on Structural Repairs and Maintenance of Heritage Architecture, July, 22–24, WIT Press, Tallinn, Estonia, 231–241 (2009)
7. Gulotta, D., Toniolo, L., Tedeschi, C., Binda, L., Commercial ready-mixed mortars for the conservation of cultural heritage: Characterization and preliminary durability test, in advanced materials research, vol. **133–134**, pp. 259–264, ISSN: 1022–6680 Trans Tech Publications, Switzerland (2010) doi:10.4028/www.scientific.net/AMR.133-134.259.978-0-87849-239-8
8. Tedeschi, C., Binda, L., Gulotta, D., Toniolo, L.: Durability to salt decay of commercial ready-mixed mortars for the conservation of cultural heritage, 2nd Historic Mortars Conference Rilem TC 203-RHM repair mortars for historic masonry, IV. 19, September, 22–24, Prague, RILEM-PRO 78, In: Valek, J., Groot, C., Hughes, J.J., (ed.) Rilem Publication, S.A.R.L., 1015–1022 (2010)
9. Rilem MS-A1, Determination of the resistance of wallettes against sulphates and chlorides (1998)
10. Garavaglia, E., Anzani, A., Binda, L., Cardani, G.: Fragility curve probabilistic model applied to durability and long term mechanical damages of masonry, Springer Ed. Mater. Struct. **41**(4), 733–749 (2008)
11. Garavaglia, E., Gianni, A., Molina, C.: Reliability of porous materials: Two stochastic approaches. J. Mater., ASCE, USA, **16**(5), 419–426 (2004)
12. Garavaglia, E., Lubelli, B., Binda, L.: Two different stochastic approaches modeling the deterioration process of masonry wall over time, Spring Ed. Mater. Struct. **35**(248), 246–256 (2002)
13. Berra, M., Binda, L., Squarcina T., Fatticcioni, A.: Laboratory and in situ measurement of the decay of masonry surfaces, 6th Int. DBMC, Japan, Vol. **2**, 834–843 (1993)
14. Binda, L., Baronio, G., Lubelli, B., Rocca, P.: Effectiveness of surface treatments of stone and brick masonry: proposal and calibration of on site control techniques, 8 DCMC, 8th International Conference on Durability of Building Materials and Components, vol. 1, pp. 538–549. Vancouver, Canada (1999)
15. Garavaglia, E., Cardani, G., Binda, L.: Application of a probabilistic model for the prediction of the decay due to salt crystallization of masonry building materials, 9 DBMC-9th International conference on durability of building materials and components, Brisbane, Queensland, Australia, CD-ROM, Paper 243 (2002)
16. Bekker, P.C.F.: Modelling of lifetime performance in building, International Journal of Architectural Heritage, Special Issue: In Honor of Professor Luigia Binda and Over 25 Years of masonry research and education **5**(4–5), 395–411 (2011)

Environmental Factors Affecting Corrosion of Steel Inserts in Ancient Masonry

L. Bertolini, M. Carsana, B. Daniotti and E. Marra

Abstract Steel inserts are often present in ancient masonry in order to improve the structural behaviour of buildings or to prevent the propagation of cracks. The corrosion of inserts embedded in the materials of the masonry and consequently expansive phenomena may be harmful for the durability of the whole structure. This chapter shows the results of a study on the corrosion behaviour of steel inserts in contact with mortars and/or brick to simulate masonry structures. The effects of temperature and moisture on corrosion rate of steel and resistivity of both aerial and hydraulic mortars and brick specimens exposed in a climatic chamber to cycles of temperature (5–20–40 °C) and relative humidity (65–80–95 % and at saturation) are discussed. The results show that corrosion rate is negligible in specimens exposed to 65–80 % RH (even at 40 °C), whereas it reaches high values in wet environments or in the presence of water suction (in particular, in gypsum mortars). A correlation between electrical resistivity of embedding materials and corrosion rate of steel has been observed. In order to extend the results obtained on small-scale specimens at case studies that should consider the effect of the real hygrothermal conditions on corrosion of steel inserts in ancient masonry, numerical simulations have been also performed. The validation of these data by means of a bi-dimensional HMT model is now in progress.

L. Bertolini (✉) · M. Carsana
CMIC Department, Politecnico di Milano, Milan, Italy
e-mail: luca.bertolini@polimi.it

M. Carsana
e-mail: maddalena.carsana@polimi.it

B. Daniotti
BEST Department, Politecnico di Milano, Milan, Italy
e-mail: bruno.daniotti@polimi.it

E. Marra
BEST/CMIC Department, Politecnico di Milano, Milan, Italy
e-mail: eleonora.marra@chem.polimi.it

V. P. de Freitas and J. M. P. Q. Delgado (eds.), *Durability of Building Materials and Components*, Building Pathology and Rehabilitation 3,
DOI: 10.1007/978-3-642-37475-3_9, © Springer-Verlag Berlin Heidelberg 2013

Keywords Corrosion · Steel inserts · Ancient brick masonry · Hygrothermal conditions · Electrical resistivity

1 Introduction

Steel inserts are often present in ancient masonry, either due to an original design choice or as the result of later restoration works. They may have different aims, e.g. improving the structural behaviour of buildings (chains and ties) or preventing the propagation of cracks, and they may be applied externally or embedded in the materials of the masonry (mortar, bricks, stone blocks, etc.). In the latter case, the presence of these elements may be harmful for the durability of ancient masonry, since their corrosion may produce deleterious expansive phenomena leading to the detachment of the embedding materials [1–5].

It is important to make a distinction between steel inserts dated to the creation of the original structure, normally embedded in the structural elements, and those fixed later during restoration phases (most often applied externally). While externally applied steel inserts are directly exposed to the action of the atmosphere (either inside or outside the building), inserts embedded in the masonry may have a complex corrosion behaviour, which depends on the material they are in contact (e.g. bricks, hydraulic mortars, gypsum) and its moisture content. Microstructure of these materials and chemical composition of the solution contained in their pores will influence the corrosion behaviour of steel. For example, the nearly neutral pH of some types of mortar can cause the corrosion process on the steel surface as soon as they are wet; conversely, in the case of alkaline mortars that promote passivity of the steel, corrosion can only take place after carbonation of the mortar cover has occurred. The corrosion rate will also depend on the availability of water and oxygen in the pores of the embedding material near the steel surface; hence it will be a function of environmental conditions. Temperature and humidity of the environment influence the electrical resistivity of the mortar or bricks and the corrosion rate of the embedded steel insert. For instance, a mortar with very high moisture content (that is near saturation, e.g. due to wetting or water condensation) is characterized by low resistivity, but, if saturation conditions are maintained for long time, the oxygen content at the steel surface may be low due to the low diffusivity through the water-filled pores. Conversely, in the case of masonry characterized by lower water content, although oxygen is available at the steel surface, the resistivity is high and it increases as the moisture content decreases.

Concerning masonries, rarely condition of permanent saturation are reached, thus oxygen is available at the steel surface and the corrosion rate depends on the resistivity of the materials in contact with steel; in general, the lower the moisture content, the higher the electrical resistivity of these materials and the lower the corrosion rate of steel. Nevertheless, a large number of parameters, related to both the environmental conditions and the materials, may have a complex role, which makes any prediction of the actual corrosion behaviour of steel inserts quite difficult. As matter of facts, the study of the dependence of resistivity of masonry and corrosion rate of steel insert on the environmental conditions is a subject of great interest in relation to optimising restoration methods and promoting a durability approach for historical buildings. Improving the understanding on corrosion of steel embedded in masonry could provide a tool for the design of repair works which are not merely aimed at the remediation of corrosion damages, but also at the control of the corrosion rate necessary to fulfil conservation requirements of preserving as much as possible the original materials.

This chapter shows the preliminary results of a study aimed at the evaluation of the corrosion behaviour of steel inserts embedded in masonry. The first part of this study deals with the assessment of the corrosion behaviour of low carbon steel embedded in mortar (based on several types of binders: gypsum or blends of gypsum-lime, lime-pozzolana and lime-cocciopesto), brick and composite brick and mortar samples, exposed to several environments. The effect of temperature and humidity on corrosion rate of steel and resistivity of specimens exposed in a climatic chamber to cycles of temperature (5–20–40 °C) and relative humidity (65–80–95 % and at saturation) are discussed. In order to extend the results obtained on small-scale specimens at case studies that should consider the effect of the real hygrothermal conditions on corrosion of steel inserts in ancient brick masonry, numerical simulations have been performed.

2 Corrosion of Low Carbon Steel Embedded in Building Materials

This part deals with the corrosion study performed on low carbon steel embedded in mortar, brick and composite brick and mortar samples. The experimental procedure and the results concerning the characterisation of the materials used in this study and the role of temperature and relative humidity on the corrosion behaviour of the lab specimens are presented in the following paragraphs.

2.1 Experimental Procedure

The experimental study has been performed at Politecnico di Milano, in the laboratories of the Department of Chemistry, Materials and Chemistry Engineering "G. Natta" at Politecnico of Milan. In the following paragraphs materials, specimens and laboratory tests are described.

2.1.1 Materials and Specimens Description

Different types of mortars have been cast in the laboratory using gypsum (G), lime and gypsum (LG) and blended lime with two different types of hydraulic additions: pozzolana (LP) and cocciopesto (LCP). Mix proportions of the mixtures studied are presented in Table 1.

Starting from these four types of mortars, three kind of specimens have been cast: $40 \times 40 \times 160$ mm^3 prisms, aimed at the evaluation of the compressive strength and the resistance to carbonation; 50 mm cubes, for the assessment of the water uptake due to capillary suction; finally, $80 \times 90 \times 30$ mm^3 reinforced prisms (three samples for each type of mortar), where two carbon steel bars (6 mm in diameter) and two stainless steel wires (AISI 304, 2 mm in diameter) have been inserted, in order to study the corrosion behaviour of steel embedded in these materials.

As far as steel bars are concerned, a low carbon steel (C 0.04 %, Si 0.041 %, Mn 0.285 %, S 0.012 % and P 0.012 %) and ferritic microstructure (grains size of about 20 μm) and pearlitic inclusions, has been used (see Fig. 1).

Moreover, in order to study the behaviour of steel embedded in masonry made of fired bricks and lime-gypsum mortar, two other sets of reinforced specimens have been manufactured: brick specimens (B) and composite brick and mortar specimens

Table 1 Mix proportions of the mortars manufactured in the laboratory (w/b = water/binder ratio, Agg/b = sand/binder ratio)

Mixture	w/b (–)	Agg/b (–)	Water (kg/m^3)	Sand (kg/m^3)	Lime (kg/m^3)	Gypsum (kg/m^3)	Pozzolana (kg/m^3)	Cocciopesto (kg/m^3)
G	0.42	–	515	–	–	1227	–	–
LG	0.63	1.20	404	769	212	430	–	–
LP	1.05	3.01	402	1148	191	–	191	–
LCP	1.05	3.01	402	1148	191	–	–	191

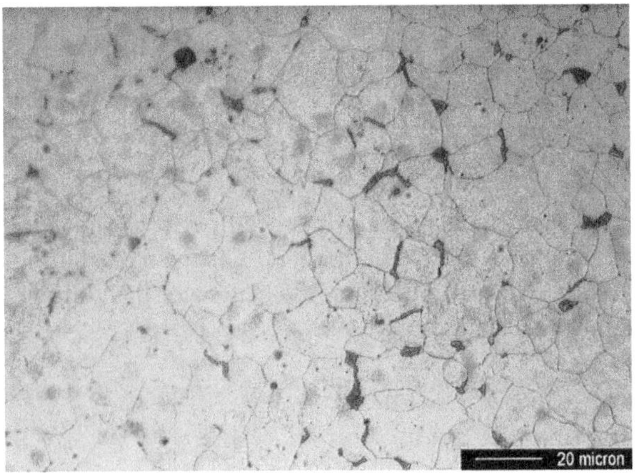

Fig. 1 Metallographic section of a steel bar used in this study

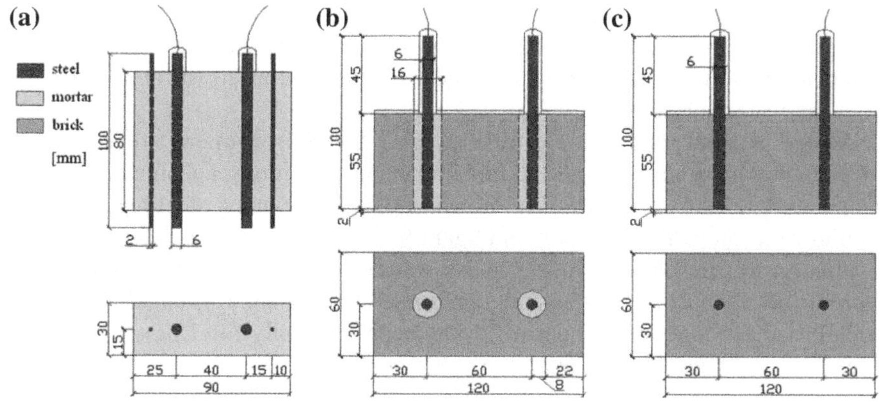

Fig. 2 Longitudinal section (*up*) and cross-section (*low*) of the specimens manufactured in the laboratory: (**a**) mortar sample (G, LG, LP and LCP) with 2 low carbon steel bars and 2 stainless steel wires; (**b**) composite brick and mortar sample (B + LG) with 2 low carbon steel bars embedded in a 5 mm mortar layer; (**c**) fired brick sample (B) with 2 low carbon steel bars. Dimensions in mm

(indicated with B + LG). Brick specimens had a size of 55 × 120 × 60 mm^3, in order to obtain several samples from the same brick; in two of these specimens two carbon steel bars (6 mm in diameter) have been inserted in two holes and the space between the bar and the brick has been filled with lime-gypsum mortar; in the other specimen bars have been forced in direct contact with the brick. A sketch of reinforced specimens geometry is shown in Fig. 2.

2.1.2 Environmental Conditions

Mortar specimens have been cast in PVC moulds and demoulded after 5 days, following the procedure described in the EN 1015-11 standard. Every sample has been cured for 28 days in a climatic chamber (7 days at 95 % RH and 21 days at 65 % RH, 20 °C T).

At the end of the curing time non-reinforced samples have been exposed both to accelerated carbonation (2 % CO_2, 65 % RH and 20 °C) and to natural conditions (in the laboratory, 65 % RH and 20 °C ca.), whereas all the reinforced specimens have been kept in the carbonation chamber until complete carbonation has been achieved (14 and 28 days respectively for the composite brick and mortar and for the mortar specimens), as shown by tests with an alcoholic solution of phenolphthalein performed on non-reinforced prisms. After the exposure in the carbonation chamber, reinforced specimens have been further exposed to several environments, in order to evaluate the role of temperature and relative humidity on corrosion behaviour of steel inserts.

Mortar specimens have been exposed to twelve different environments, varying relative humidity (65–80–95 % and at saturation) and temperature (5–20–40 °C).

Each condition has been kept for at least 28 days, apart from the immersion in water (1 day). In the latter condition, only 10 mm of the samples has been put in water, to allow water uptake towards steel bars placed at 20 (b) and 80 (a) mm from the water surface.

As far as brick and composite brick and mortar specimens are concerned, instead, they have been exposed to three moisture conditions, keeping constant the temperature at 20 °C: 95 % RH (C1), water (C2, maintaining steel bars at 20 and 60 mm form the water surface), and again 95 % RH (C3).

Finally, for the blended lime-gypsum mortar, one of the three specimens has been further exposed to the moisture conditions (C1, C2 and C3) undertaken by the brick-based specimens (storage in C2 has been kept only for 1 day, maintaining steel bars at 20 and 80 mm from the water surface).

2.1.3 Tests on Mortar Specimens

Mortar specimens have been characterized by measuring consistence (wokability), density, compressive strength, resistance to carbonation and water uptake.

As far as fresh mortars are concerned, both density and consistence have been measured. In particular, the consistence of the mortars has been measured using a flow table test, according to EN 1015-3 standard.

The compressive test has been performed on $40 \times 40 \times 160\ mm^3$ prisms, previously divided into two parts, after 28 days of curing. A preload of 0.5 kN and a load increased at the rate of 2400 N/s has been applied until the failure of the specimen, according to EN 196-1 standard.

The same specimens, exposed both under natural condition (in the laboratory) and accelerated conditions (2 % CO_2), have been used to evaluate the resistance to the penetration of carbonation. After different exposure times, the minimum and the maximum carbonation depth have been measured on the fracture surface of each specimens by means of an alcoholic solution of phenolphthalein test and an average value has been calculated. As far as specimens exposed to accelerated carbonation are concerned, measurements have been performed after 7, 14 and 28 days of exposure; specimens exposed to natural carbonation have been tested after 28 and 112 days of exposure.

2.1.4 Corrosion Measurements

Corrosion behaviour of reinforced specimens has been investigated by monitoring electrical resistivity of the mortar or brick (ρ, Ωm), corrosion potential (E_{corr}, mV vs Ag/AgCl) and corrosion rate (i_{corr}, mA/m^2) on the steel bars.

Resistivity measurements could provide information about the moisture content of specimens and their porosity. Electrical conductance (G, mS) has been measured with a conductivity meter and converted in resistivity by using the relationship: $\rho = K/G$ (where K is the cell constant that considers the specimen geometry, which was determined with a FEM model); these measurements have been carried out using pairs of stainless steel wires (in the case of mortar samples) or carbon steel bars (in the case of B and B + LG specimens), connected to the conductivity meter.

Corrosion potential has been measured versus an external reference electrode (Ag/AgCl) placed on the surface of the mortar (through a wet sponge), using a high impedance voltmeter.

Finally, corrosion rate has been measured with the polarisation resistance R_p method, by imposing potential steps of $\Delta E = \pm 10$ mV versus the free corrosion potential and measuring the resulting current density (i, mA/m^2); corrosion rate has been calculated as: $i_{corr} = B/(\Delta E/i)$, considering B = 26 mV [6]. With regard to the measurement of corrosion rate on reinforced mortar samples, stainless steel wires have been used as counter-electrodes; carbon steel bars have been utilized in turn as counter-electrode or working-electrode in the case of B and B + LG specimens.

2.2 Characterization of Mortars

Table 2 reports density and consistence of each freshly mixed mortar, as well as compressive strength and density measured at the end of the curing time. Consistence corresponds to almost 55 % for all the mortars manufactured; only slightly lower for the LG mortar, which shows a value of 42.5 %. In relation to the possible use of these mortars also as integration materials in the repair works, a comparable workability is required. As far as density is concerned, values of 1810 ÷ 1930 kg/m^3 have been achieved in all fresh mortars, whereas at the end of the 28 days of curing, density decreased to 1440 ÷ 1620 kg/m^3 due to drying; measurements have been performed at 20 °C, 65 % RH. Compressive strength has been tested on 40 × 40 × 160 mm^3 mortar prisms after 28 days of curing. The highest values are reached in gypsum-based mortars (G, 14 MPa; LG, 10 MPa), whereas the hydraulic mortars showed very low values, lower than 1 MPa (LCP mortar shows slightly lower values if compared to LP mortar). Differences in terms of compressive strength between the gypsum-based mortars and those with pozzolana and cocciopesto are also due to the different water/binder ratio, equal to 0.42–0.63 for the former and 1.05 for the latter.

Table 2 Fresh mortars: density and consistence. Mortars after 28 days of curing: density and compressive strength (R_c)

Mortars	Consistence(%)	$Density_{fresh}(kg/m^3)$	$R_{c\_28\ days}(MPa)$	$Density_{28\ days}(kg/m^3)$
G	55.0	1811	13.91	1442
LG	42.5	1878	10.24	1619
LP	56.9	1858	0.94	1614
LCP	53.1	1933	0.77	1526

Fig. 3 Carbonation depth as a function of the time of exposure and of the environment (*bold symbols* refer to accelerated environment, *filled ones* to natural conditions)

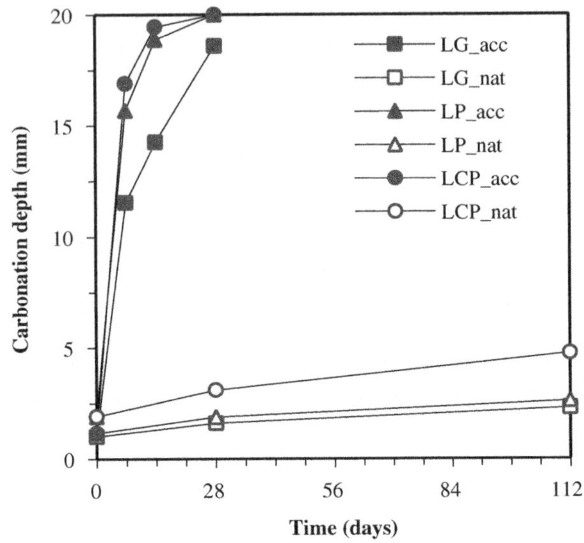

Figure 3 shows the average penetration depth calculated on the fracture surface at different times of exposure, on LG, LP and LCP specimens. Results show the carbonation depth on three sets of specimens, respectively after 7, 14 and 28 days of exposure in a chamber with a constant flow of 2 % of carbon dioxide (bold symbols), and of other two sets of specimens, respectively after 28 and 112 days of exposure in natural conditions (in the laboratory, filled symbols). All specimens have been exposed to carbonation at the end of the 28 days curing time.

Whatever the time of exposure is, the greater carbonation depths in accelerated conditions are reached in the LCP and LP specimens, followed by LG samples. The lower resistance to carbonation of the hydraulic mortars (with pozzolana and cocciopesto) could be explained considering higher porosity of these mortars than that of lime-gypsum specimens characterized by a lower water/binder ratio (0.63 for LG and 1.05 for LP and LPC). Furthermore, the dry conditions of exposure (20 °C, 65 % RH) of the samples during the carbonation might have also stopped the hydration process of mortars with pozzolana and cocciopesto which is slower than the hydration of gypsum. After 28 days, the carbonation depth of the LP and LCP specimens exposed to 2 % of CO_2 reached 20 mm, equal to half thickness of the specimens, while in LP specimens it reached only 18 mm. Then, the exposure in the carbonation chamber has been interrupted, since 15 mm are the cover depth of the reinforced specimens. Therefore it can be assumed that the carbonation front has reached the bars embedded in the mortar samples, even though it was not uniform, as underlined by the presence of the so called Liesgang Patterns (typical phenomenon of the carbonated lime-based mortars, [7]).

An example of Liesgang Patterns can be observed in Fig. 4, where the fracture surface of a LG sample after 7, 14 and 28 days of exposure to accelerated environment is shown.

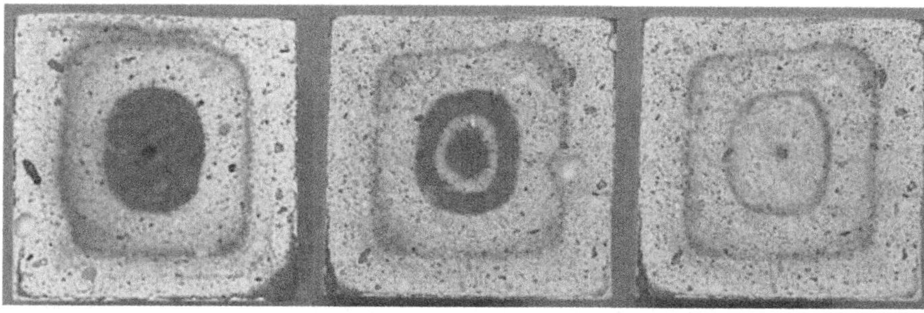

Fig. 4 Fracture surfaces of LG specimens, tested with an alcoholic solution of phenolphthalein: from *left*, carbonation depth at 7, 14 and 28 days of exposure to accelerated environment

Figure 3 shows also the carbonation depth measured at different time on specimens exposed in laboratory to simulate natural condition and compare these results with those obtained on specimens exposed in carbonation chamber.

Clearly the carbon dioxide content in the atmosphere influences the rate of penetration of carbonation; it can be observed that the exposure to an indoor environment determines a slower propagation of carbonation. In fact, also after 4 months of exposure in the laboratory, the carbonation depth of the different types of mortars varies from 2 to 5 mm (see Fig. 3).

2.3 Effect of Relative Humidity and Temperature on Corrosion Behaviour of Steel Inserts

The effect of the environmental conditions (i.e. temperature and relative humidity) on steel inserts in mortars and brick is presented. The first part deals with the mortar specimens, whereas in the second part composite specimens are considered (the discussion has been limited to the LG based samples).

2.3.1 Mortar Specimens

Figure 5 shows, as an example, results of corrosion potential (E_{corr}) and corrosion rate (i_{corr}) and electrical resistivity (ρ), as a function of time, of an LG specimen, in three different moisture conditions (65–80–95 % RH) at 20 °C. As far as corrosion potential and corrosion rate are concerned, grey lines refer to measurements performed respectively on each steel bar of the sample, while the black line represents an average of the measurements performed on the two bars embedded into the specimen. Potential values of about −220 mV vs Ag/AgCl are reached only in the first hygrometric condition (65 % RH); as relative humidity increases, a steady decrease of the corrosion potential can be observed (−380 mV at 80 % RH and −420 mV vs Ag/AgCl at 95 % RH).

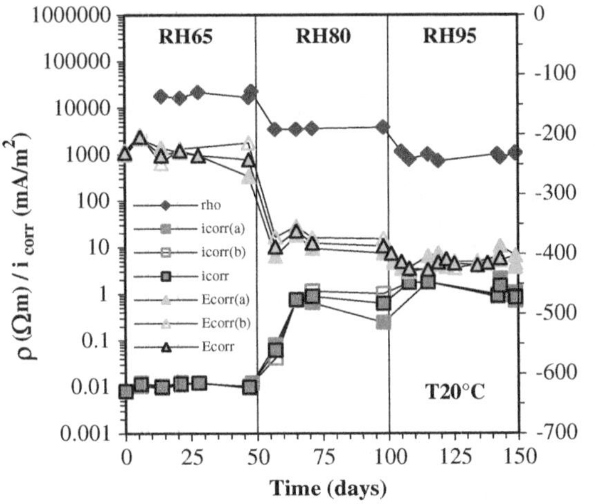

Fig. 5 Electrical resistivity ρ (rho), corrosion potential E_{corr} and corrosion rate i_{corr} as a function of time, of steel bars (*a* and *b*) embedded in a lime and gypsum mortar specimen, in three moisture conditions (65, 80 and 95 % R) at 20 °C. *Grey lines* refer to values measured on each bar; *black line* refers to average values

Only in the dry condition negligible corrosion rate has been measured (being the threshold conventionally fixed at 1 mA/m^2, which is equal to 1.17 μm/year, assuming uniform corrosion); in the other cases corrosion rates near to the threshold have been observed (0.8 mA/m^2 at 80 % RH and 1 ÷ 2 mA/m^2 at 95 % RH). Also the electrical resistivity of the mortar is influenced by the relative humidity: very high values of resistivity have been measured in the 65 % RH condition, whilst as relative humidity increased values drop of one order of magnitude.

Replicated LG samples have been exposed to the same hygrometric conditions, at 5 °C and 40 °C. Moreover, the same exposure conditions have been undertaken by G, LP and LCP mortar samples (each of them has been stored at a fixed relative humidity, just varying the temperature).

For every type of mortar (G, LG, LP and LCP) Figs. 6, 7, 8, 9, 10, 11, 12, 13, 14 show the average, the minimum and the maximum values of steady state measurements of electrical resistivity (Figs. 6, 7, 8), corrosion potential (Figs. 9, 10, 11) and corrosion rate (Figs. 12, 13, 14) in each hygrometric condition (at 5 °C, 20 °C and 40 °C). As far as saturation condition is concerned (C2), separate values are reported for corrosion potential and corrosion rate of bars at different distance from the water uptake surface (bar *b*, 20 mm and bar *a*, 60 mm).

Results of Figs. 6, 7, 8 suggest that electrical resistivity decreases as the humidity of environmental increases; a similar relationship was found between electrical resistivity and relative humidity for all mortars studied. At constant relative humidity, Fig. 6 shows variations due to the type of mortars; higher values of electrical resistivity have been measured in the hydraulic mortars (LP and LCP) than those gypsum-based mortars (G and LG) although the latter have shown lower porosity (§ 2.2). These differences could be explained probably by the possible differences in the pore solution of the mortars. As for concrete [8, 9], electrical resistivity is associated with the electrical migration of ions in the aqueous solution contained in the capillary pore of mortar. Gypsum-based mortars

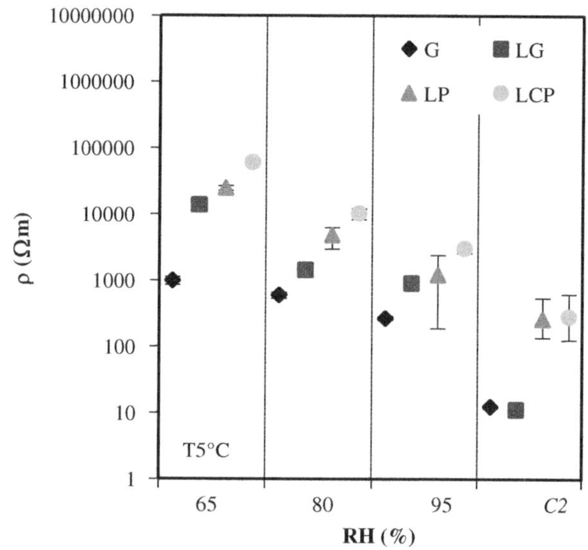

Fig. 6 Electrical resistivity ρ of G, LG, LP and LCP mortars as a function of different moisture conditions at 5 °C of temperature. Average values of replicated specimens

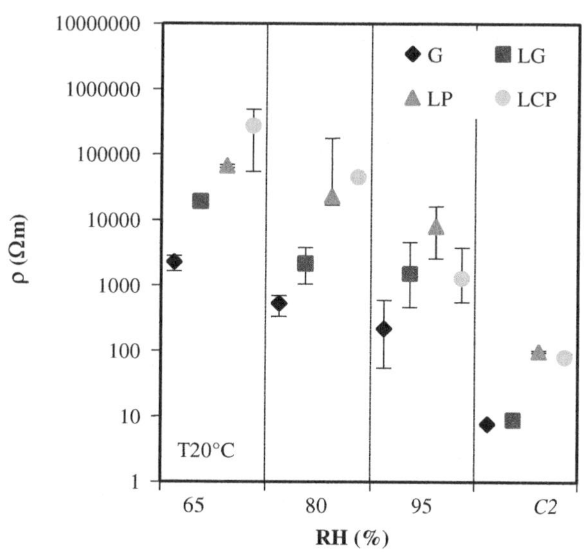

Fig. 7 Electrical resistivity ρ of G, LG, LP and LCP mortars as a function of different moisture conditions at 20 °C of temperature. Average values of replicated specimens

contain in their pore sulphate ions which could favorite ionic movement much more than that presented in carbonated hydraulic mortars. The comparison of results obtained at different temperatures (see Figs. 6, 7, 8) does not show remarkable variations of relationship.

Figures 9, 10, 11, 12, 13, 14 show that also corrosion potential and corrosion rate are influenced by the exposure conditions.

As far as corrosion potential is concerned, the higher values are reached at 5 °C if compared with those achieved at 20 °C and 40 °C. This effect, almost negligible

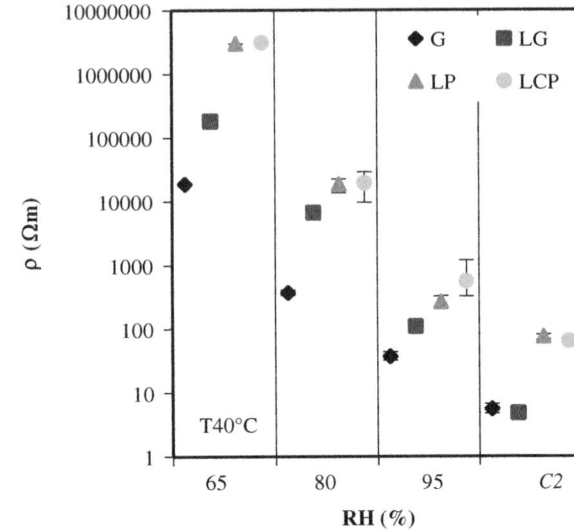

Fig. 8 Electrical resistivity ρ of G, LG, LP and LCP mortars as a function of different moisture conditions at 40 °C of temperature. Average values of replicated specimens

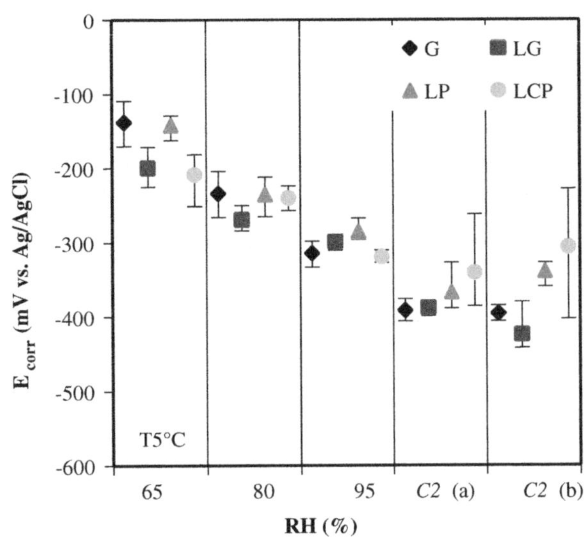

Fig. 9 Corrosion potential E_{corr} of steel embedded in G, LG, LP and LCP mortars as a function of different moisture conditions at 5 °C of temperature. Average values of replicated specimens

at 65 % RH, turns to be more remarkable in the other hygrometric conditions (at saturation, potential drops from −350 mV vs Ag/AgCl at 5 °C to −500 mV vs Ag/AgCl at 20 °C and 40 °C). The type of mortar slightly affects the measurements (differences of about 50 mV in each condition).

Finally, corrosion rate increases as relative humidity increases. Corrosion rate is below 1 ÷ 2 mA/m^2 for samples stored at both 65 % RH and 80 % RH, in every kind of mortar considered, both at lower and higher temperature (0.01 ÷ 0.1 mA/m^2 at 65 % RH). At 95 % RH, instead, corrosion rate increases up to 10 mA/m^2; at

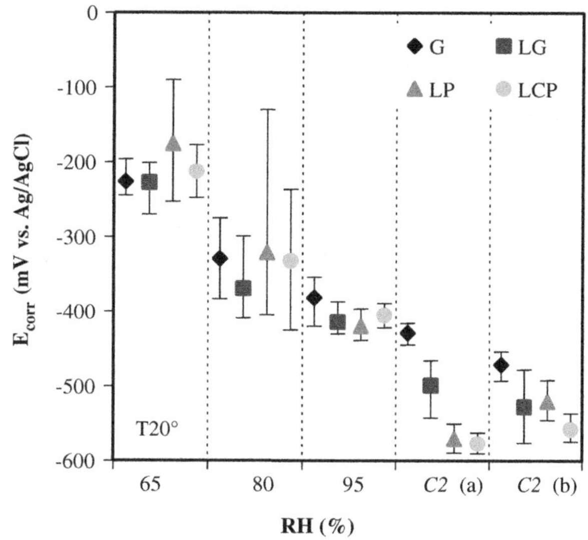

Fig. 10 Corrosion potential E_{corr} of steel embedded in G, LG, LP and LCP mortars as a function of different moisture conditions at 20 °C of temperature. Average values of replicated specimens

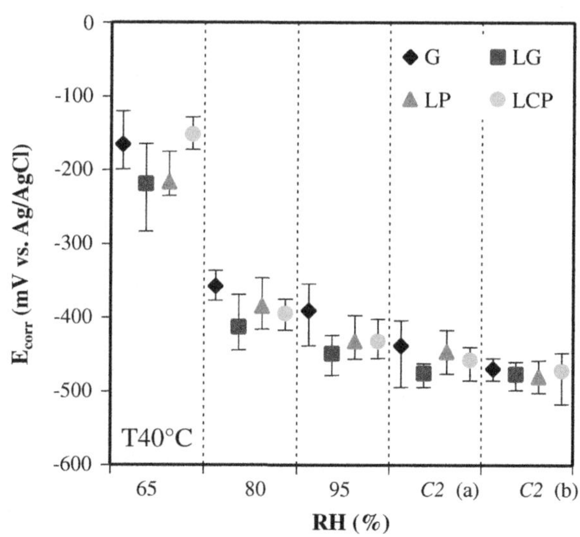

Fig. 11 Corrosion potential E_{corr} of steel embedded in G, LG, LP and LCP mortars as a function of different moisture conditions at 40 °C of temperature. Average values of replicated specimens

saturation (C2), corrosion rate is higher than 10 mA/m^2 and in the case of G and LG mortars it reaches 100 mA/m^2. Concerning the role of temperature, it can be also observed that in dry environment (65 % RH) an increase in temperature enhances drying out hence reducing the corrosion rate of bars (below 0.01 mA/m^2); as relative humidity raises (80 % RH), this effect is reduced to the point of being negligible. Indeed, at 95 % RH the increase in the kinetic of the corrosion process is the key factor, so corrosion rate is higher at 40 °C (10 mA/m^2) than at 5 °C and 20 °C (1 ÷ 3 mA/m^2). As far as the role of materials is concerned, in every exposure condition, the maximum corrosion rate is achieved by the gypsum-based mortars,

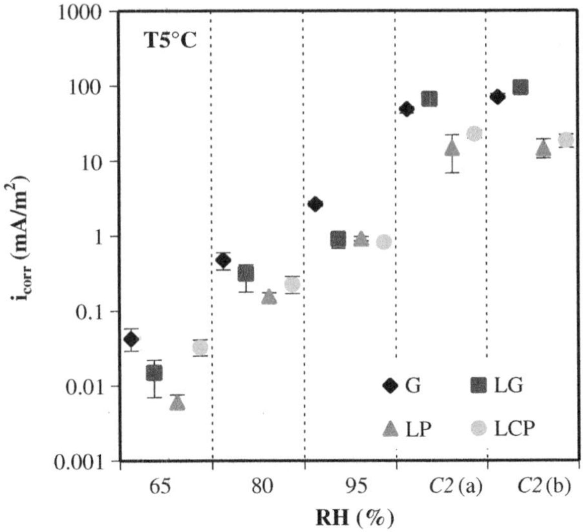

Fig. 12 Corrosion rate i_{corr} of steel embedded in G, LG, LP and LCP mortars as a function of different moisture conditions at 5 °C of temperature. Average values of replicated specimens

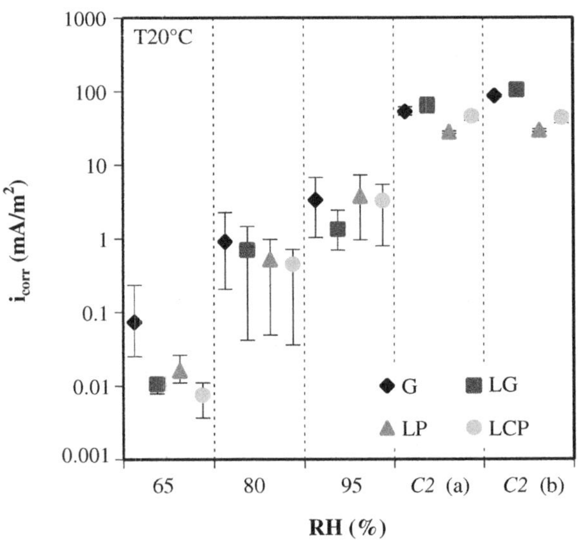

Fig. 13 Corrosion rate i_{corr} of steel embedded in G, LG, LP and LCP mortars as a function of different moisture conditions at 20 °C of temperature. Average values of replicated specimens

which also had the lower values of electrical resistivity. Moreover, the higher values of corrosion rates reached in the G and LG mortars, especially in humid or saturated environment, could be explained considering that the water solution in the pores of these mortars is slightly acidic, being more aggressive for the bars.

The results of tests carried out on mortar specimens exposed to different environmental conditions allow an evaluation of the role of temperature and relative humidity, in steady state conditions, on the corrosion behaviour of steel inserts embedded in carbonated mortars or in gypsum mortars. In general, it can be

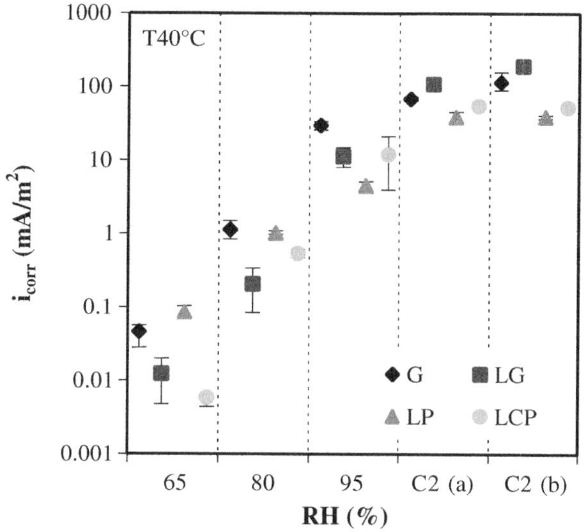

Fig. 14 Corrosion rate i_{corr} of steel embedded in G, LG, LP and LCP mortars as a function of different moisture conditions at 40 °C of temperature. Average values of replicated specimens

assumed that the higher the corrosion potential of the inserts and the lower the resistivity of the embedding material, the higher the corrosion rate of inserts.

This can be observed in Fig. 15, where the relationship corrosion potential—corrosion rate (Fig. 15a) and electrical resistivity—corrosion rate (Fig. 15b) in the LG specimens is presented as example, hence allowing an evaluation of the influence of different hygrothermal conditions on such parameters. Light gray and gray line and dark line refer respectively to constant temperature of 5, 20 °C and 40 °C; increasing degree of filling of the symbols indicates an increase in the moisture condition (65 %, 80 %, 95 % and at saturation C2). Concerning the latter condition (C2), it should be observed that only the measurements taken on the *b* bars, i.e. the nearest to the water uptake surface (20 mm), have been considered.

Figure 15a shows a linear relationship between corrosion potential and corrosion rate of steel embedded in lime-gypsum mortars; it can be observed that in dry conditions, steel has potential values more positive than −250 mV vs Ag/AgCl and the corrosion rate is lower than 0.11 mA/m^2. Conversely, corrosion potential (Fig. 15a) decreases by increasing the environmental relative humidity: the higher values, indeed, have been observed at 65 % RH (about −200 ÷ −250 mV vs Ag/AgCl), whereas the lower are reported at saturation (about −400 ÷ −600 mV vs Ag/AgCl). The results obtained for different temperatures show linear trend with slope slightly different.

The absence of a generalized correlation show that the simple measure of corrosion potential of steel embedded in mortar cannot be used to evaluated the corrosion rate (although such an approach would be very useful, since potential measurements are easier to carry out than corrosion rate measurements.

Also, the electrical resistivity (Fig. 15b) can be correlated to corrosion rate with a linear relationship. Figure 15b shows that the affect of the relative humidity are greater than that of temperature on the corrosion rate of steel inserts embedded in

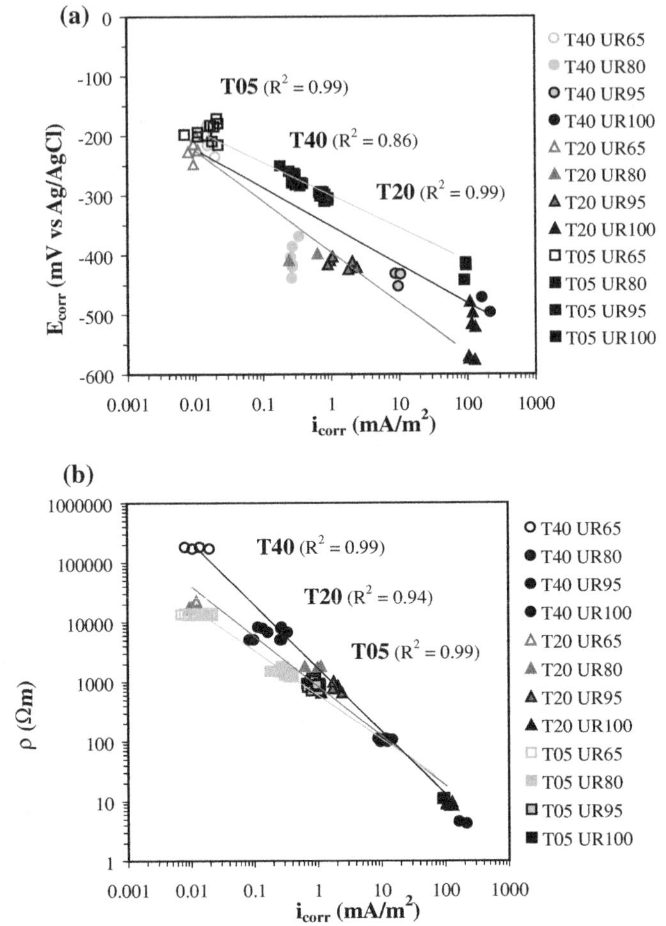

Fig. 15 **a** Relationship between corrosion potential, E_{corr} and corrosion rate, i_{corr} and **b** relationship between electrical resistivity, ρ and corrosion rate, i_{corr} in LG mortar during the exposure to various RH conditions. C2 refers to saturation.

LG mortars, especially for critical range of corrosion rate. Similar trends have been confirmed also for the other types of mortars studied. These results suggest that the evaluation of corrosion rate through resistivity measurements might be successful to investigate in details the corrosion behaviour of steel insert in masonry.

2.3.2 Composite Specimens

Figure 16a shows corrosion potential (empty symbols) and corrosion rate (filled symbols), as a function of time, of different specimens (LG, B and B + LG) in three different moisture conditions (at 20 °C): humid environment at 95 % RH (C1), wet (C2), and again 95 % RH (C3). Each curve represents an average of the measurements carried out respectively on different replicated bars. Corrosion potential values of about −200 mV vs Ag/AgCl (suggesting negligible corrosion) have been registered only on brick-based specimens (B and B + LG) in the 95 % RH condition (C1), while in all the other cases corrosion potential falls to −400 ÷ −600 mV vs Ag/AgCl showing active corrosion. These observations are supported by average values of the corrosion rate measurements: indeed, only in brick-based samples (B and B + LG) in humid condition (C1) negligible corrosion rate have been measured; in the other specimens, moderate (2 mA/m^2 on LG sample, in environment C1) or even high (10 ÷ 100 mA/m^2 on every other sample, in C2) corrosion rates have been observed. Table 3 shows the average values and standard deviations of steady state measurements in each exposure condition (C3 condition has been excluded since steady values have been not reached). In this table, for condition C2, separate values are reported for corrosion potential and corrosion rate of bars at different distance from the water uptake surface (20 and 60 ÷ 80 mm). Lower bars at 20 mm, that experienced a higher water content have shown higher corrosion rate, though, LG specimen was more affected by this phenomenon.

Corrosion behaviour of steel can be correlated to results of electrical resistivity measurements. Indeed, Fig. 16b shows the resistivity, as a function of time, of the specimens. Extremely high values of resistivity have been achieved in C1 condition by B + LG specimens (grey curve represents an average of two replicated specimens). This could be attributed to the presence of a discontinuity between the brick and the mortar layer; further investigations are required for a full comprehension of such an aspect. Lower resistivity is shown by LG (about 1100 Ωm) and especially B (black line, about 500 Ωm) specimens, the latter being more porous than the former. In wet condition (C2) these differences tend to become negligible

Table 3 Mean value μ and standard deviation σ of electrical resistivity ρ, corrosion potential E_{corr} and corrosion rate i_{corr} measured on brick (B), brick and blended lime-gypsum mortar (B + LG), blended lime-gypsum mortar (LG), exposed to different environmental conditions

Environ.condition	Material	ρ (Ωm)		E_{corr} (mV vs Ag/AgCl)		i_{corr} (mA/m^2)	
		μ	σ	μ	σ	μ	σ
T 20 °C RH 95 % (C1)	B + LG	36532	14001	−197	77	0.03	0.01
	B	482	51	−201	32	0.97	0.22
	LG	1072	535	−415	10	2.06	1.65
T 20 °C in H_2O (C2)	B + LG	39	2	−510[a]	76[a]	3.53[a]	0.55[a]
				−587[b]	9[b]	9.95[b]	1.15[b]
	B	16	3	−545[a]	20[a]	33.7[a]	1.65[a]
				−579[b]	55[b]	35.2[b]	5.01[b]
	LG	9	0	−472[a]	54[a]	64.9[a]	9.84[a]
				−494[b]	60[b]	118[b]	16.6[b]

[a] bars at 60 ÷ 80 mm, [b] bars at 20 mm from the water uptake surface

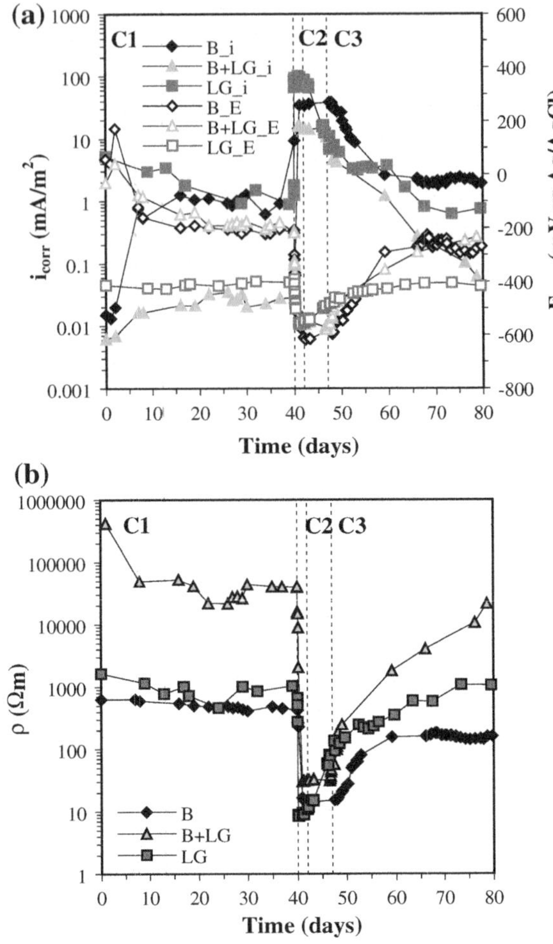

Fig. 16 **a** Corrosion potential E_{corr} and corrosion rate i_{corr} of steel bars embedded in various materials as a function of time exposure in three different conditions at 20 °C (*C1* and *C3* refer to 95 % RH, while *C2* refers to wet environment). *Black, grey* and *light grey symbols* refer respectively to brick (*B*), composite mortar and brick (*B + LG*), mortar samples (*LG*); *empty symbols* refer to E_{corr} values, filled ones to i_{corr} values. **b** Electrical resistivity ρ, as a function of time (same specimens and exposure conditions). Average values of replicated specimens (4 bars for LG, and 2 bars for B and B + LG)

(values roughly 10 ÷ 40 Ωm), whereas in the final condition (C3) there is a progressive recover of the initial differences.

Comparing the wetting process with the drying one, it is possible to observe the differences in the kinetic of the hygrometric pattern: in the former case (C1-C2 boundary) the transient is faster (constant values are reached in few hours), while in the latter (C2-C3 boundary) is slower (resistivity returns to its first range value in more than 10 days). Moreover, the drying process seems to be quicker in the B + LG specimens, followed by the B sample and by the LG one.

Since electrical resistivity measurements can provide information about the water content of specimens and their porosity [10], it could be used as a corrosivity index of steel inserts in active conditions when exposed at different environments. In general, the lower the resistivity of the embedding material, the higher the corrosion rate of inserts.

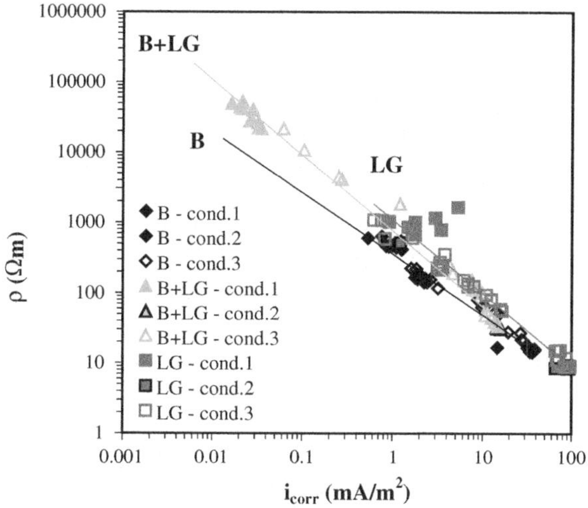

Fig. 17 Relationship between electrical resistivity ρ and corrosion rate i_{corr} in different materials during the exposure to various RH conditions (T constant at 20 °C). *Black, light grey* and *grey line symbols* refer respectively to brick (*B*), composite mortar and brick (*B + LG*), mortar samples (*LG*). *Filled symbols* refer to 95 % RH (*C1*), *bold ones* to water uptake (*C2*), *empty ones* to drying at 95 % RH (*C3*)

This can be observed in Fig. 17, where the relationship between electrical resistivity and corrosion rate in the composite specimens is presented, hence allowing an evaluation of the influence of different hygrothermal conditions on such parameters. Similar relationships can be found for specimens made with the other types of mortars studied.

3 Hygrothermal Simulations

Laboratory trials have highlighted the key role of temperature, relative humidity and water (rising damp, condensation or rainwater) on the corrosion behaviour of steel bars embedded in masonry materials. Though moisture and temperature conditions within a brick masonry wall consist of a complex set of parameters (e.g. material properties, indoor and outdoor environmental conditions), taking into account for all these variables could be done by means of hygrothermal models. Indeed, heat and moisture transfer (HMT) models and tools have had a great spread in the last two decades [11], and have shown a good aptitude in predicting moisture and temperature conditions with a reasonable degree of accuracy [12].

Therefore, in order to extend the results obtained in the laboratory on small-scale specimens, numerical simulations have been performed at the BEST Department of Politecnico of Milan: temperature, relative humidity and water conditions at the interface between the metal inserts and the materials in which they are embedded have to be analysed.

3.1 Simulation Models

As far as preliminary assessment is concerned, two series of simulations have been performed, both considering mono-dimensional geometries and implementing a finite difference algorithm: the first one studies only the temperature field; the second one analyses the simultaneous effects of temperature and relative humidity. In particular, concerning temperature and relative humidity variations, the behaviour of a typical ancient brick works (consisting of non-hydraulic mortar and fired bricks, width 380 mm) has been studied by means of different hygrothermal boundary conditions, which are steady state ones (20 °C T, 50 % RH) and dynamic sinusoidal ones (48, 24, 12, 6 and 3 h periods, mean value and maximum variation respectively of 20 °C and 5 °C for T, 50 and 15 % for RH).

Moreover, two further series of situations have been modelled. The first one simulates again traditional brick masonries (width 380 mm), assuming as external boundary condition a reference year in the geographic area of Milan, whereas as internal boundary condition constant temperature and relative humidity (20 °C T, 50 % and 65 % RH) has been considered.

In order to validate the experimental data obtained in the laboratory, instead, the second one reproduces brick blocks (width 120 mm) subject to the hygro-thermal fluctuations laboratory tests conditions: temperature steps have been imposed (5–20–40 °C), keeping constant the relative humidity (65–80–95 %); in addition, relative humidity steps have been fixed (65–80–95 %), keeping constant the temperature (5–20–40 °C).

Also a bi-dimensional model has been implemented, in order to validate the experimental data obtained in the laboratory. Brick, mortar and composite brick and mortar samples have been modelled, hence considering the cross-section of each kind of specimen (squared-section bars has been considered). Concerning materials, six different kind of fired masonry bricks and two types of mortars (both hydraulic and aerial) have been studied; the main properties of each material are reported in Table 4. As boundary conditions, the same temperature (5–20–40 °C) and humidity (65–80–95 %) steps have been kept.

3.2 Thermal and Hygrothermal Response

The simulations results have shown that an unsteady state analysis is needed for a consistent evaluation of the thermal behaviour of the brick masonries considered (Fig. 18a), whereas a steady state approach is enough to describe the hygrometric behaviour inside the domain, apart from the boundary zone (Fig. 18b): while the thermal fluctuations involve the whole samples thickness, the relative humidity variations are limited to the very first centimetres from the interface with the environment.

Table 4 Main hygrothermal properties of the materials used for the bi-dimensional simulations

ID	Material	ρ (kg/m^3)	ε (m^3/m^3)	C (J/kgK)	λ (W/mK)	μ_d (−)	A_w (kg/m^{2}s$^{0.5}$)
1[a]	Fired masonry brick	1650	0.41	850	0.60	10	0.400
2[a]		1725	0.38	850	0.60	15	0.300
3[a]		1800	0.31	850	0.60	17	0.360
4[b]		1630	0.35	850	0.60	10	0.267
5[b]		1780	0.32	850	0.60	10	0.283
6[b]		1560	0.38	850	0.60	15	0.583
1[b]	Hydraulic mortar	1700	0.35	850	0.80	15	0.087
1[b]	Aerial mortar	1785	0.28	850	0.70	15	0.153
1[a]	Steel	7800	0.00	450	60.0	15000	0.000

Legend ε = porosity, ρ = bulk density, C = thermal capacity, λ= thermal conductivity, μ_d = vapour diffusion resistance factor, A_w = water uptake coefficient
References [a] *IBP* Fraunhofer Institute for Building Physics, [b] *TUV* Technical University Vienna, [c] *NRC* National Research Council Canada

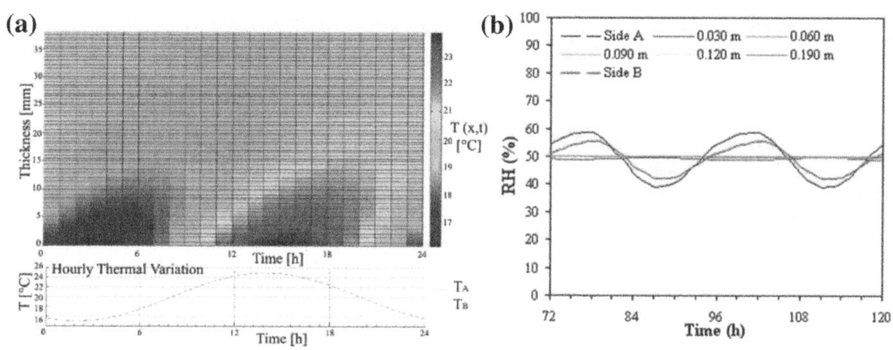

Fig. 18 **a** Thermal response of a 380 mm brick masonry wall to a sinusoidal variation, 24 h period on side A (*down*), constant on side B (*up*). **b** Hygrometric response of the same wall to a sinusoidal variation, 24 h period on side A (*black curve*), constant on side B (*grey curve*)

The simulations of the experimental laboratory tests, confirming such results, have highlighted a very fast thermal response (24 h are enough to reach the equilibrium with the environment) and a slower hygrometric one (asymptotic values are not reached in the core region). Finally, since these results have to be validated by the experimental laboratory data concerning the corrosion behaviour of steel inserts, an attempt to consider bi-dimensional HMT models is in progress.

In this respect, all samples modelled have achieved hygrothermal equilibrium with the environment after 180 ÷ 240 h (Figs. 19, 20, 21, 22), as observed on laboratory specimens. Comparison among the brick-based samples has shown their consistent variation in hygrometric behaviour (Fig. 19) and no great differences in thermal behaviour (Fig. 20). As far as mortars are concerned, instead, no consistent differences in hygrothermal behaviour have been detected between them

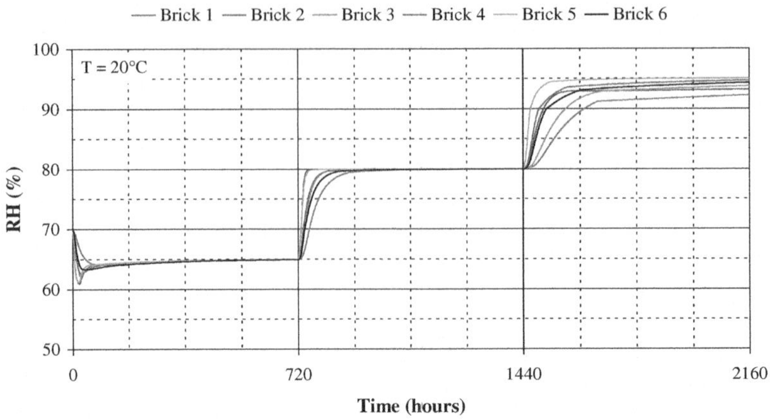

Fig. 19 Hygrometric response of brick specimens to relative humidity steps (65–80–95 %), at a constant temperature of 20 °C

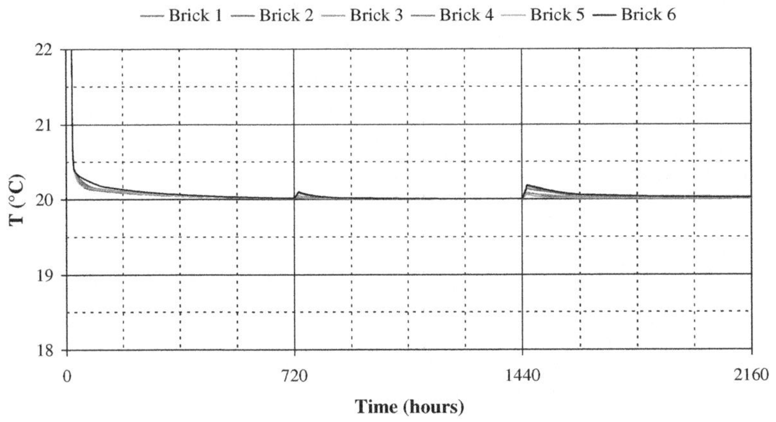

Fig. 20 Thermal response of brick specimens to relative humidity steps (65–80–95 %), at a constant temperature of 20 °C

(Figs. 21, 22). Brick samples show higher hygrometric inertia than mortar ones. These results could be correlated to the corrosion rate measurements performed in the laboratory.

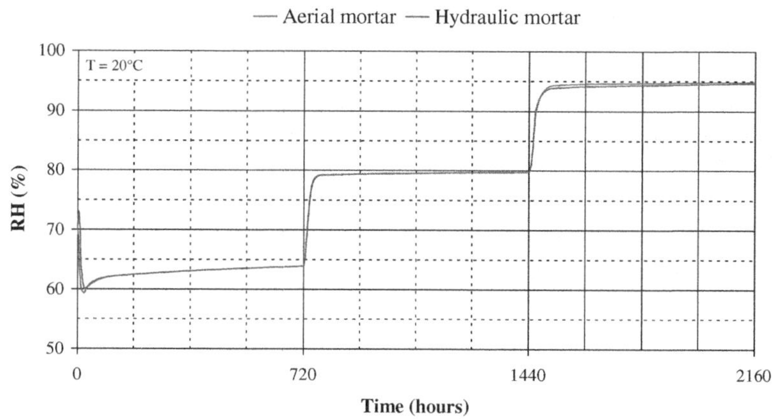

Fig. 21 Hygrometric response of mortar specimens to relative humidity steps (65–80–95 %), at a constant temperature of 20 °C

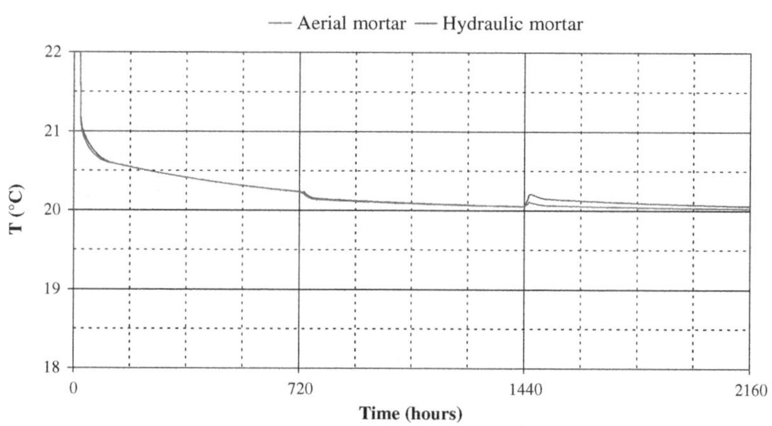

Fig. 22 Thermal response of mortar specimens to relative humidity steps (65–80–95 %), at a constant temperature of 20 °C

4 Conclusions

This study has shown the key role of temperature, relative humidity and water on the corrosion behaviour of low carbon steel inserts embedded in masonry materials. Tests in specimens with four different kinds of mortars and fired brick have shown that corrosion rate is strongly related to the moisture content of the embedding material. Corrosion rate was negligible in specimens exposed to relative humidity of 80 % and 65 %, even at 40 °C. Conversely, in wet environments or in presence of water suction, corrosion rate has reached high values.

Correlations between electrical resistivity of embedding material and corrosion rate of steel, and corrosion potential and corrosion rate of steel have been observed.

With the aim to extend the results obtained on small-scale laboratory specimens to case studies that should consider the effect of the real hygrothermal conditions on corrosion of steel inserts in ancient brick masonry, numerical simulations have been performed. Concerning this, an attempt to define a bi-dimensional HMT model, which can predict corrosion behaviour of inserts as a function of the hygrothermal conditions, is now in progress.

References

1. Bertolini, L., Carsana, M., Marra, E.: Degradation of mortars and steel inserts from the ciborium of the medieval abbey of San Pietro al Monte. In: Special Topics on Materials Science and Technology: An Italian Panorama. Brill Publisher, Leiden (2009)
2. Bertolini, L., Carsana, M., Gastaldi, M., Lollini, F., Redaelli, E.: Corrosion assessment and restoration strategies of reinforced concrete buildings of the cultural heritage. Mater. Corros. **62**(2), 146–154 (2011)
3. Fort Gonzalez, R., De Alvarez Buergo, M., Mingarro Martìn, F., De Lopez Azcona, M.C.: Stone decay in 18th century monuments due to iron corrosion. The Royal Palace, Madrid (Spain). Build. Environ. **39**, 357–364 (2004)
4. Lourenco, P.B.: Recommendations for restoration of ancient buildings and the survival of a masonry chimney. Constr. Build. Mater. **20**, 239–251 (2006)
5. Straube, J., Schumacher, C.: Assessing the durability impacts of energy efficient enclosure upgrades using hygrothermal modelling. WTA-J. **2**, 197–222 (2006)
6. Bertolini, L., Elsener, B., Pedeferri, P., Polder, R.: Corrosion of steel in concrete. Wiley-VCH, New York (2004)
7. Rodriguez-Navarro, C., Cazalla, O., Elert, K., Sebastian, E.: Liesegang pattern development in carbonating traditional lime mortars. R. Soc. **458**, 2261–2273 (2002)
8. Bertolini, L., Polder, R.B.: Concrete resistivity and reinforced corrosion rate as a function of temperature and humidity of the environment, TNO-Report, 97-BT-R0574, (1997)
9. Pedeferri, P.: Corrosione e protezione dei materiali metallici. Polipress, Milano (2007)
10. Bürchler, D., Elsener, B., Böhni, H.: Electrical resistivity and dielectric properties of hardened cement paste and mortar. In: Gerhardt, R.A., Taylor, S.R., Garborczi E.J. (eds.) Electrically based micro-structural characterization. Mat. Res. Soc. Symp. Proc. **411**, 407 (1996)
11. Straube, J., Burnett, E.F.P.: Overview of hygrothermal (HAM) analysis methods. In: Trechesel, H.R. (ed.) ASTM Manual Series MNL 40: Moisture analysis and condensation control in building envelopes, pp. 81–89. ASTM International, Philadelphia (2001)
12. Künzel, H.M.: Simultaneous Heat and Moisture Transfer in Building Components. IBP Verlag, Stuttgart (1995)

Corrosion of Reinforcement in Existing Concrete Façades

Jukka Lahdensivu, Hanna Mäkelä and Pentti Pirinen

Abstract The most common degradation mechanism causing the need to repair concrete façades in Finland, is the corrosion of reinforcement due to carbonation and the small cover depths of the reinforcement. During the last almost 20 years the repair of concrete façades and balconies has been very active and this work has produced a great deal of information about the actual condition of those repaired buildings. This active condition assessment and repair work have produced an extensive and growing information base, which can be used, for example, for anticipating the upcoming repair need of concrete structures. For this purpose the data is gathered from 947 concrete buildings, which includes the distribution of different kinds of states of damage and the interdependence between the damage and other factors. The carbonation of concrete has widely advanced in the façades made in 1970s or earlier. The corrosion of reinforcement has been possible approximately last 20–30 years in those façades. Despite insufficient cover depths of concrete and far advanced carbonation of concrete, visually seen corrosion damage are relative rare. Corrosion damage appear mostly on façades, which get more rain. In Finland those are upper parts of southern and western façades.

Keywords Corrosion of reinforcement · Carbonation · Concrete façade · Annual rainfall · Condition assessment · Durability

J. Lahdensivu (✉)
Tampere University of Technology, Tekniikankatu 1233101 Tampere, Finland
e-mail: jukka.lahdensivu@tut.fi

H. Mäkelä · P. Pirinen
Finnish Meteorological Institute, Erik Palmenin aukio 100560 Helsinki, Finland
e-mail: hanna.tietavainen@fmi.fi

P. Pirinen
e-mail: pentti.pirien@fmi.fi

V. P. de Freitas and J. M. P. Q. Delgado (eds.), *Durability of Building Materials and Components*, Building Pathology and Rehabilitation 3,
DOI: 10.1007/978-3-642-37475-3_10, © Springer-Verlag Berlin Heidelberg 2013

1 Introduction

The growth of European suburban areas was fast in the 1960s and 1970s. Migration from the countryside into towns and changes in social structure created demand for fast and massive housing production. Large suburbs were built which changed the former pre-war townscape remarkably.

Due to the massive need for residential buildings and the rapid development of prefabrication techniques of precast concrete panels in the 1960s and 1970s, concrete soon became the dominant material of façades and balconies in multi-storey residential and office buildings in Finland [16].

Since the 1960s a total of about 44 million square metres of precast concrete panel façades have been built in Finland as well as 900,000 precast concrete balconies [28]. As a matter of fact, more than 60 % of the Finnish building stock has been built in the 1960s or later [23]. Compared to the rest of Europe, the Finnish building stock is quite young.

Despite the rather young age of the Finnish concrete building stock, several problems have been encountered in their maintenance and repair. The structures have deteriorated due to several different deterioration mechanisms whose progress depends on many structural, exposure and material factors. Thus the service lives of structures vary widely. In some cases the structures have required remarkable and often unexpected, technically significant and costly repairs less than 10 years after their completion. For this reason many new methods have been developed in Finland for maintaining and repairing these concrete structures during the last 20 years. The methods include a condition investigation practice and its extensive utilization, rational repair methods and their selection, as well as first-rate repair products and appropriate instructions for managing repair projects.

Concrete structures have been repaired extensively in Finland since the early 1990s. During the almost 20-year period, approximately 10 % of the stock built in 1960 until 1980 has been repaired once. It is estimated that the total annual value of building repair work in Finland is approximately 5.500 million Euros, of which approximately 30 % involves external structures (walls, balconies, roofs, windows, etc.). The total annual volume of façade renovation is approximately 15 million m^2. In addition, 40,000 balconies are repaired annually and 4,500 new balconies are added to old buildings. It is estimated that the volume of façade renovation will grow 2 % annually [27, 28].

Because of the great amount of these existing concrete structures, it is very important to solve their incident repair need economically and in a technically durable way. This means that the most suitable repair methods have to be used for each case and it is also important to be able to determine the optimal time of the repairs.

1.1 Structures of Concrete Façade Panel

The concrete panels used in exterior walls of multi-storey residential buildings were, and still are, chiefly prefabricated sandwich-type panels with thermal insulation placed between two concrete layers. Façade panels are made up of two relatively thin reinforced concrete layers connected to each other by steel trusses. The thermal insulation between the layers is most often mineral wool of 60–145 mm nominal thickness depending on the building regulations in force at the time of design and construction.

The usual nominal thickness of the outer layer has been 40–70 mm while 50 and 60 mm are the most usual values [19]. The layers are most typically reinforced with steel mesh of a wire diameter of 3 mm and spacing of 150 mm (Fig. 1). Rebars 6–8 mm in diameter are typically used as so-called edge bars and often also diagonally at the corners of windows and other major openings in the layers. The bars are spliced by lap splices, which increase the overall thickness of the reinforcement. The compression strength of concrete is typically near C20/25.

The outer layer is generally supported by the inner layer. Sandwich façade panels are connected to the building frame by the inner layer, usually by means of cast concrete joints and reinforcement ties. Panels are typically equipped with lifting straps of 16 mm steel rod for installing them in place. Lifting straps of non-bearing panels are anchored to both the inner and outer layer. After a panel is in place, the straps are to be cut to avoid cold bridges. In the case of a bearing panel, the lifting strap is anchored only to the inner layer and is used to fix the panel to the building frame.

In the Finnish prefabricated concrete building system, the inner layer of end façade panels is load-bearing while that of long façade panels is non-bearing. The thickness of the inner layer is normally from 150 to 160 mm (load bearing panels) or 70 mm (non-load bearing panels). The outer layers of both panel types always have the same dimensions and reinforcement. All vertical and horizontal joints between outer layers are elastic, made primarily with polymer sealants to allow

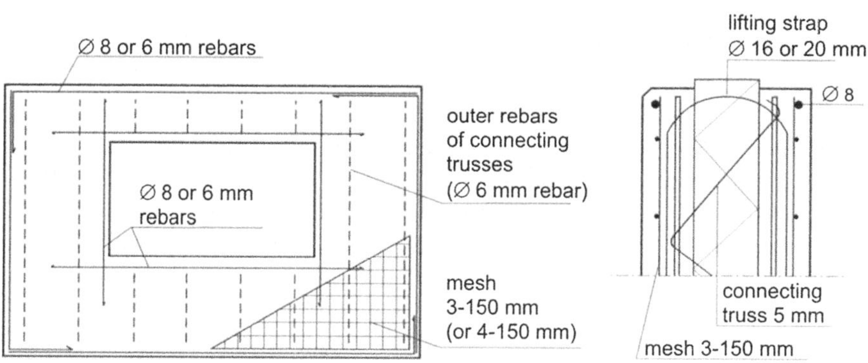

Fig. 1 Typical reinforcement of outer layer of a concrete façade panel

thermal as well as other movement of the layers. It should also be noted that usually there is no ventilation gap behind the outer layers of precast exterior wall panels. Thus, if the thermal insulation gets wet e.g., due to leakage through the joints, the structure dries slowly. The drying of the outer layers is also slow because of the relatively efficient thermal insulation that limits the drying heat flow through the wall. This means that the concrete may remain moist for long periods.

1.2 Objective

The objective of this chapter was to study the factors that have actually had an impact to the existence and progress of corrosion of reinforcement in existing Finnish concrete façades.

This chapter is based on the author's experiences from about 150 condition assessments of concrete structures, long-continued development of condition investigation systematic, and Lahdensivu's dissertation Durability properties and actual deterioration of Finnish concrete façades and balconies [12].

2 Corrosion of Reinforcement

The most common deterioration mechanisms causing the need to repair concrete facades in Finland, and concrete structures in general, are the corrosion of reinforcement due to carbonation or chlorides as well as the insufficient frost resistance of concrete which leads to, for instance, frost damage [19]. Reinforcing bars within concrete are normally well protected from corrosion due to the high alkalinity of concrete pore water. Corrosion may start when the passivity is destroyed, either by chloride penetration or due to the lowering of the pH in the carbonated concrete [17].

2.1 Carbonation of Concrete

Corrosion protection of mild steel in concrete is based solely on the high alkalinity of concrete, or rather of the pore solution in the concrete at the steel surface, deriving from small quantities of readily soluble alkali hydroxides, NaOH and KOH, and a large proportion of less soluble lime, $Ca(OH)_2$. These are mainly responsible for the buffering action of concrete [4, 8]. The pH of solid concrete is usually above 13. Such high alkalinity forms a thin and dense oxide layer on a steel surface [17], which very efficiently protects all embedded steel from corrosion.

Passivation of a steel surface is very efficient corrosion protection, because the passive film is usually self-healing as long as pH remains high and chloride content remains sufficiently low. According to Parrott [18], the critical pH level of non-chloride-contaminated concrete is somewhere between 11 and 11.5, which cannot be preserved under the passive layer.

The corrosion protection of steel is electrochemically based primarily on alkalinity. On the other hand, physical protection in the form of a sufficiently dense, thick and uniform layer of concrete on top of the steel is needed to prevent harmful substances, like chlorides and acids, from penetrating into the concrete surrounding the steel [7].

Carbonation is a chemical reaction where atmospheric carbon dioxide diffuses into the pore system of concrete and reacts with the alkaline hydroxides of concrete. Due to carbonation, the pH of the concrete decreases from the level of 13 to the level of 8.5 and the passivity of the reinforcing steel loses its passivity. Carbonation begins at the surface of a newly constructed member and propagates slowly as a front at a decelerating rate deeper into the structure. The speed of propagation is influenced mainly by the quality of concrete (amount of cement and porosity of concrete) as well as the moisture exposure. Heavy moisture exposure, for example due to rainfall, slows down carbonation because water blocks the pores from CO_2 [18, 20, 26].

2.2 Chloride Contamination

The passivity of steel may be destroyed also by the presence of chlorine ions derived either from the environment or from the use of contaminated constituents of concrete. Chlorides act as catalysts to corrosion and, therefore, the corrosion process may proceed rather quickly. Whereas carbonation-induced corrosion leads to more or less uniform corrosion, chloride attack usually causes localised corrosion, which can result in severe reduction in the steel diameter. Chloride-induced corrosion becomes highly accelerated when carbonation reaches the reinforcement depth. This means that the extent of visible corrosion damage may increase strongly in a short time. Steel corrosion in chloride migrated concrete has been widely studied by Parrott [18], Richardson [20], Schiessl [22], Broomfield [5] etc.

Many different limit values for critical chloride content are presented in literature ranging from 0.17 to 2.5 % of the weight of cement either as acid-soluble or free water-soluble chlorides [25]. Alonso et al. [1] suggested a critical total chloride limit for Ordinary Portland Cement of 1.24–3.98 % by weight of cement where the corresponding share of free water-soluble chlorides is 0.39–1.16 % by weight of cement. Finnish guidelines consider 0.03–0.07 % acid-soluble chloride by weight of concrete critical [6]. To avoid confusion, it should be noted that critical chloride content is usually expressed as a proportion of the weight of cement.

2.3 Active Corrosion

Once the passivity is destroyed either by carbonation or by chloride contamination, active corrosion may start in the presence of moisture and oxygen [17]. Corrosion may run for a long time before it can be noticed on the surface of the structure. Because corrosion products are not water soluble, they accumulate on the surface of steel nearby the anodic area [13]. This generates an internal pressure, because the volume of the corrosion products induced by carbonation is four to six times bigger than in the original steel bars [18, 26].

The internal pressure caused by corrosion products leads to cracking or spalling of the concrete cover. Visible damage appears first on the spots where the concrete cover is the smallest and the moisture content of concrete is the highest. According to Tuutti [26], the relative humidity of concrete has a strong influence on the corrosion speed. The rate of corrosion proceeds significantly only if the relative humidity of concrete exceeds 80 %. In the typical Finnish outdoor climate between September and April, the relative humidity in concrete façades is between 90 and 95 %. This means corrosion current of 0.1–1 $\mu A/cm^2$ respectively.

The corrosion of steels embedded in concrete structures generally affects their appearance more than their load-bearing capacity. According to Andrade [3], at the rapid corrosion rate of 1 $\mu A/cm^2$ steel corrodes 10 μm/a. Thus, it takes several years of severe corrosion for a steel cross-section to corrode enough to have a significant impact on the load-bearing capacity of a structure. The corrosion of steel causes cracks and chipping in the concrete cover when the cross-section has decreased about 15–50 μm [2]. Loss of reinforcement anchorage or setting due to cracking or spalling of concrete may affect structural integrity and performance much faster than a reduction in the cross-section of steels.

Chloride-induced corrosion is typically pitting corrosion, where at least part of the corrosion products are water soluble, which allows corrosion of steels to propagate considerably before visual damage can be detected on the concrete cover.

3 Research Material

The research material consists of the database of deterioration and material properties of existing Finnish concrete façade panels built up between 1961 and 1996, and weather observations since 1961 made by the Finnish Meteorological Institute (FMI).

3.1 Condition Assessment of Concrete Façades

The condition assessment of concrete façade is a systematic method to determine the condition of an existing structure, future propagation of deterioration and recommended repair measures. The content of a condition investigation is based on the properties of the target building, such as used structure types, materials, environmental stress conditions, already visible damage, and the goals set for the assessment. The contents of a condition assessment must always be planned separately for each target [6].

In a condition assessment, the condition and performance of a structural member or group of them is determined systematically in terms of different degradation mechanisms using various research methods including review of design documents, visual inspection of the target, various field measurements and surveys as well as sampling and laboratory analyses. The aim is to find out the causes, extent and impacts of the damage existing at the time of investigation as well as to anticipate future damage at the stage when visible damage does not yet exist. The data are collected as samples while the properties and condition of the structure vary in its different parts. Thus, the condition assessment of an old structure always involves uncertainty, which is why an effort is made to reduce it by using parallel methods in investigating degradation mechanisms and by collecting data from as many sources as possible.

3.2 Database

The database consists of data derived from condition assessment reports on concrete façades and balconies. Condition assessment reports have been collected from owners of rental residential properties and consulting companies conducting condition investigations as well as the Laboratory of Structural Engineering at Tampere University of Technology. It contains condition assessment report data on 422 targets. A condition investigation report on a single target comprises the condition assessment data of 1–30 buildings. The database contains the assessment results on a total of 947 precast concrete buildings since many assessments looked at several buildings at the same time.

The block of flat buildings in the database was built in 1960–1996; mostly in the 1970s and early 1980s. They have been divided into three different climatic regions by geographic location: the southern coastal area, inland and northern Finland. Moreover, the metropolitan area has been examined as a separate entity due to its large building stock.

Inland areas constitute the largest examined area, which also has a lot of precast concrete buildings. Northern Finland has considerably fewer such buildings than the rest of Finland, which means that relatively few condition assessment reports on its buildings were available. These buildings have been made part of the inland building stock in order to be able to include also these targets of condition assessment in the study.

3.3 Meteorological Observations

The meteorological observations have been compiled from observations of the Finnish Meteorological Institute (FMI) starting from 1961. Climate conditions vary across Finland, which is why observation data have been collected from several localities. The majority of the Finnish multi-storey residential building stock is located in the southern coastal areas and elsewhere in southern Finland. Based on that fact and FMI's meteorological observatory network, the following stations were selected:

- Turku Airport (on the southern coast);
- Helsinki Kaisaniemi (on the southern coast);
- Helsinki-Vantaa Airport (in southern coastal area);
- Jyväskylä Airport (inland);
- Oulu Airport (only wind speed and direction during rain in northern coastal area);
- Rovaniemi Airport (only wind speed and direction during rain and sleet in northern inland area).

Besides total monthly rainfall amounts, snowfall in the form of water or sleet has been classified since only liquid rain can be absorbed by capillary suction into the pore system of a porous material. Precipitation days have been recorded into five categories by amount so that 0.1 mm represents lightest precipitation. The other categories are ≥ 1, ≥ 3, ≥ 5 and ≥ 10 mm.

Wind directions and velocities during rain or sleet have been compiled as part of precipitation data, as well as continuous September-April and corresponding yearly rain-wind observations that also include summer rains. Such observations have been organised by periods of five observation years.

4 Results and Discussion

Most of the blocks of flats stock in Finland is situated in the southern part of Finland. In the database the building stock has been divided into three different group based on their geographical position: coastal area, inland and North Finland. Approximately a third of those buildings is situated on the coastal area and two-thirds inland. Only a few of the buildings in the database are situated in North Finland.

Fig. 2 The most typical surface types in Finnish façade panels. **a** Brushed painted concrete. **b** Exposed aggregate concrete. **c** Painted plain concrete. **d** Brick tile finishing

All prefabricated concrete panels have basically been constructed in the same way, but there are many differences in their surface finishing and manufacturing. Those have a strong influence, for instance, on the situation of rebars and the quality of concrete. In the database the most common surface finishing is brushed painted concrete, exposed aggregate concrete and painted plain concrete, see Fig. 2. The distribution of the surface finishing on the database is shown in Fig. 3.

4.1 Structural and Material Properties

This chapter looks at the durability properties of concrete façades on the basis of realised dimensions and properties of structures. The analysis is based on observations and examinations of concrete core samples taken from structures and field measurements.

4.1.1 Thickness of Outer Layer

The façades of precast multi-storey residential buildings consist of sandwich panels that are typically non-bearing panels with window openings along the sides and solid load-bearing panels at the ends. The thickness of the outer layer of the

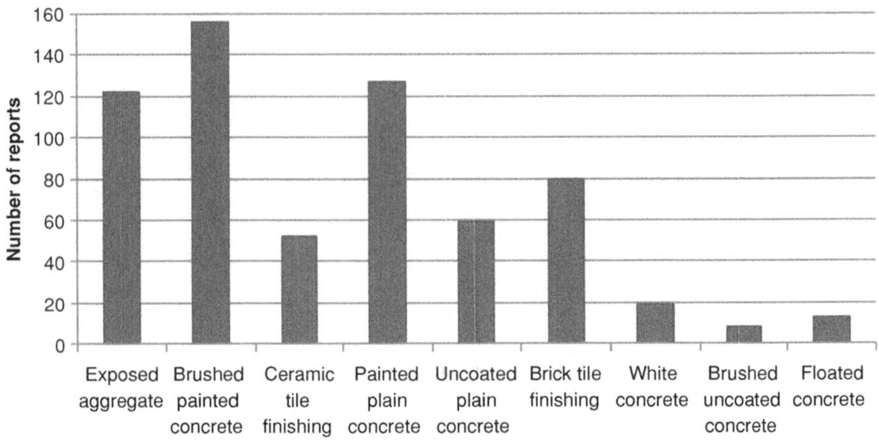

Fig. 3 Distribution of surface finishing in the database (n = 947 single buildings)

concrete sandwich panel has been measured as the length of the samples taken during condition assessment. Measured outer layer thicknesses have been compiled in Table 1 according to different surface types of façades.

The outer layer of a sandwich panel must be thick enough to ensure that a sufficient concrete cover can be placed on top of the centrally installed reinforcing mesh and edge bars, and that the bonding of trusses embedded in the inner surface of the outer layer is sufficient and reliable. Some of the panel's lifting lugs are also installed in the outer layer and also require sufficient cover thickness and anchoring. The design thickness of outer layers has varied in different periods

Table 1 Thickness of outer layer of precast sandwich panels based on length of samples of condition assessments

	Bearing panel					Non-bearing panel				
	Length [mm]				No.	Length [mm]				No.
	Min.	Max.	Av.	Std. dev.		Min.	Max.	Av.	Std. dev.	
Exp. aggregate	26	100	62.2	11.1	552	37.5	95	63.2	9.7	339
Exp. agg. layer	2	79	26.6	10.3	335	9	85	27.2	11.4	224
Backing con. layer	0	75	35.1	13.1	335	0	71.5	36.2	14.2	223
Brushed painted	31	175	59.5	13.2	736	32	109	58.9	11.2	527
Clinker-clad	34	175	67.2	17.6	214	44	105	67.3	11.3	70
Painted form-finish	32	135	65.3	19.1	204	27	111	62.7	13.0	135
Unpainted form finish	44	153	78.0	21.2	155					
Brick panel-clad	44	126	82.3	11.6	350	33	107.5	78.6	13.5	131
Brushed unpainted	36	84	54.8	11.8	34	44	57	50.1	4.9	7
White concrete	46	118	69.0	14.9	48	59	89	69.5	8.9	31
White con. layer	11	55	24.7	11.2	20	11	62	33.1	16.2	15
Backing con. layer	21	58	39.6	10.3	20	0	62	37.6	21.5	15
Floated unpainted	50.5	75	62.7	7.5	13		60			1

according to façade surface. Most typically it has been 60 mm, but in the early years of prefabricated construction even 40–50 mm [19].

On average, façade panel outer layer thicknesses are quite close to each other ranging from 55 to 70 mm. Exceptions in the thicker direction are the unpainted form-finish panel, generally used in the plinth storey of a building, and the brick panel-clad panel, where the around 20–30 mm brick panel increases the total thickness of the outer layer. The thickness of the concrete section in the case of these brick panel-clad panels is nevertheless of the same magnitude as in façade panels in general. Thicknesses of the outer layer vary considerably in all surface types. No significant differences in the average thickness of the outer layer and variation of thickness have been observed between bearing and non-bearing panels with the same façade panel surface type.

The concrete of the outer layer of exposed aggregate and white concrete panels is generally a combination of two layers of different concrete types cast together. The thickness of the surface layer varies considerably from a few millimetres to the thickness of the entire outer layer. The average exposed aggregate and white concrete layers are 25–33 mm.

Although average outer layer thicknesses are close to design values, quite small outer layer thicknesses also occur with all façade surface types. Yet, just one sample has been reported to have broken (length 10 mm). A total of 43 (1.2 %) under 40 mm outer layer thicknesses were discovered, mainly in bearing panels, whose façade surface was either exposed aggregate (11 samples) or brushed painted (14 samples; six of the samples were from non-bearing panels). In the case of other panel and façade surface types less than 40 mm thicknesses occurred only in isolated cases. On the other hand, a total of 101 unusually thick, over 100 mm, outer layer thicknesses were measured. They were most common in form-finish (45) and brick panel-clad panels (26). The greatest individual outer layer thicknesses measured were 175 mm in the case of a brushed painted and a clinker-clad panel.

Providing sufficient reinforcement cover depth for corrosion protection is impossible with outer layers less than 40 mm thick. By force of circumstances reinforcements remain either too close to the outer surface, whereby the protective concrete layer requirement is not met, or too close to thermal insulations, which may compromise bonding. In theory, it is possible to install the planned reinforcement with splices and lifting lugs only in a 65 mm thick outer layer, so as to meet the cover depth requirement of 20 mm. A cover depth of 25 mm requires an outer layer of 85 mm considering installation tolerances. Reinforcements of outer layers will not function optimally also in the case of excessively thick outer layers. When reinforcements lie too deep, they cannot prevent cracking due to the shrinking of concrete.

4.1.2 Cover Depth of Reinforcement

The concrete cover depth distributions of steels in concrete samples are shown in Table 2. Only a small share of less than 5 mm cover depths measured from the outer surface of samples were found in painted form-finish and brushed painted panels. In clinker-clad panel samples the share of less than 10 mm cover depths was already significant at 6.2 %. With other facade surface types, the share of less than 10 mm cover depths is at the most 3.6 %. Only isolated samples were taken from floated unpainted facades, which make them incomparable.

The share of small reinforcement cover depths deviates totally from other façade surface types in the case of brick panel-clad façades. The share of less than 10 mm cover depths is 15 %. However, it should be noted that cover depths have not been measured from the outer surface of samples, but from the interface between concrete and brick panel, thereby ignoring the about 20 mm contribution of the brick panel. On the other hand, the carbon dioxide diffusion resistance of brick panel is so low that it does not retard the migration of carbon dioxide into the backing concrete [24], which means that the share of the brick panel should actually be ignored in cover depth measurement.

The cover depths measured from samples taken from the inner surface of the outer layer are considerably smaller than those on the outer surface. Small cover depths occur in especially large numbers in façades cast with the façade surface

Table 2 Shares of reinforcement cover depths by classes in concrete samples drilled from façade panels

Surface type		Share of cover depth classes [%]				No.
		0–4 mm	5–9 mm	10–14 mm	15–19 mm	
Exposed aggregate	Outer surf.	0.00	0.64	3.81	13.60	472
	Inner surf.	2.11	4.68	12.41	13.30	427
Brushed painted	Outer surf.	0.47	1.24	4.81	11.32	645
	Inner surf.	9.69	13.44	19.73	16.67	588
Clinker-clad	Outer surf.	0.00	6.21	14.29	19.25	161
	Inner surf.	2.29	3.05	6.10	15.30	131
Painted form-finish	Outer surf.	1.06	1.60	13.30	13.30	188
	Inner surf.	5.52	9.20	9.20	17.18	163
Unpainted form-finish	Outer surf.	0.00	3.61	4.82	16.87	83
	Inner surf.	1.52	7.58	9.09	10.61	66
Brick-panel clad	Outer surf.	8.86	6.27	9.23	4.42	271
	Inner surf.	4.17	3.13	7.29	9.90	192
Brushed unpainted	Outer surf.	0.00	0.00	0.00	18.18	22
	Inner surf.	6.67	13.33	20.00	13.33	15
White concrete	Outer surf.	0.00	0.00	5.36	12.50	56
	Inner surf.	2.33	2.33	2.33	4.65	43
Floated unpainted	Outer surf.	0.00	16.67	0.00	0.00	6
	Inner surf.	11.11	33.33	11.11	11.11	9

up. It means that the support of reinforcement for concrete placement has been quite inadequate allowing the reinforcement to sink close to the thermal insulation.

In field investigations the cover thicknesses of façade panels have been measured widely from the outer surface of façade using a non-destructive cover depth meter, which provides a considerably larger sample than merely measuring cover depths from samples. The cover depth distributions of the reinforcements of different façade types are shown in Table 3 separately for reinforcement mesh and edge bars.

The shares of small, under 5 and under 10 mm cover depths, are generally smaller than the corresponding outer surface cover depths measured directly from samples, see Table 2. Exceptions are brick panel- and clinker-clad façades. In the case of clinker-clad façades, at least in some measurements, the reading of the cover depth meter has clearly been reduced—typically by 20 mm—which is the only possible explanation for the less than 20 mm cover depths. The share of less than 10 mm cover depths in clinker-clad façades is clearly larger based on sample measurements than field measurements. If the typically about 5 mm (often also 8 and 12 mm) reduction due to the thickness of the clinker tile is taken into account in field measurements, the cover depth distributions based on sample and field measurements are quite close.

The surface of most façade panels, such as exposed aggregate and brushed panels, is rough, which means that cover depth meters may indicate slightly exaggerated cover depths. Surface roughness is 3–5 mm depending on façade

Table 3 Shares of cover depths of meshes and edge bars by cover depth categories measured from the outer surface of façade panels

Surface type		Share by cover depth categories [%]				No.
		0–4 mm	5–9 mm	10–14 mm	15–19 mm	
Exposed aggregate	Mesh	0.00	0.47	2.60	10.70	39 473
	Edge bars	0.00	0.34	1.14	5.62	21 167
Brushed painted	Mesh	0.01	0.57	1.90	8.62	68 804
	Edge bars	0.01	1.10	2.55	8.17	36 154
Clinker-clad	Mesh	0.00	1.97	6.65	22.38	8 582
	Edge bars	0.00	1.10	6.16	18.49	4 690
Painted form- finish	Mesh	0.13	3.22	4.11	13.09	18 183
	Edge bars	0.26	2.44	3.83	11.51	7 991
Unpainted form- finish	Mesh	0.01	2.07	4.70	14.72	7 056
	Edge bars	0.31	0.89	3.01	11.69	2 265
Brick panel-clad	Mesh	1.41	7.42	5.16	2.97	14 988
	Edge bars	0.02	3.78	4.32	3.31	9 759
Brushed unpainted	Mesh	0.00	1.38	5.01	11.97	1 653
	Edge bars	0.00	6.38	10.79	15.09	1 332
White concrete	Mesh	0.00	0.10	1.80	11.64	3 869
	Edge bars	0.00	0.00	1.61	15.28	2 193
Floated unpainted	Mesh	0.00	0.57	1.78	6.75	1 578
	Edge bars	0.00	2.11	1.74	4.78	520

surface type and roughness, which has not generally been considered when measuring and reporting cover depths. Cover depths of areas suffering from already visual damage are not measured in field investigations either. This is another reason why the shares of the smallest cover depths indicated by field measurements are smaller than in reality.

The cover depth of reinforcements can be measured considerably more accurately directly from concrete samples than by using non-destructive methods. Thus, a cover depth distribution measured directly from samples may be considered more reliable although the share of visible corrosion damage is ignored also in that case. On the other hand, the sampling required is much smaller than when using a cover depth meter during field investigations.

Although the cover depths of reinforcement to a quite large extent do not meet the minimum requirements set in codes and durability guidelines, it can be stated that the shares of small, under 5 and under 10 mm cover depths, are on the whole quite small in the case of facade panels.

4.2 Carbonation Depth of Concrete Compared with Cover Depths of Reinforcement

Carbonation of concrete cannot be seen visually on façades. The determination of the carbonation depth of concrete always requires sampling and laboratory testing. The most typical carbonation depths of concrete façades constructed in the 1970s or earlier is nowadays around 10 and 20 mm if the concrete quality is normal. Concrete façades covered with ceramic or brick tiles can be carbonated only through the cement-based pointing of tiles. Furthermore paintings used on concrete façades have resistance against the diffusion of carbon dioxide. The average carbonation speed of different concrete façades is shown in Fig. 4.

According to Fig. 4, the average carbonation depth is in the concrete façade built in the year 1970 now between 6 and 17 mm depending on the surface

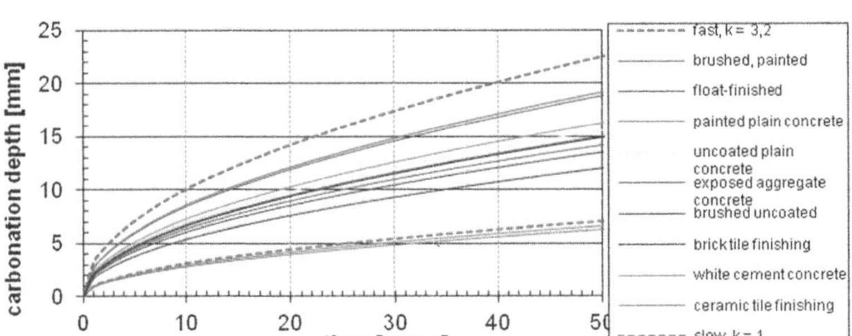

Fig. 4 Average carbonation speed of concrete has a great deal of variation according to the surface finishing

finishing of concrete. However, there is a great deal of scattering in the carbonation rate also in each group of surface finishing.

Average carbonation of the outer surface is clearly slowest with clinker- and brick panel-clad and white concrete surfaced panels. Clinker tiles prevent effectively the diffusion of carbon dioxide into concrete meaning that carbonation may propagate only through the joints between tiles or in areas where clinker tiles have not completely bonded to the substrate. Brick panels improve the water-cement ratio of concrete and thus the density of concrete at the panel/concrete interface, which has been found to clearly slow down the carbonation of concrete. The slow carbonation of white concrete is most likely due to the higher grade of concrete containing more cement than the concrete used in so-called more conventional facade types.

Carbonation is clearly fastest in brushed and floated panels. Their façade surfaces have been treated while fresh. The carbonation rates of other façade surface types fall fairly evenly between the rate of the above-mentioned façade surface types.

There is wide deviation between concrete outer surface carbonation coefficients as can be seen from Fig. 5. Distribution of carbonation was wide also between samples drilled from different panels of the same building, which points to wide variation in concrete quality between different production batches. The concrete carbonation coefficient exceeds $k = 6\ mm/a^{0.5}$ with a few samples per façade surface type. The database includes only four samples of through-carbonated outer layers. With clinker-clad, unpainted brushed and floated, and white concrete surfaced panels the carbonation coefficient exceeds $k = 3\ mm/a^{0.5}$ only in a few isolated cases. Compared to earlier Finnish studies [9, 10, 15], the average carbonation coefficients of the material of this study are slightly smaller, but slightly larger than in Irish precast concrete buildings [21].

The depth of the concrete cover on reinforcement varies a lot, depending on the manufacturing of concrete panels and the quality of labour work. The cover depth of reinforcement varies between 0 and over 50 mm. When determining the suitable repair method for a concrete façade, the amount of cover depths less than

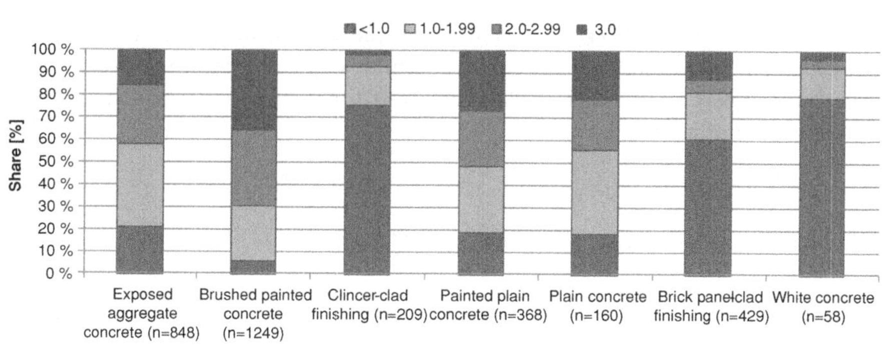

Fig. 5 Carbonation coefficients, k, of outer surfaces of façades by façade surface types

10 mm is critical in most of the cases. A typical amount of small cover depths is less than 5 % of all reinforcement in façades. In most cases the smallest cover depths are in ceramic tile finished façades, where the reinforcement is situated just behind the tiles.

The carbonation of concrete has widely achieved the reinforcement of all concrete façades built during the 1960s and the 1970s. Furthermore, corrosion has been possible for 20–30 years for now.

4.3 Visual Corrosion Damage

According to the database, 59.2 % of Finnish concrete façades showed visible corrosion damage during the condition assessment. 53.5 % of those were local, while 5.7 % were wide spread. The corrosion of reinforcement was induced by the carbonation of concrete and the small cover depths of reinforcement. Much more visual corrosion damage was seen in the coastal area than inland. Despite the small cover depths of reinforcement in carbonated concrete, 40.8 % of façades did not demonstrated any visual corrosion damage.

The corrosion rate has strong dependence on the moisture content in carbonated concrete. According to Tuutti [26], the corrosion rate increases sharply as relative humidity of concrete exceeds 85 %. The moisture stress on a carbonated structure does thus essentially affect the rate of reinforcement. According to Mattila [14], the density of corrosion current is high in uncovered concrete if the annual amount of rain and sleet is approximately 480 mm or higher. In Fig. 6 the annual liquid precipitation, i.e., rain or wet snow, in Finland is shown.

In the coastal area (Turku) and southern Finland (Vantaa) the amount of annual rainfall is compared to inland (Jyväskylä). In the coastal area and southern Finland the annual rainfall also exceeds the critical amount of 480 mm per year most years, especially after the 1980s. In the 1980s also the carbonation of concrete achieved the near surface remained reinforcement. Thus the higher rainfall amounts would be a reasonable explanation for more corrosion damage in the coastal area than inland.

The weather observations of FMI also indicate that the amount of rain and sleet is the highest in autumn when the relative humidity of outdoor air is typically high. These two factors keep concrete structures wet for long periods, which keeps the corrosion rate at a high level.

4.4 Influence of Prevailing Wind Direction

The annual amount of rain and sleet does not fall uniformly across all façades. The distribution depends on the height of the building and the prevailing wind directions during rain and sleet. Prevailing wind directions and wind speeds largely

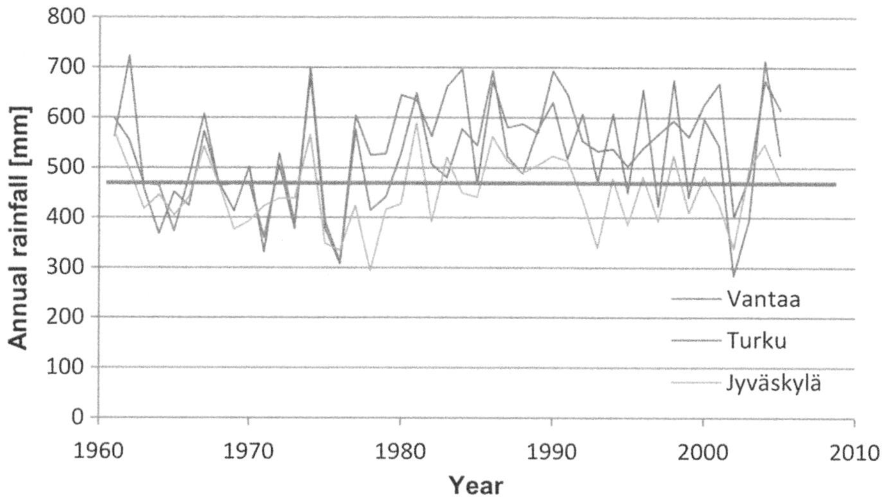

Fig. 6 Annual liquid precipitation in Jyväskylä (inland), Vantaa (south costal area) and Turku (*coast line*) between 1961 and 2005. *The horizontal line* expresses the annual rainfall that leads to a faster corrosion rate

determine on the distribution of rain and sleet on a building. Most of the liquid precipitation comes with southerly to westerly winds in all parts of Finland. Rain events with wind from other directions have been rare; see Fig. 7.

The amount of rain and sleet falling on façades depends on the drop-size distribution of rain, the velocity of falling drops, wind speed, building height, protection offered by the environment, structures guiding wind, etc. According to FMI, the average velocity of falling rain drops is 4–8 m/s, which means that an average of 40–60 % of the rainfall hits façades. During high winds, which occur more frequently in southern Finland, a larger portion of the rainwater hits the vertical surfaces of façades. According to Jerling and Schechninger [11], the upper parts and corners of façades receive more rainfall than lower and central parts.

Hardly any corrosion damage can be detected on the northern or eastern façades despite the reinforcement being embedded in carbonated concrete as on the southern and western façades. The high outdoor relative humidity in Finland during winter is not high enough to increase the humidity of carbonated concrete to the extent that corrosion initiates or propagates relatively fast, which is evidenced e.g., by the relatively slow corrosion of the soffits of concrete balcony slabs and the minor corrosion of the reinforcements of façades that are protected from rain and sleet or receive very little of them. Thus, a high corrosion rate is not possible without rain and sleet. In Finland, the prevailing wind directions during rainfall are southerly to westerly. Snowfall, which usually is accompanied by northerly winds, cannot be absorbed in the pore structure of concrete. That is a reasonable explanation for more corrosion damage occurring on southern and western façades than northern and eastern ones.

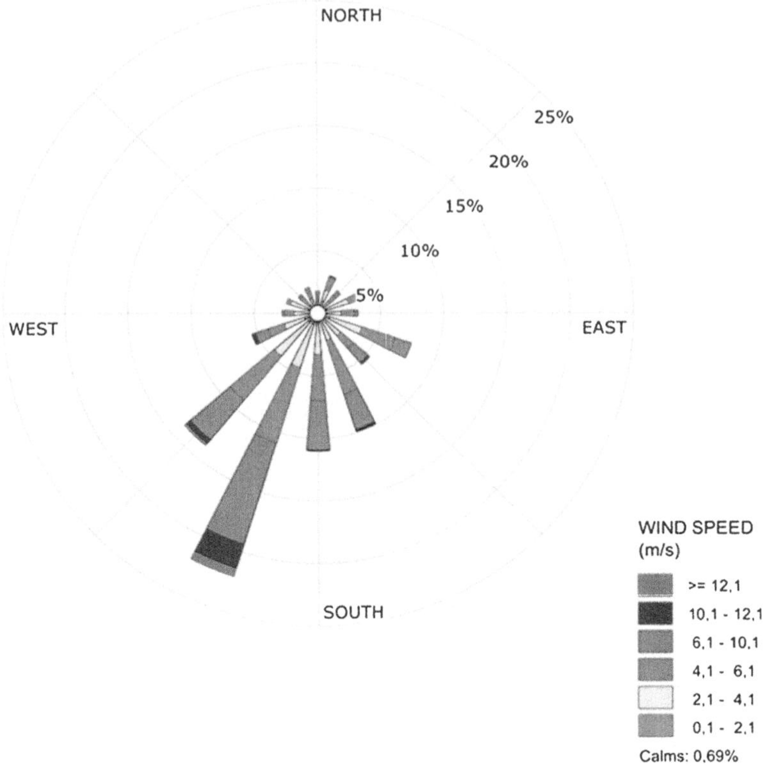

Fig. 7 Prevailing wind directions in wintertime during rain and sleet events. Weather data is measured at Helsinki-Vantaa airport during Sept. 1975 to Apr. 1980

4.5 Corrosion Induced by Chlorides

The chloride content of concrete of a total of 496 facade samples has been determined. The Condition investigation manual for concrete façade panels [6] considers 0.03–0.07 w % of the weight of concrete a critical chloride content for corrosion of reinforcements.

Seven samples from five spots in concrete façades exceed the above lower limit. The chloride contents of the samples are 0.04–0.09 w %, the average being 0.06 w %. The buildings are located in different parts of southern Finland. Three were completed in the 1960s and the other two in 1974 and 1979.

All in all, the critical chloride content required for initiation of reinforcement corrosion was exceeded in only a few isolated buildings completed mainly before 1976. The use of chlorides as a concrete-hydration accelerator has been considered primarily a problem of on-site prefabrication plants of the 1960s. Thus, it was surprising to discover high chloride contents also in concrete panels manufactured in the 1970s. Chlorides cannot have penetrated into the concrete after completion of any of the buildings, but must have entered into the concrete in connection with fabrication of panels intentionally or accidentally.

Although 1.4 % of the samples from façades exceed the critical chloride content, reinforcement corrosion initiated by chlorides is on the whole extremely rare and is not regionally concentrated.

5 Conclusions

5.1 Thickness of Outer Layer

The average thickness of the outer layer of sandwich panels is 55–70 mm depending on the façade surface type, which is close to the design values of both bearing and non-bearing panels. In theory, it is only possible to install the designed reinforcement with its splices and lifting lugs in an outer layer 65 mm thick to meet the 20 mm cover depth requirement. A 25 mm cover depth requires, considering the installation tolerances, an 85 mm outer layer. From the viewpoint of the durability of a concrete structure, it has thus been impossible even theoretically, to provide steels the 25 mm cover depth set in requirements.

However, local less than 40 mm and over 80 mm outer layer thicknesses occur in all façade surface types. In the case of under 40 mm outer layers, it is impossible to place the reinforcement at a proper depth from the viewpoint of corrosion protection.

5.2 Cover Depths of Reinforcement

Reinforcement cover depths can be measured most reliably from concrete samples since the measurements of field investigations generally do not eliminate the effect of the surface structure of the panel or the finishing products. If the effect of surface roughness or, for instance, the thickness of a clinker tile is deducted from the reading of the cover depth meter, the reinforcement cover depth distributions measured from samples and those determined by the cover depth meter are quite consistent.

Reinforcement cover depths do not meet in all respects the minimum requirements set for them in design and durability guidelines of various periods. Yet, it can be stated that on the whole the shares of small, under 5 and under 10 mm, cover depths are quite small in all facade surface types—less than 3.6 %. In clinker-clad panels the share of less than 10 mm cover depths is 6.2 %, which is significant from the viewpoint of the economy of cementitious patch repairs.

5.3 Presence of Chlorides

The critical chloride content of concrete necessary for the initiation of reinforcement corrosion is exceeded in only a few individual cases—in buildings mostly completed before 1976. Use of chlorides as a hydration accelerator has been considered primarily a problem of on-site prefabrication plants of the 1960s. Therefore, it was surprising to discover high chloride contents also in nine precast panels of buildings completed in the 1970s. It was impossible for chlorides to have penetrated into the concrete of any of the buildings after their completion. They must have entered it in connection with the manufacture of the precast panels either intentionally or accidentally.

Although 1.4 % of the samples drilled from façades exceed the critical chloride content, corrosion of reinforcements initiated by chlorides is on the whole extremely rare and is not regionally concentrated.

5.4 Annual Rainfall

Both annual and winter-time precipitation measurements show that more rain and sleet is received in the coastal area than inland. Thus, the higher rainfall could be a reasonable explanation for more corrosion damage in the coastal area than inland. It has also been discovered that rainfall is the highest in autumn when the relative humidity of outdoor air is typically also high. These factors together keep concrete structures wet for long periods. As a result, the corrosion rate of steels remains high for long, and water collects deep inside the pore system of concrete before the beginning of the frost period.

5.5 Prevailing Wind Directions

Rain and sleet do not fall evenly on all façades of a building, but are largely driven about by prevailing winds. During the autumn and winter rain and sleet, the prevailing winds on the southern and south-western coast blow from between east and south-west, and inland from between south-east and south-west. Compared to year-round wind directions, the distribution of wind directions during the survey period in question is quite different. Wind direction and speed during snowfall have no impact on the damage to concrete structures since snow cannot be absorbed into the pore system of concrete. During summer rains, the winds blow most often from between east and south-west, yet the southerly winds dominate.

Thus, prevailing winds during rain and sleet are a quite clear reason for the fact that less corrosion damage of concrete occurs on façades facing north to east than on those facing south to west.

Due to the stronger winds prevailing in the coastal area during rain/sleet, a larger portion of it hits the façades than inland. With the contribution of the higher precipitation of the coastal area, the moisture stress received by façades is considerably higher in the coastal area than inland, which also has a clear causal relationship with the more prevalent corrosion of reinforcements of concrete in the coastal area. Since winds are stronger at the level of the higher sections of buildings, it is natural that the upper parts of high buildings receive more precipitation stress than lower buildings, and the lower parts or buildings in general.

References

1. Alonso, C., Andrade, C., Castellote, M., Castro, P.: Chloride threshold values to depassivate reinforcing bars embedded in a standardized OPC mortar. Cem. Concr. Res. **30**, 1047–1055 (2000)
2. Alonso, C., Andrade, C., Rodriguez, J., Diez, J.M.: Factors controlling cracking of concrete affected by reinforcement corrosion. Mater. Struct. **31**, 435–441 (1998)
3. Andrade, C.: Measurement of R_p on-site. In: Weydert, R. (ed.) Corrosion of Steel in Reinforced Concrete Structures, pp. 88–104. Final reports of single projects 1997–2002. Luxembourg University of Applied Sciences (2002)
4. Bakker, R.: Initiation period. In: Schiessl, P. (ed.) Corrosion of Steel in Concrete, pp. 22–55. Chapman and Hall, London (1988)
5. Broomfield, J.: Corrosion of Steel in Concrete: Understanding, Investigation and Repair, p. 240. E & FN Spon, London (1997)
6. Condition investigation manual for concrete facade panels: BY 42. Concrete Association of Finland, Helsinki, p. 178 (in Finnish) (2002)
7. Durable Concrete Structures: Design Guide. Thomas Telford, Somerset, p. 112 (1992)
8. Gjørv, O.E.: Durability Design of Concrete Structures in Severe Environments, p. 220. Taylor & Francis, London (2009)
9. Heimala, A., Punakallio, E.: Facade study. Helsinki. Asuntohallitus, Tutkimus- ja suunnitteluosasto. Asuntotutkimuksia 5/1993, 174 p. + app. 8 p. (in Finnish) (1993)
10. Huopainen, J.: Carbonation of concrete facades—field study. Tampere University of Technology, Tampere. Structural engineering, Master's Thesis. 104 p. + app. 51 p. (in Finnish) (1997)
11. Jerling, A., Schechninger, B.: Fogars beständighet. Byggforskningrådet. Rapport R89:1083. Stockholm, p. 172 (in Swedish) (1983)
12. Lahdensivu, J.: Durability properties and actual deterioration of Finnish concrete facades and balconies. Tampere University of Technology, Tampere, Publication **1028**, 117 p + app. 37 p. (2012)
13. Mattila, J.: Realkalisation of concrete by cement-based coatings. Tampere University of Technology, Tampere. Structural Engineering. Licentiatés Thesis, p. 161 (In Finnish) (1995)
14. Mattila, J.: On the durability of cement-based patch repairs on Finnish concrete facades and balconies. Tampere University of Technology, Tampere, Publication **450**, p. 111 (2003)
15. Mehto, L., Pentti, M., Käkönen, H.: Carbonation of concrete facades. Tampere University of Technology, Tampere. Structural Engineering, Research report **41**, 149 p. + app. 37 p. (in Finnish) (1990)
16. Mäkiö, E., et al.: Apartment buildings 1960–1975. Rakennustieto Oy, Tampere, p. 288 (in Finnish) (1994)
17. Page, C.L.: Basic principles of corrosion. In: Schiessl, P. (ed.) Corrosion of Steel in Concrete, pp. 3–21. Chapman and Hall, London (1988)

18. Parrott, L.J.: Review of Carbonation in Reinforced Concrete, p. 42. Cement and Concrete Association, Wexham Springs (1987)
19. Pentti, M., Mattila, J., Wahlman, J.: Repair of concrete façades and balconies. Part 1: Structures, degradation and condition investigation. Tampere University of Technology, Structural Engineering, Tampere, Publication **87**, p. 156 (In Finnish) (1998)
20. Richardson, M.: Carbonation of Reinforced Concrete, p. 203. Dublin, CITIS (1988)
21. Richardson, M.: Parameters affecting the rate of carbonation: A survey of concrete in Ireland. In: Page, C.L., Treadaway, K.W.J., Bamforth, P.B. (eds.) Corrosion of Reinforcement in Concrete, p. 612. Elsevier Science Publisher Ltd, Essex (1990)
22. Schiessl, P., Breit, W., Raupach, M.: Durability of local repair measures on concrete structures damaged by reinforcement corrosion. In: Mathora, V.M. (ed.) Durability of Concrete, pp 1195–1215. American Concrete Institute, Detroit (1994)
23. Statistics Finland: www.tilastokeskus.fi. Reference date 9.8.2010. (in Finnish) (2010)
24. Sulankivi, H.: Carbonation of concrete coverer by brick plates. Tampere University of Technology, Tampere. Structural Engineering, Masteŕs Thesis. 134 p. + app 7 p. (in Finnish) (1993)
25. Taylor, P.C., Nagi, M.A., Whiting, D.A.: Threshold chloride content for corrosion of steel in concrete: A literature review. Illinois, Portland Cement Association. PCA R& D Serial No. 2169, p. 32 (1999)
26. Tuutti, K.: Corrosion of Steel in Concrete. Swedish Cement and Concrete Research Institute, CBI Research, Stockholm **4**(82), 304 (1982)
27. Vainio, T., et al.: Repair, maintenance and improvement work in Finland 2000–2010. Espoo 2002. VTT Research notes 2154. 60 p. + app. 25 p. (In Finnish) (2002)
28. Vainio, T., Lehtinen, E., Nuuttila, H.: Building and renovation of façades. Tampere. VTT 26 p. + app. 13 (In Finnish) (2005)

The manufacturer's authorised representative in the EU is Springer Nature Customer Service Centre GmbH, Europaplatz 3, 69115 Heidelberg, Germany. If you have any concerns regarding our products, please contact ProductSafety@springernature.com

Printed and bound by CPI Group (UK) Ltd, Croydon, CR0 4YY
22/09/2026
02221064-0013